European flora
of the desmid genera
Staurastrum and Staurodesmus

European flora
of the desmid genera
Staurastrum and Staurodesmus

by Peter F. M. Coesel

and Koos (J.) Meesters

with 120 plates and 5 figures in the text

Contents

Preface

The genus *Staurastrum* stands as one of the largest and taxonomically most compli-
cated desmid genera. Indeed, after the authorative flora by W. and G.S. West dating
from the beginning of the last century no critical flora dealing with this genus has
appeared. In the absence of objective, reliable criteria for delimitation of desmid
species it is not surprising that attempts to make an up-to-date taxonomic revision
of this large genus have failed so far. Unfortunately, Teiling's (1948) splitting off
of the genus *Staurodesmus* led to more complications rather than diminishing the
taxonomic and nomenclatural chaos.

The present flora does not pretend to solve all problems of identification. As a
matter of fact, assessing mutual affinities in this predominantly asexually reproducing group of microalgae is only possible after taking into account DNA data. On a
species level, however, this information is fragmentary. Yet, there is an increasing
need to label the microorganisms encountered when sampling the aquatic environ-
ment for assessment of water quality or conservation value. Among the desmids,
representatives of the genera *Staurastrum* and *Staurodesmus* are particularly well-
known for their occurrence in the euplankton of freshwaters, both oligotrophic and
eutrophic. Hopefully, our flora will contribute to more consistent species identification and a more reliable comparison of data sets.

We are much indebted to our colleague, Chris F. Carter, for linguistic screening
of the introductory sections.

Peter F.M. Coesel
Koos (J.) Meesters
Amsterdam, January 2013

Introduction

The present manual aims to be a critical flora of the European *Staurastrum* and *Staurodesmus* species. For pragmatic reasons Iceland and the Azores are taken as the western limit; the western frontiers of the former Soviet Union (exclusive of the Baltic States) and Turkey are taken as the eastern border of the geographical region covered by this flora.

All European species recorded in the literature under the genus names of *Staurastrum*, *Staurodesmus* and *Arthrodesmus* were screened, but only records accompanied by an illustration were considered for inclusion. Species judged to be either highly questionable or to be identical to others are noted in a list at the end. Taxa described as varieties and formae of those questionable species are not dealt with at all, nor is any attempt made to classify all varieties and formae that are considered not to belong to the species under the name of which they were originally described.

Following the critical flora by Růžička (1977, 1981) the monothetic taxon concept is adopted, i.e., each taxon is considered the sum of all subordinate taxa. Consequently, the diagnosis of a species may not be contrary to the diagnosis of any of its subordinate varieties or formae.

In writing the present flora it appeared that many taxon names published after 1953 should formally be considered invalid, not being in accordance with one or more articles of the International Code of Botanical Nomenclature (ICBN): usually by want of a Latin diagnosis, indication of a nomenclatural type, or adequate reference to a basionym. For instance, most of the recombinations made by Teiling (1967) when transferring *Staurastrum* and *Arthrodesmus* species to his newly established genus *Staurodesmus,* suffer from an inadequate or even complete lack of reference to the basionym. In the present flora, for pragmatic reasons, invalidly published names are not indicated as such. In a number of cases the validity of the name may be subject to interpretation of the ICBN articles in question. Apart from that, the intention of our flora is to provide a useful identification manual, not a compilation of taxonomic revisions.

Because of the huge difficulties in tracing and examining type specimens of desmid taxa in the original samples and it being infeasible to preserve a single cell as a type specimen according to art. 37.5 of the Vienna Code (McNeill & al. 2006), names of newly described taxa are connected to a given illustration as holotype.

Taxon description

Taxon diagnoses are confined to the essential discriminating morphological characteristics. More details may often be found in the original description referred to. Cell dimensions are inclusive of all projections (arm-like processes, spines and such-like) unless explicitly stated otherwise. Taxon names including their authorities are in accordance with the International Code of Botanical Nomenclature (McNeil & al. 2006) taking Ralfs (1848) as starting point for desmid nomenclature. As a consequence, all taxon names (and author names belonging to them) published before 1848 are ignored. Taxon-linked author names are not abbreviated. If necessary, initials are added according to Brummitt & Powell (1992).

Each taxon dealt with is referred to the paper in which the taxon name in question was originally published, as well to a possible earlier description under another combination of names (basionym). As for synonyms (other than possible basionyms), reference is only made to recently used, current combinations, holding the epitheton under discussion. Parts of other synonyms may be traced from the various lists of excluded taxa.

Taxa mentioned somewhere in the European literature but not worked in the present flora because of reasonable doubt about their identity, either because of an unclear picture, incongruity with the original diagnosis, or possible synonymy, are listed as 'excluded'.

All references to taxon-linked author names are specified in the list of references at the end, except those of excluded taxa.

Infraspecific taxa

Where infraspecific taxa are distinguished in a given species, the nominate taxon is dealt with separately. To assess this need, also infraspecific taxa exclusively known from non-European literature were considered. If those non-European taxa were indeed found to deserve an infraspecific status, they are separately dealt with in a smaller font. When, however, such taxa are considered to be identical to one of the European taxa they are mentioned as a synonym under the taxon name they are assigned to. Questionable non-European taxa are mentioned without additional comments. If a non-European infraspecific taxon is believed not to belong to the species under discussion it is labeled 'taxon excluded'.

Ecology and distribution

A rough indication of the taxon-specific habitat conditions in terms of acidity and trophic status is given. Where known, preference for a euplanktonic, tychoplanktonic (living among submerse aquatic weeds along the shore and from there often entrained into the open water body), benthic or subatmophytic (living on wet substrates) way of life is added.

An indication of the global distribution is given by listing the continents from which reliable records (checked from illustrations) are known. In this context New Zealand is accounted as the Australian continent and Greenland as the North American one. Cosmopolitan species are known from all continents except Antarctica (where conditions for the vast majority of desmid species are too extreme).

Illustrations

For each taxon, where meaningful, the original illustration is represented next to the others. In a number of taxa, however, the quality of the original illustration is so poor that no relevant information is provided. Usually this relates to sketchy and/or fragmentary figures in old publications. In those cases, taxonomic concept and illustration of the species in question are mostly based on those given in the well-known flora series by W. & G.S. West (1912, 1923). Next to the original figure a variable number of illustrations made by other authors are given, roughly representing the morphological variability of the taxon in question. These latter figures exclusively refer to European material.

To ensure uniformity in style of representation, all figures have been redrawn. For simplification, inner cell walls as well as cell wall punctations — if at all present in the original illustrations — have not been copied. Where original illustrations show cell contents that mask any of the cell wall ornamentation, our version replaces this with a uniform grey shade. All illustrations are approximately on the same scale of magnification, i.e., 600x. Apical cell views linked to frontal ones by means of a dotted line do not per se originate from one and the same real cell.

Identification keys

Identification keys are dichotomous. Where successive key numbers are physically far apart the preceding key number is given between brackets. Identification keys are based exclusively on vegetative cell characteristics. Statement of cell length and cell breadth includes possible processes. Where the ratio between length and breadth of the semicell body is an issue, processes are excluded. In the case of a gradual transition between the semicell body and the arm-like processes, the breadth of the semicell body is taken at the base (just above the cell isthmus).

Problems in species delimitation

The genera *Staurastrum* and *Staurodesmus* are notorious for their problematic species delimitation and, consequently, difficult identification of those species (Mollenhauer 1988). One of the main causes for this is the infrequent occurrence of sexual reproduction, a process of gene exchange resulting in reduced morphological variation. This phenomenon is not only characteristic of *Staurastrum* and *Staurodesmus* species but also of other desmid genera and many other groups of microalgae. Predominantly asexual (clonal) reproduction in those unicellular organisms results in the formation of microspecies, mutually differing in only minor characteristics, just as in apomictically reproducing macrophytes (Coesel & Krienitz 2008). In addition to this genotypically based morphological variation, many *Staurastrum* and *Staurodesmus* species exhibit a remarkable phenotypic variation. Often, the extremes in such morphological ranges seem to represent quite different species (e.g., Grönblad & Růžička 1959, Péterfi 1972, Brook 1959, Ling & Tyler 1995). In connection with the above-described phenomena, there are two contrary tendencies in desmid taxonomy, i.e. "splitting" versus "lumping" of traditional species. Unfortunately, in most cases, the nature of morphological variability (whether genotypic or phenotypic) is unknown and not seldom one gets the impression that 'species' are interconnected by transitional forms rendering taxon delimitation rather arbitrary. Without objective criteria for ranking species, subspecies, varieties and forms, traditional desmid taxonomy has suffered from nomenclatural chaos (e.g., Kouwets 2008) and numerous synonyms only add to the confusion.

For pragmatic reasons, in the present manual the following starting-points are chosen:

1) The number of infraspecific taxa is highly limited. Most of the *Staurastrum* and *Staurodesmus* forms described as infraspecific taxa are either considered to represent species of their own or to be a part of the taxonomically irrelevant variability of the species in question. It should be stressed that the choice between these two possibilities is quite subjective, so a matter of best professional judgement.

2) The only infraspecific level maintained is that of the variety. In general, the variety level is used for those forms that deviate too little from the nominate form to justify (by intuition) distinction of a separate species but are to be linked to a somewhat deviating ecological or geographical distribution.

3) In cases where alternative taxonomic options are being considered, traditional species and variety delimitations have been maintained as far as possible. This approach does not complicate in a needless way any comparison with existing literature or data bases.

Delimitation of the genera *Staurastrum* and *Staurodesmus*

The desmid genus *Staurastrum*, erected in 1828 by Meyen and validated by Ralfs in 1848, is exceptionally polymorphic and should be considered a highly artificial one. Unfortunately, phylogenetic relationships obtained by DNA analyses so far do not correlate with well-defined morphological features (Gontcharov & Melkonian 2005, 2011). Therefore, as yet, for species identification we must retain current classification schemes.

Staurastrum paradoxum Meyen ex Ralfs, considered the type species of this genus, is characterized by four- or triradiate cells provided with arm-like processes. Ralfs (1848), however, under the genus name of *Staurastrum* represents not only such armed cell forms but also compact, rounded cell forms, both smooth-walled and furnished with granules, warts or spines. Radiation of cells may range from bi- to pluriradiate. Actually, the genus *Staurastrum* as conceived by Ralfs (1848) incorporates all unicellular tri- to pluriradiate desmid forms as well as all biradiate ones the cell body of which is provided with arm-like processes.

As radiation in desmids appears to be a relatively unstable characteristic that may vary within a single species or may even differ between the two semicells of a single cell (e.g., Teiling 1950) it will be clear that boundaries between *Staurastrum* and a number of other desmid genera are far from sharp. In practice, triradiate forms of for example, *Cosmarium* and *Xanthidium* species cannot reliably be distinguished from *Staurastrum*. In those species it is the predominant radiation that is decisive. See, for example, triradiate forms of normally biradiate *Xanthidium armatum* (Skuja 1964), *Cosmarium speciosum* (Růžička 1957), *Cosmarium margaritiferum* (Messikommer 1951), or *Cosmarium bulliferum* (Růžička 1962) and biradiate forms of normally triradiate *Staurastrum lunatum* (Grönblad 1942) or *Staurastrum forficulatum* (West & al. 1923).

In the case of those species with a characteristic cell outline or cell wall sculpturing as in the above-mentioned ones, identification of such anomalous forms is relatively easy. Not seldom, however, species morphology is less precise, giving rise to the possibility that bi- and triradiate forms of the same species have been described under different genera. For instance, *Staurastrum circulare* Schmidle (1896), especially its var. *americanum* Grönblad (1962) might refer to a triradiate form of *Cosmarium margaritatum* (P. Lundell) J. Roy et Bisset; *Staurastrum enontekiense* Grönblad (1942) might refer to a triradiate form of *Cosmarium furcatospermum* W. et G.S. West; *Staurastrum trigonum* (Boldt) Kossinskaja (1936) most likely refers to a triradiate form of *Xanthidium bifidum* (Brébisson) Deflandre.

The genus *Staurodesmus* was established by Teiling (1948) for a group of monospinous desmid species (semicells with one spine per radius). Up to then, those species were partly placed in the genus *Staurastrum* Meyen, and partly in the genus *Arthrodesmus* Ehrenberg ex Ralfs. The reasons for doing this stemmed from the occasional observations of co-occurring triradiate cell forms (formerly *Staurastrum*) and biradiate forms (formerly *Arthrodesmus*) but obviously belonging to one and the same species. These relationships were reinforced by the occurrence of dichotypic cells (so-called Janus forms) consisting of one triradiate and one biradiate semicell.

It is true that with the introduction of *Staurodesmus* the problem of the artificial and unsatisfactory delimitation between *Staurastrum* and *Arthrodesmus* had been solved. On the other hand, another problematic delimitation had now been created, i.e., that between *Staurodesmus* and *Staurastrum*. For, in Teiling's concept, the expression of the angular processes in *Staurodesmus* is highly variable, ranging from long spines to minute, often hardly distinguishable papillae or just somewhat thickened semicell angles. In view of this and also because of the unclear, confusing taxonomic treatment in Teiling's (1967) monograph of the *Staurodesmus* species, it is understandable that Teiling's concept of *Staurodesmus* is not generally accepted. One of the reasons that, in the present flora, we do distinguish *Staurodesmus* as a separate genus, is that the genus name

Arthrodesmus Ralfs 1848, being a homonym of the chlorococcalean genus *Arthrodesmus* Ehrenberg 1838, has not been validly published (Compère 1976a). However, as for the delimitation of the genus *Staurodesmus*, we do not fully share Teiling's (1967) view. Only species characterized by distinct spines or acute mucros are accounted as *Staurodesmus*. Species marked by blunt papillae or gradually changing angular cell wall thickenings are designated *Staurastrum* (or *Cosmarium*). It has to be stressed, for that matter, that intrinsically spinous species may incidentally develop anomalous forms in which the cell wall spines are largely or even completely reduced.

Notes on reproduction, ecology and distribution

As in other desmid genera, sexual reproduction in *Staurastrum* and *Staurodesmus* species is predominantly vegetative, by cell division. Sexual reproduction, by conjugation, is only occasionally observed and in many species unknown. Particularly in species with a planktonic way of life, circulating in the open water of large lakes, the chances of cells meeting for sexual reproduction are almost nil. Successful conjugation is only possible under conditions of undisturbed cell contact for the time needed to induce pairing of conjugants and amoeboid migration of gametes. Such conditions are to be expected in shallow, stagnant water bodies, in the shelter of submerged vegetation, but not in more turbulent open water.

Most of the *Staurastrum* and *Staurodesmus* species occur in (slightly) acidic oligo-mesotrophic habitats, but especially among euplanktonic species quite a number are known that flourish under eutrophic conditions. Incidentally, those species may even reproduce so abundantly that they cause a water bloom (e.g. Lenzenweger 1980, Brook 1981, Andersen & al. 1987). Species living in deep lakes or reservoirs are often characterized by long processes increasing their cell surface/volume ratio and hence reducing their sinking velocity. On the other hand, cells of species living semiatmophytically on wet substrates or in ephemeral puddles that periodically dry out are marked by more cylindric cells with a relatively smaller surface area.

Many species exhibit distinct geographical distribution patterns within Europe. As well as atlantic species, confined to western and nordic Europe, arctic-alpine species are also known (Coesel 1996b, Coesel & Krienitz 2008). Remarkably, a number of atlantic-arctic flagship species, such as *Staurastrum arctiscon*, *S. ophiura*, *S. elongatum* and *S. cerastes* disappeared from the Netherlands about 60 years ago, concomitant with rising average annual temperatures.

Genus *Staurodesmus*

The genus *Staurodesmus* was erected by Teiling (1948) but he omitted to designate a type species of this genus. This omission was corrected by Compère (1977). His choice of *Staurodesmus triangularis* (Lagerheim) Teiling as type species was, among other things, dictated by the consideration that in this species all of 2-, 3- and 4-radiate forms occur as well as Janus forms, thus illustrating Teiling's motives for the erection of the genus *Staurodesmus* (see Introduction).

Type species: *Staurodesmus triangularis* (Lagerheim) Teiling 1948, Bot. Not. 1948: 68, fig. 63-68.
Type specimen: *Arthrodesmus triangularis* Lagerheim 1886, Öfver. Förh. Kongl. Svenska Vetensk.-Akad. 42 (7): 244, pl. 27, fig. 22.

Diagnosis of genus:
Staurodesmus Teiling 1948
Cells bi- to pluriradiate, usually with a (moderately) deep sinus. Semicells smooth-walled with one spine (or mucro) per radius. Chloroplasts axial with radiating ridges.

Essential to the genus *Staurodesmus* is the occurrence, at the periphery of a smooth-walled semicell body, of one single spine per semicell lobe. However, incidentally one or more of those projections may be completely reduced or, instead of one spine per radius two may occur.

Valuation of diagnostic features in species identification

The most relevant morphological characteristic in species identification is the shape of the semicell body and, in connection with that, the shape of the sinus. Also significant is the place of insertion of the spines: on the apical, basal or lateral side of the semicell body (Figs 1, 2). Projection (Fig. 3) and length of the spines in some of the species are relatively consistent, in other species, on the contrary, they may be quite variable (Pl. 7). The same holds for the degree of radiation: in some species it is consistent, in others it is variable.

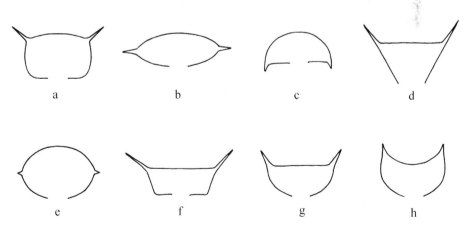

Fig. 1. Basic types of semicell shape in *Staurodesmus*: a = subrectangular, b = fusiform; c = semicircular; d = cup-shaped (triangular); e = elliptic; f = (obversely) trapeziform; g = bowl-shaped; h = semilunate.

Fig. 2. Variation in triangular semicell shape: a = with elevated apex; b = with indented lateral sides

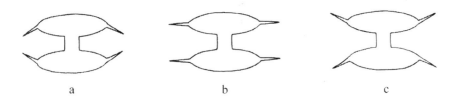

Fig. 3. Projection of spines in *Staurodesmus*: a = convergent; b = parallel; c = divergent

Key to the species

Unless explicitly stated otherwise, descriptions refer to cells in frontal view and to cell dimensions inclusive of spines.

1	Cells as a rule biradiate	2
-	Cells as a rule tri-pluriradiate	27
2	Semicells with very short spines (often almost papillate), semicell body about cuneate in outline	*Std. controversus*
-	Semicells (if not reduced) furnished with marked spines, semicell body various in outline	3
3	Cells with an elongate isthmus	4
-	Cell isthmus not elongate	11
4	Semicell body (above the isthmus) hexagonal with distinctly convergent spines	*Std. subhexagonus*
-	Semicell body shaped otherwise, spines convergent or divergent	5
5	Semicell body above the isthmus trapeziform, more than twice as broad as long	*Std. extensus*
-	Semicell body shaped otherwise	6
6	Semicell body with an elevated apex	10
-	Apex of semicell body not distinctly elevated	7
7	Spines usually convergent, relatively short (< breadth of semicell body)	*Std. ralfsii*
-	Spines usually divergent, relatively long (> breadth of semicell body)	8

8	Apex of semicell deeply concave	*Std. phimus*
-	Apex of semicell about straight	9
9	Isthmus distinctly elongate	*Std. longispinus*
-	Isthmus hardly or not elongate	*Std. incus*
10	(6) Semicell body (above the isthmus) triangular in outline	*Std. triangularis*
-	Semicell body elliptic-rhomboid in outline	*Std. subtriangularis*
11	(3) Cells about rectangular in outline	12
-	Cells shaped otherwise	13
12	Sinus deep, closed for the greater part	*Std. bulnheimii*
-	Sinus very shallow, widely open	*Std. subquadratus*
13	Semicells elliptic in outline, spines laterally inserted	14
-	Semicells triangular or cuneate in outline, spines apically inserted	15
14	Spines convergent (often curved)	*Std. convergens*
-	Spines about parallel, straight	*Std. subulatus*
15	Spines usually convergent	16
-	Spines divergent or parallel	18
16	Semicell body relatively broad (at least twice the semicell length)	*Std. depressus*
-	Semicell body but little broader than long	17
17	Cells somewhat longer than broad exclusive of spines. Isthmus often slightly elongate	*Std. ralfsii*
-	Cells about as long as broad exclusive of spines. Isthmus not elongate	*Std. glaber*
18	Semicells with convex apex	19
-	Semicell apex, at least in its mid region, straight or concave	20
19	Semicell apex only convex in its very middle, so actually a bit undulate	*Std. hebridarus*
-	Semicell apex evenly convex	*Std. subulatus* var. *nordstedtii*
20	Apex of semicells (slightly) elevated	21
-	Apex of semicells not elevated	22
21	Semicell body (above the isthmus) triangular in outline	*Std. triangularis*
-	Semicell body elliptic-rhomboid in outline	*Std. subtriangularis*
22	Apex of semicell body deeply concave	23
-	Apex of semicell body, if concave at all, at best slightly concave	24
23	Apical angles of semicells inflated, abruptly passing into the spines	*Std. phimus*
-	Apical angles not inflated, gradually passing into the spines	*Std. semilunaris*
24	Cells relatively large (exclusive of spines usually longer than 20 μm) and furnished with long spines (> 14 μm)	25
-	Cells small (≤ 20 μm in length) with short spines (≤ 11 μm)	26

25	Semicells bowl-shaped	*Std. validus*
-	Semicells cup-shaped (about triangular)	*Std. incus*
26	Apical angles of the semicell body abruptly attenuated into the spines	*Std. pterosporus*
-	Apical angles gradually passing into the spines	*Std. omearae*
27	(1) Cells usually 5-radiate	*Std. wandae*
-	Cells at best 4-radiate	28
28	Lateral sides of the semicell body (slightly) indented	29
-	Lateral sides of the semicell body not indented	32
29	Apical angles remarkably inflated and abruptly passing into a minute spine	*Std. andrzejowskii*
-	Apical angles less inflated and more gradually passing into the spines	30
30	Apical angles furnished with rather long spines (> 10 μm)	*Std. aristiferus*
-	Apical angles furnished with minute spines (< 5 μm) that are often completely reduced	31
31	Semicell body triangular with convex apex	*Std. leptodermus*
-	Semicell body cuneate with concave apex	*Std. corniculatus*
32	(28) Semicell body shaped like a burst pomegranate, the radii being curved in a lunate way	*Std. unguiferus*
-	Semicells shaped otherwise	33
33	Semicell body bowl-shaped. Apical spines projected upwards, largely following curvation of the lateral sides	34
-	Semicells shaped otherwise	36
34	Isthmus slightly elongate	*Std. dejectus*
-	Isthmus not elongate	35
35	Semicell apex concave or straight	*Std. connatus*
-	Semicell apex convex	*Std. patens*
36	Semicells cuneate with concave apex and slightly convex lateral sides, spines very short and projected upwards	*Std. corniculatus*
-	Semicells shaped otherwise	37
37	Semicell body elliptic, rhomboid to semicircular	38
-	Semicell body triangular to cup-shaped	42
38	Sinus widely open, strongly undulate	*Std. recurvus*
-	Sinus not remarkably undulate	39
39	Isthmus highly elongate	*Std. cuspidatus*
-	Isthmus not elongate	40
40	Cells with convergent spines	*Std. dickiei*
-	Cells with about parallel or divergent spines	41

41	Cells relatively large (usually more than 40 µm in length) with stout spines (more than 10 µm in length)	*Std. megacanthus*
-	Cells smaller (usually less than 30 µm in length) with short spines (up to 6µm)	*Std. mucronatus*
42	(37) Isthmus elongate	43
-	Isthmus not elongate	44
43	Apical angles rather abruptly passing into the spines	*Std. cuspidatus*
-	Apical angles gradually passing into the spines	*Std. cuspidicurvatus*
44	Cell body without spines distinctly longer than broad	*Std. incus*
-	Cell body without spines about as broad as long or broader than long	45
45	Cells with distinctly convergent spines	*Std. glaber*
-	Cells with about parallel or divergent spines	46
46.	Relatively large cells (inclusive of spines broader than 50 µm) with stout spines (longer than 10 µm)	47
-	Smaller cells (inclusive of spines less than 50 µm in breadth) with slender spines (usually less than 10 µm in length)	48
47	Sinus acute-angled	*Std. megacanthus* var. *scoticus*
-	Sinus obtuse-angled, rounded	*Std. cuspidicurvatus*
48	Semicell body without spines broader than 20 µm, spines shorter than 6 µm	*Std. mucronatus* var. *subtriangularis*
-	Semicell body without spines less than 20 µm in breadth, spines usually longer than 6 µm	49
49	Apical angles of the semicell body abruptly attenuated into the spines	*Std. pterosporus*
-	Apical angles gradually passing into the spines	*Std. omearae*

Staurodesmus andrzejowskii (Woloszynska) Teiling   Plate 16: 14

Teiling 1967, p. 550

Basionym: *Staurastrum andrzejowskii* Wolozynska 1921, p. 139, fig. 8, 9

Cells 3-radiate, about as long as broad, deeply constricted. Sinus open and acute-angled. Semicells cup-shaped with an indentation at the transition of the semicell body towards the radiate parts. Apical angles distinctly inflated, produced obliquely upwards and abruptly passing into a minute, upwardly projected spine. Semicells in apical view triangular with concave sides and inflated angles. Zygospore unknown. Cell length (?20-)35-38 µm, cell breadth (?20-)35 µm.

Ecology: in acidic, oligotrophic water.

Distribution: only known from a site in Poland.

Std. andrzejowskii with its single record is a bit questionable species showing resemblance with given forms of *Std. leptodermus*, especially its tropical var. *ikapoae* (Schmidle) Teiling 1967, a taxon that be better considered a species of its own.

Staurodesmus aristiferus (Ralfs) Teiling

Cells (3-)4-radiate, about as long as broad, deeply constricted. Sinus widely open, acute- to obtuse-angled at the apex. Semicells triangular in outline with an indentation at the transition of the central semicell body towards the radiate parts. Apical angles somewhat inflated, produced obliquely upwards and abruptly passing into long, divergent spines. Semicells in apical view tri-or quadrangular with slightly inflated angles ending in a long spine. Zygospore unknown. Cell length including spines 35-60 µm, cell breadth including spines 35-60 µm.
Ecology: in plankton of acidic, oligotrophic water bodies.
Distribution: Europe, North America. In Europe, *Std. aristiferus* is a rare species with an atlantic distribution, known from the United Kingdom, France, the Netherlands and Sweden.

var. ***aristiferus***   Plate 10: 10-13
Teiling 1950, p. 311
Basionym: *Staurastrum aristiferum* Ralfs 1848, p. 123, pl. 21: 2
Cells 4-radiate, in apical view with concave lateral sides.

var. ***protuberans*** (W. et G.S. West) Teiling   Plate 10: 14-15
Teiling 1967, p. 561
Basionym: *Staurastrum aristiferum* var. *protuberans* W. et G.S. West 1903, p. 544, pl. 14: 5
Cells 3-radiate, in apical view with lateral sides that are convex in the middle.

Taxa excluded:
Staurastrum aristiferum var. *indentatum* G.M. Smith 1924
S. aristiferum var. *gracile* Lütkemüller 1900
S. aristiferum var. *parallelum* W. et G.S. West 1896
S. aristiferum forma *parvulum* Skuja 1949
S. aristiferum var. *planum* W.B. Turner 1892
S. aristiferum var. *prescottii* Irénée-Marie 1952
S. aristiferum var. *projectum* Jao 1949

Staurodesmus bulnheimii (Raciborski) Round et Brook   Plate 3: 1-5
Round & Brook 1959, p. 184
Basionym: *Arthrodesmus bulnheimii* Raciborski 1889, p. 95, pl. 6: 17
Cells 2-radiate, exclusive of spines about as long as broad, or slightly longer than broad, deeply constricted. Sinus closed for the greater part. Semicells subrectangular in outline, the apical angles furnished with long, divergent spines. Semicells in apical view elliptic with a long spine at each pole. Zygospore unknown. Cell length including spines 40-60(-70) µm, cell breadth including spines 40-60(-80) µm.
Ecology: in (tycho)plankton of acidic, oligotrophic water bodies.
Distribution: Eurasia, North America. In Europe, *Std bulnheimii* is rare and mainly confined to atlantic regions.

For validation of species name see Coesel 1993. Eichler 1896 and Skuja 1964 depict distinct cell wall scrobiculae.

Accounted *Staurodesmus bulnheimii* (var. *bulnheimii*):
Std. bulnheimii var. *subrotundatum* (Printz) Teiling 1967
Arthrodesmus bulnheimii var. *subrotundatum* Printz 1915

Taxa excluded:
Std. bulnheimii var. *huitfeldtii* (Strøm) Teiling 1967
Std. bulnheimii var. *subincus* (W. et G.S. West) Thomasson 1966 (= *Std. validus* var. *subincus*)

Staurodesmus connatus (P. Lundell) Thomasson   Plate 16: 15-18
Thomasson 1960, p. 34
Basionym: *Staurastrum dejectum* var. *connatum* P. Lundell 1871, p. 60, pl. 3: 28
Synonym: *Staurastrum connatum* (P. Lundell) Roy et Bisset 1886, p. 237

Cells 3-radiate, exclusive of spines slightly longer than broad. Sinus deep, widely open from an acute-angled apex. Semicells bowl-shaped with strongly convex sides and long, almost vertical spines at the apical angles. Semicells in apical view triangular with slightly concave sides and broadly rounded angles furnished with a spine. Zygospore unknown. Cell length including spines 37-50 μm, cell breadth including spines 20-35 μm.
Ecology: in acidic, oligotrophic water bodies.
Distribution: possibly cosmopolitan. In Europe, *Std. connatus* is widely distributed but not common.

For validation of species name, see Coesel 1993.

Accounted *Staurodesmus connatus* (var. *connatus*):
Std. connatus var. *intermedius* (B. Eichler et Gutwiński) Teiling 1967

Taxa excluded:
Staurodesmus connatus var. *africanus* (Bourrelly) Thomasson 1960
Std. connatus var. *americanus* (W. et G.S. West) Teiling 1967
Std. connatus var. *isthmosus* (A.M. Scott et Grönblad) Thomasson 1966
Staurastrum connatum var. *muticum* Playfair 1912
S. connatum var. *pseudoamericanum* Grönblad 1920 (cf *Std. unguiferus*)
S. connatum var. *rectangulum* J. Roy et Bisset 1886
S. connatum var. *spencerianum* Maskell 1889 (= *Std. spencerianus*)
S. connatum var. *warmbadianum* Claassen 1961

Staurodesmus controversus (W. et G.S. West) Teiling
Cells 2(-4)-radiate, about as long as broad or slightly longer than broad, moderately constricted. Sinus widely open, obtuse-angled. Semicells cuneate with slightly convex sides. Apical angles furnished with a minute, acute spine. Semicells in apical view elliptic-fusiform, sometimes triangular or quadrangular, the angles furnished with a minute spine. Zygospore globose, furnished with simple spines. Cell length (8-)10-24 μm, cell breadth 10-25 μm.
Ecology: in (tycho)plankton of oligotrophic, acidic waters.
Distribution: Eurasia, American continents. In Europe, *Std. controversus* is pretty rare, most finds are from atlantic regions.
Relevant literature: Teiling 1967.

Std. controversus may be confused with small reduction forms of other species.

var. *controversus*   Plate 19: 14-17
Teiling 1967, p. 504
Basionym: *Arthrodesmus controversus* W. et G.S. West 1894, p. 9
Original illustration: W. West 1892, pl. 22: 10 (as *Arthrodesmus glaucescens* forma *convexa* W. West)
Cells slightly longer than broad, less than 15 μm in length.

19

Accounted *Staurodesmus controversus* var. *controversus*:
Staurodesmus controversus var. *brasiliensibus* (Borge) Teiling 1967

var. *crassus* (W. et G.S. West) Coesel et Meesters stat. et comb. nov.   Pl. 19: 18-26
Basionym: *Arthrodesmus crassus* W. et G.S. West 1903, p. 541, pl. 14: 8, 9
Synonym: *Staurodesmus crassus* (W. et G.S. West) M. Florin 1957, p. 130
Cells about as long as broad, at least 15 µm in length.

Accounted *Staurodesmus controversus* var. *crassus*:
Staurodesmus controversus var. *zachariasii* (Schröder) Teiling 1967

Taxa excluded:
Staurodesmus controversus var. *alaskensis* (Irénée-Marie) Teiling 1967
Std. crassus var. *productus* (Skuja) Teiling 1967 (cf *Std. omearae*)

Staurodesmus convergens (Ralfs) S. Lillieroth
Cells 2(-3?)-radiate, exclusive of spines somewhat broader than long to about as broad as long, deeply constricted. Sinus widely open to its extremity. Semicells ellipsoid in outline, at the poles with a shorter or longer, downwardly projected spine. Semicells in apical view ellipsoid (rarely triangular?). Zygospores, known from the nominate variety, globose and smooth-walled. Cell length 30-55 µm, cell breadth including spines 40-90 µm.
Ecology: in benthos and plankton of acidic, oligo-mesotrophic water bodies.
Distribution: cosmopolitan. In Europe, the nominate variety of *Std. convergens* is widely distributed and locally common, var. *deplanatus* is rather rare.
Relevant literature: Teiling 1967.

As for its cell dimensions, site of insertion of the spines, length of the spines and shape of the sinus, *Std. convergens* is a polymorphic taxon. Spineless forms may be easily confused with given *Cosmarium* species (in particular *C. depressum*). Ralfs (1848) depicted three, rather different forms under the name of *Arthrodesmus convergens*. Contrary to Teiling (1967) we like to select Ralfs' plate 20: 3b, representing the most common and widely distributed form, as type of the nominate variety of this species.

var. *convergens*   Plate 1: 1-11
Lillieroth 1950, p. 264
Basionym: *Arthrodesmus convergens* Ehrenberg ex Ralfs 1848, p. 118, pl. 20: 3
Semicells elliptic to semicircular in outline.

Accounted *Staurodesmus convergens* var. *convergens*:
Std. convergens var. *laportei* Teiling 1967
Std. convergens var. *pumilus* (Nordstedt) Teiling 1967
Std. convergens var. *ralfsii* Teiling 1967
Arthrodesmus convergens forma *curta* W.B. Turner 1892
A. convergens forma *exaltata* Cedergren 1926
A. convergens var. *incrassatus* Gutwiński 1892
A. convergens forma *inermius* (Jacobsen) Schmidle 1898a
A. convergens var. *xanthidioides* Grönblad 1920

var. **deplanatus** (Deflandre) Coesel et Meesters comb. nov.   Plate 1: 12-13
Basionym: *A. convergens* forma *deplanata* Deflandre 1926, p. 994, fig. 8
Semicells more or less cuneate in outline.

Accounted *Staurodesmus convergens* var. *deplanatus*:
Arthrodesmus convergens var. *deplanatus* (Deflandre) Laporte 1931

Taxa excluded:
Staurodesmus convergens var. *depressus* (Woloszynska) Teiling 1967 (= *Std. depressus*)
Std. convergens var. *wollei* (Irénée-Marie) Teiling 1967
Arthrodesmus convergens var. *divaricatus* Maskell 1889
A. convergens var. *mucronatus* Borge 1896
A. convergens var. *obesus* W. et G.S. West 1896

Staurodesmus corniculatus (P. Lundell) Teiling   Plate 16: 1-9
Teiling 1948, p. 76
Basionym: *Staurastrum corniculatum* P. Lundell 1871, p. 57, pl. 3: 23
Cells 3(-4)-radiate, about as long as broad or somewhat longer than broad, rather deeply constricted. Sinus widely open, acute-angled to subrectangled at the apex. Semicells (sub)cuneate
in shape with a (slightly) concave apex. Apical angles upwards produced, often ending in a short
spine but not seldom obtuse. Semicells in apical view usually triangular with about straight sides
and obtusely rounded angles often tipped with a minute spine. Zygospore unknown. Cell length
20-45 μm, cell breadth 20-35 μm.
Ecology: in benthos and tychoplankton of acidic, oligotrophic water bodies.
Distribution: cosmopolitan. In Europe, *Std. corniculatus* is of rare occurrence. Most of the records
originate from boreal-(sub)alpine regions.

Accounted *Staurodesmus corniculatus* (var. *corniculatus*):
Staurodesmus corniculatus var. *spinigerus* (W. West) Croasdale 1957
Staurastrum corniculatum var. *americanum* Scott et Grönblad 1957
S. corniculatum var. *pelagicum* Teiling 1946
S. corniculatum var. *spinigerum* W. West 1892a

Taxa excluded:
Staurodesmus corniculatus var. *subspinigerus* (Kurt Förster) Teiling 1967
Staurastrum corniculatum var. *australis* Raciborski 1892 (cf *Std. leptodermus*)
S. corniculatum forma *maior* Grönblad 1936 (cf *Std. leptodermus*)
S. corniculatum var. *pseudoconnatum* Playfair 1908
S. corniculatum var. *variabile* Nordstedt 1888a (cf *Std. leptodermus*)

Staurodesmus cuspidatus (Ralfs) Teiling 1948   Plate 11: 1-20
Teiling 1948, p. 60
Basionym: *Staurastrum cuspidatum* Brébisson ex Ralfs 1848, p. 122, pl. 21: 1, pl. 33: 10
Cells (2-)3(-4)-radiate, exclusive of spines about as long as broad or slightly longer than broad,
deeply constricted. Isthmus (highly) elongate. Semicell body (above the isthmus) elliptic to trian-
gular in outline, at the angles fairly abruptly passing into a (rather) long spine. Semicells in apical
view usually triangular with slightly to deeply concave sides, the angles ending in a (long) spine.
Zygospores globose, furnished with long, simple spines that are inflated at the base. Cell length
including spines 20-55 μm, cell breadth including spines 25-75 μm.
Ecology: in (tycho)plankton of oligo- to mesotrophic waters, both acidic and slightly alkaline.

Distribution: cosmopolitan. In Europe, *Std. cuspidatus* is widely distributed and locally common.

Following Teiling (1967), in literature given forms of *Std. cuspidatus* are often identified as *Std. mamillatus* (Nordstedt) Teiling 1967. In our opinion, however, *Staurastrum mamillatum* as originally described by Nordstedt (1870) from Brazil is characterized by distinctly capitate semicell angles and in its distribution confined to the American continents.

Accounted *Staurodesmus cuspidatus* (var. *cuspidatus*):
Std. cuspidatus forma *alaskanus* Croasdale 1957
Std. cuspidatus subsp. *constrictus* (G.M. Smith) Teiling 1948
Std. cuspidatus var. *divergens* (Nordstedt) Coesel 2007
Std. cuspidatus forma *granulatus* Compère 1967
Std. cuspidatus subsp. *tetragonus* (K.E. Hirn) Teiling 1948
Std. cuspidatus subsp. *tricuspidatus* (Brébisson) Teiling 1948
Staurastrum cuspidatum var. *acuminatum* Nygaard 1949
S. cuspidatum var. *alaskense* Irénée et Hilliard 1963
S. cuspidatum var. *canadense* G.M. Smith 1922
S. cuspidatum var. *compactum* Messikommer 1957
S. cuspidatum var. *coronulatum* Gutwiński 1890
S. cuspidatum var. *divergens* Nordstedt 1870
S. cuspidatum var. *elegans* Tarnogradsky 1960
S. cuspidatum forma *incurva* Heimerl 1891
S. cuspidatum var. *inflexum* Raciborski 1889
S. cuspidatum var. *maximum* W. West 1891a
S. cuspidatum var. *robustum* Messikommer 1928

Questionable non-European taxon:
Staurodesmus cuspidatus var. *longispinus* (Grönblad) Compère 1967

Taxa excluded:
Staurodesmus cuspidatus var. *groenbladii* Kurt Förster 1969 (cf *Sd. cuspidicurvatus*)
Std. cuspidatus var. *pseudogroenbladii* Kurt Förster 1974
Std. cuspidatus var. *subexcavatus* (W. et G.S. West) Kurt Förster 1969 (cf *Std. cuspidicurvatus*)
Staurastrum cuspidatum var. *columbianum* G.S. West 1914
S. cuspidatum var. *delpontei* Irénée-Marie 1952

Staurodesmus cuspidicurvatus Coesel et Meesters nom. nov.   Plate 12: 1-7
Synonym: *Staurastrum curvatum* W. West 1892, p. 172, pl. 13
Cells (2-)3(-4)-radiate, exclusive of spines about as long as broad or slightly longer than broad, deeply constricted. Isthmus (slightly) elongate. Semicell body (above the isthmus) triangular in outline, at the angles gradually passing into (rather) long, divergent spines. Semicells in apical view usually triangular with slightly to deeply concave sides, the angles ending in a (long) spine. Zygospore unknown. Cell length including spines 30-60 μm, cell breadth including spines 55-75 (-95) μm.
Ecology: in plankton of acidic, oligotrophic lakes.
Distribution: cosmopolitan? In Europe, *Std. cuspidicurvatus* is locally common in atlantic lakes but very rare in more central parts of the continent.

Std. cuspidicurvatus (= *S. curvatum* W. West) should not be confused with the non-European species *Std. curvatus* (W.B. Turner) Thomasson 1965 (invalidly published, so incorrectly represented under that name in Teiling 1967) that was originally described by Turner (1872) as *Arthrodesmus curvatus*.

Accounted *Staurodesmus cuspidicurvatus* (var. *cuspidicurvatus*):
Staurodesmus cuspidatus var. *curvatus* (W. West) Teiling 1967
Staurastrum curvatum forma *elliptica* Borge 1930
S. curvatum var. *elongatum* G.M. Smith 1924
S. curvatum var. *inflatum* E.M. Lind et Pearsall 1945

Taxa excluded:
Staurastrum curvatum forma *biradiata* Willi Krieger 1932
S. curvatum forma *brevispina* Nygaard 1949
S. curvatum var. *cruciatum* Willi Krieger 1932
S. curvatum var. *variabile* Skuja 1976 (= *Staurodesmus skujae* Croasdale et Flint 1994)

Staurodesmus dejectus (Ralfs) Teiling

Cells 3(-4)-radiate, about as long as broad exclusive of spines, deeply constricted. Isthmus (slightly) elongate. Semicell body (above the isthmus) bowl-shaped with an upwardly projected rather short spine at each apical angle. Semicells in apical view tri- or quadrangular with concave sides, the inflated angles abruptly passing into a (short) spine. Zygospores, known from var. *dejectus* and var. *apiculatus*, globose and furnished with stout thorn-like processes. Cell length including spines (15-)20-35(-40) μm, cell breadth including spines (15)20-35(-40) μm.
Ecology: in benthos and plankton of oligo-mesotrophic water bodies.
Distribution: cosmopolitan. In Europe, both the nominate variety and var. *apiculatus* are widely distributed and locally common; var. *robustus* is less common.
Relevant literature: Teiling 1967.

var. *dejectus*   Plate 17: 1-10
Teiling 1954, p. 128
Basionym: *Staurastrum dejectum* Brébisson ex Ralfs 1848, p. 121
Original figure: Teiling 1954, fig. 1 (after Brébisson in litt., as *Staurastrum dejectum*)
Semicell body with flat or only slightly convex apex. Apical spines projected obliquely upwards.

Accounted *Staurodesmus dejectus* var. *dejectus*:
Staurodesmus dejectus var. *brevispinus* (Nygaard) Coesel 1993

var. *apiculatus* (Brébisson) Croasdale   Plate 17: 11-15
Croasdale 1957, p. 128
Basionym: *Staurastrum apiculatum* Brébisson 1856, p. 142, pl. 1: 23
Synonym: *Staurastrum dejectum* var. *apiculatum* (Brébisson) P. Lundell 1871
Semicell body with flat or only slightly convex apex. Apical spines (almost) vertically directed or even recurved. Cells relatively small.

var. *robustus* (Messikommer) Coesel   Plate 17: 16-20
Coesel 1993, p. 110
Basionym: *Staurastrum cuspidatum* var. *robustum* Messikommer 1928, p. 208, pl. 8: 11
Semicell body with highly convex apex. Apical spines projected obliquely upwards. Cells relatively stout.

23

Accounted *Staurodesmus dejectus* var. *robustus*:
Staurodesmus dejectus var. *borealis* Croasdale 1965

Taxa excluded:
Staurastrum dejectum var. *biundulatum* Noda et Skvortzov 1969
S. dejectum var. *convergens* Wolle 1884
S. dejectum var. *debaryanum* Nordstedt 1889 (cf. *Std. glaber*)
S. dejectum var. *decumbens* Sampaio 1944
S. dejectum forma *falcatum* Kurt Förster 1964
S. dejectum var. *inflatum* W. West 1892 (cf *Std. patens*)
S. dejectum var. *latius* C. Bernard 1908
S. dejectum forma *longispina* Nygaard 1949 (cf *Std. omearae*)
S. dejectum forma *major* W. et G.S. West 1906
S. dejectum forma *mediocris* Nygaard 1949 (cf *Std. omearae*)
S. dejectum var. *mucronatum* (Ralfs ex Brébisson 1856) Wolle 1884 (= *Std. mucronatus*)
S. dejectum var. *patens* Nordstedt 1887 (= *Std. patens*)
S. dejectum var. *subapplanatum* Lobik 1916
S. dejectum var. *subglabrum* Grönblad 1920 (cf *Std. glaber*)
S. dejectum subsp. *tellamii* W. et G.S. West 1894
S. dejectum var. *triangulatum* Willi Krieger 1932

Staurodesmus depressus (Woloszynska) Coesel et Meesters stat. nov.   Plate 1: 14-16

Basionym: *Arthrodesmus convergens* var. *depressum* Woloszynska 1921, p. 137, fig. 13, 14
Synonym: *Staurodesmus convergens* var. *depressus* (Woloszynska) Teiling 1967
Cells 2-radiate, exclusive of spines broader than long, deeply constricted. Sinus widely open, acute-angled at the apex. Semicells broadly cuneate with slightly convex lateral sides. Apical angles ending in a stout, downwards projected spine. Semicells in apical view ellipsoid. Zygospore unknown. Cell length 25-32 µm, cell breadth including spines 53-62 µm.
Ecology: in tychoplankton of acidic, oligotrophic water.
Distribution: Europe. Only known from a few sites in Poland and Finland.

Staurodesmus dickiei (Ralfs) S. Lillieroth

Cells 3(-4)-radiate, exclusive of spines about as long as broad, deeply constricted. Sinus closed or slightly open at the apex, widely open to the extremity. Semicell body elliptic, semicircular or rhomboid in outline, with a shorter or longer spine at each of the poles/basal angles. Spines projected downwards and often curved. Semicells in apical view usually triangular with concave sides, the angles ending in a spine (spines sometimes bent sideways). Zygospores, known from the nominate variety, globose and furnished with simple spines. Cell length (15-)25-45(-55) µm, cell breadth including spines (20-)30-60(-80) µm.
Ecology: in benthos and plankton of acidic, oligo-mesotrophic waters.
Distribution: cosmopolitan. In Europe, both the nominate variety and var. *circularis* are widespread and locally common. Var. *rhomboideus* is rare, being predominantly known from some lakes in Scotland and Ireland.

Both in var. *dickiei* and var. *circularis* forms with a densely punctate cell wall have been described, sometimes as a separate variety or forma. As yet, the taxonomic relevance of this cell wall feature is questionable.

var. ***dickiei*** Plate 20: 1-13
Lillieroth 1950, p. 264
Basionym: *Staurastrum dickiei* Ralfs 1848, p. 123, pl. 21: 3
Semicells elliptic in outline, spines inserted at the poles.

Accounted *Staurodesmus dickiei* var. *dickiei*:
Staurodesmus dickiei var. *groenlandicus* (Børgesen) Teiling 1967
Std. dickiei var. *maximus* (W. et G.S. West) Thomasson 1963
Std. dickiei var. *vanoyei* Teiling 1967
Staurastrum dickiei var. *gedanense* P. Schulz 1922
S. dickiei forma *longispina* F.E. Fritsch et F. Rich 1937
S. dickiei var. *minutum* G.S. West 1914
S. dickiei var. *punctatum* W. West 1892

var. ***circularis*** (W.B. Turner) Croasdale Plate 21: 1- 4
Croasdale 1957, p. 130
Basionym: *Staurastrum dickiei* var. *circulare* W.B. Turner 1892, p. 105, pl. 16: 5
Semicells semicircular in outline, spines inserted at the basal angles.

Accounted *Staurodesmus dickiei* var. *circularis*:
Staurodesmus dickiei var. *microspinus* Hinode 1971

var. ***rhomboideus*** (W. et G.S. West) S. Lillieroth Plate 21: 5-6
Lillieroth 1950, p. 264
Basionym: *Staurastrum dickiei* var. *rhomboideum* W. et G.S. West 1903, p. 545, pl. 16: 9
Semicells rhomboid in outline, spines inserted at the lateral angles.

Accounted *Staurodesmus dickiei* var. *rhomboideus*:
Staurastrum dickiei var. *rhomboideum* forma *depressa* Irénée-Marie 1938

Questionable non-European taxon:
Staurastrum dickiei var. *polypyrenoideum* Kanetsuna 2002 (polyploid form?)

Taxa excluded:
Staurodesmus dickiei var. *bourrellyi* Teiling 1967
Std. dickiei var. *conicus* (W.B. Turner) Teiling 1967
Std. dickiei var. *crassicornutus* (Kurt Förster) Teiling 1967
Std. dickiei var. *curvispinus* (Irénée-Marie) Teiling 1967
Std. dickiei var. *denticulatus* (Nordstedt) Teiling 1967
Std. dickiei var. *galeatus* (W. et G.S. West) Teiling 1967
Std. dickiei var. *giganteus* (W. et G.S. West) Teiling 1967
Std. dickiei var. *productus* Kurt Förster 1972
Std. dickiei var. *willei* (Kurt Förster) Teiling 1967
Staurastrum dickiei var. *alaskense* Irénée-Marie et Dilliard 1963 (cf *Std. mucronatus*)
S. dickiei forma *isthmosa* Cosandey 1934
S. dickiei var. *latum* Messikommer 1960 (cf *Std. mucronatus*)
S. dickiei var. *longispinum* E. M. Lind et Pearsall 1945 (cf *Std. subulatus*)
S. dickiei var. *parallelum* Borge 1895 (cf *Std. mucronatus*)
S. dickiei forma *parva* Schmidle 1896 (cf *S. lanceolatum*)
S. dickiei var. *planum* Tarnogradsky 1960 (cf *Std. glaber*)

Staurodesmus extensus (Andersson)Teiling

Cells 2 (-3)-radiate, exclusive of spines as long as broad or slightly longer than broad, with a deep, widely open, rounded sinus. Isthmus distinctly elongate. Semicell body (above the isthmus) obversely trapeziform, the apical angles furnished with a shorter or longer spine. Semicells in apical view narrowly elliptic or triangular. Zygospores, known from var. *extensus* and var. *rectus*, globose and furnished with simple, solid spines gradually broadening towards their base. Cell length including spines (15-)20-40 µm, cell breadth including spines (15-)20-40(-50) µm.
Ecology: in benthos and plankton of acidic, oligo- or mesotrophic water bodies.
Distribution: cosmopolitan. In Europe, both var. *extensus*, var. *isthmosus* and var. *rectus* are widely distributed and locally common whereas var. *joshuae* is somewhat less common.

Of *Std. extensus* highly anomalous cell forms are known (e.g., Kouwets 1988). In addition to that, often small, hard to identify reduction forms are encountered (e.g., Coesel & Meesters 2007).

var. *extensus*   Plate 13: 1-5
Teiling 1948, p. 67
Basionym: *Arthrodesmus incus* var. *extensus* Andersson (=Borge) 1890, p. 13, fig. 7 (see also Borge 1913, p. 25, pl. 2: 23)
Spines shorter than the breadth of the semicell body, divergent, rather abruptly passing into the semicell body.

var. *isthmosus* (Heimerl) Coesel   Plate 13: 14-19
Coesel 1993, p. 110
Basionym: *Arthrodesmus incus* forma *isthmosa* Heimerl 1891, p. 603, pl. 5: 18
Spines shorter than the breadth of the semicell body, divergent to parallel, gradually passing into the semicell body.

var. *joshuae* (Gutwiński) Teiling   Plate 13: 20-24
Teiling 1967, p. 515
Basionym: *Arthrodesmus incus* forma *joshuae* ("Joshua'ii") Gutwiński 1892, p. 64, pl. 3: 6
Spines about as long as the breadth of the semicell body, convergent to parallel, abruptly passing into the semicell body.

For validation of variety name, see Coesel 1993.

var. *rectus* (B. Eichler et Raciborski) Coesel et Meesters stat. et comb. nov.   Plate 13: 6-13
Basionym: *Arthrodesmus incus* var. *vulgaris* forma *recta* B. Eichler et Raciborski 1893, p.120, pl. 3: 22, 24
Spines shorter than the breadth of the semicell body, parallel (rarely convergent), abruptly passing into the semicell body.

In literature this taxon usually is incorrrectly named *Std. extensus* var. *vulgaris* (B. Eichler et Raciborski) Croasdale 1957.

Taxa excluded:
Staurodesmus extensus var. *longispinus* (W. et G.S. West) Teiling 1967 (= *Std. longispinus*)
Std. extensus var. *malaccensis* (C. Bernard) Coesel 2007 (cf *Std. triangularis*)
Std. extensus var. *retusus* (Hirano) Tomaszewicz 1988 (cf *Std. ralfsii*)

Staurodesmus glaber (Ralfs) Teiling
Cells (2-)3(-4)-radiate, exclusive of spines about as long as broad with a deep, widely open sinus. Semicell body triangular with convergent to almost parallel spines at the apical angles. Semicells in apical view usually triangular with concave sides, the angles ending in a spine (spines sometimes bent sidewards). Zygospores, known from the nominate variety, globose and furnished with simple spines. Cell length (15-)20-25(-30), cell breadth including spines (25-)30-45(-90) μm.
Ecology: in benthos and plankton of acidic, oligo-mesotrophic waters.
Distribution: cosmopolitan. In Europe, the nominate variety is widespread and locally common, var. *hirundinella* and var. *limnophilus* are less common.
Literature: Teiling 1967.

var. **glaber**   Plate 9: 4-14
Teiling 1948, p. 69
Basionym: *Staurastrum glabrum* Ehrenberg ex Ralfs 1848, p. 217
Semicell body with a straight or slightly convex apex, spines distinctly convergent.

Accounted *Staurodesmus glaber* var. *glaber*:
Staurodesmus glabrus subsp. *brebissonii* (Raciborski) Teiling 1948
Std. glaber var. *flexispinum* (Kurt Förster) Teiling 1967

var. **hirundinella** (Messikommer) Teiling   Plate 10: 1-3
Teiling 1967, p. 559
Basionym: *Staurastrum glabrum* var. *hirundinella* Messikommer 1949, p. 246, pl. 1: 14
Semicell body with a deeply concave apex, spines distinctly convergent.

Accounted *Staurodesmus glaber* var. *hirundinella*:
Staurodesmus glaber var. *nauwerckii* Teiling 1967
Std. glabrus forma *subglabrus* (Grönblad) Teiling 1948

var. **limnophilus** Teiling   Plate 10: 4-8
Teiling 1967, p. 559
Basionym: *Staurodesmus glabrus* subsp. *brebissonii* forma *limnophilus* Teiling 1948, p. 69, fig. 18-24
Semicell body with a straight or convex apex; spines relatively long, convergent to almost parallel.

Questionable non-European taxon:
Staurodesmus glaber var. *recurvatus* Yacubson 1980

Taxa excluded:
Staurodesmus glaber var. *debaryanus* (Nordstedt) Teiling1967 (cf *Std. dickiei*)
Std. glabrus var. *ralfsii* (W. West) Teiling 1948 (= *Std. ralfsii*)
Staurastrum glabrum var. *incurvum* Skuja 1964 (= *Std. incurvus*)

Staurodesmus hebridarus (W. et G.S. West) Kurt Förster   Plate 15: 17-18
Förster 1974, p. 169
Basionym: *Arthrodesmus phimus* var. *hebridarum* W. et G.S. West 1912, p. 105, pl. 117: 22
Cells 2(-3)-radiate, exclusive of spines about as long as broad with a deep, about rectangled sinus. Semicell body triangular with a slight elevation at the middle of the apex and somewhat produced apical angles. Apical angles furnished with short, parallel or slightly divergent spines. Semicells in apical view elliptic, or triangular with about straight sides that are slightly inflated in the middle portion. Zygospore unknown. Cell length including spines 15-20 μm, cell breadth including spines 20-30 μm.

27

Ecology: in tychoplankton and benthos of acidic, oligotrophic water bodies.
Distribution: Europe, Brazil. As for Europe, only some records from Great Britain and Swedish Lappland are known.

Accounted *Staurodesmus hebridarus* (var. *hebridarus*):
Staurodesmus phimus var. *hebridarus* (W. et G.S. West) Teiling 1967

Taxon excluded:
Staurodesmus hebridarus var. *brasiliensibus* Kurt Förster 1974

Staurodesmus incus (Ralfs) Teiling

Cells 2-4-radiate, exclusive of spines (slightly) longer than broad with a deep, obtuse-angled or V-shaped, widely open sinus. Isthmus sometimes slightly elongate. Semicell body cup-shaped. Apical angles provided with stout spines. Semicells in apical view elliptic, triangular or quadrangular. Zygospores, known from the nominate variety, globose and furnished with simple spines. Cell length including spines (15-)25-60 µm, cell breadth including spines (25-)30-80 µm.
Ecology: in plankton of acidic, oligotrophic waters.
Distribution: Eurasia, American continents. In Europe, *Std. incus* has its main point of distribution in atlantic regions, particularly in the plankton of larger lakes, elsewhere it is of rare occurrence.
Literature: Teiling 1967.

var. *incus*   Plate 4: 1-9
Teiling 1967, p. 511
Basionym: *Arthrodesmus incus* Brébisson ex Ralfs 1848, p. 118, pl. 20: 4d
Cells usually 2-radiate. Semicells with about straight apex and straight or slightly convex lateral sides. Spines divergent or parallel.

The zygospore depicted in Lenzenweger (1991a) most likely does not refer to *Std. incus* but to *Std. omearae*.

Accounted *Staurodesmus incus* var. *incus*:
Arthrodesmus incus forma *perforata* Schmidle 1898
A. incus forma *scrobiculata* Schmidle 1898

var. *indentatus* (W. et G.S. West) Coesel et Meesters comb. nov.   Plate 4: 10-12
Basionym: *Arthrodesmus incus* var. *indentatus* W. et G.S. West 1912, p. 94, pl. 113: 20-24
Cells usually 2-radiate. Semicells with about straight apex and indented lateral sides making the semicell body trapeziform in its upper portion and cup-shaped in its basal portion. Spines divergent or parallel.

Accounted *Staurodesmus incus* var. *indentatus*:
Staurodesmus indentatus (W. et G.S. West) Teiling 1948
Std. indentatus subsp. *triradiatus* Teiling 1948

var. *jaculiferus* (W. West) Coesel et Meesters stat. et comb. nov.   Plate 5: 1-7
Basionym: *Staurastrum jaculiferum* W. West 1892a, p. 172, pl. 22: 14
Synonym: *Staurodesmus jaculiferus* (W. West) Teiling 1948, p. 58
Cells usually 3-radiate. Semicells with slightly convex apex and straight or slightly convex lateral sides. Spines divergent, very long.

Accounted *Staurodesmus incus* var. *jaculiferus*:
Std. jaculiferus var. *laticollis* Nygaard 1979
Std. jaculiferus var. *stroemii* Teiling 1967

Taxa excluded:
Staurodesmus incus var. *primigenius* Teiling 1967 (cf *Std. validus* var. *subincus*)
Std. incus var. *ralfsii* (W. West) Teiling 1967 (= *Std. ralfsii*)
Std. indentatus var. *rectangularis* (A.M. Scott et Prescott 1957) Teiling 1967 (cf *Std. triangularis*)
Arthrodesmus incus var. *huitfeldtii* Strøm 1921 (cf *Std. extensus*)
A. incus var. *intermedius* Wittrock 1869 (cf *Std. omearae*)
A. incus var. *malaccensis* Bernard 1909 (cf *Std. triangularis*)
A. incus forma *minor* W. et G.S. West 1912 (cf *Std. omearae*)
A. incus var. *pinguis* A.M. Scott et Grönblad 1957
A. incus var. *praelongus* G.M. Smith 1924
A. incus forma *quadrata* Schmidle 1896
A. incus var. *ralfsii* W. West 1892a (= *Std. ralfsii*)
A. incus var. *ralfsii* forma *latiuscula* W. et G.S. West 1912 (cf *Std. glaber*)
Staurastrum jaculiferum var. *excavatum* W. et G.S. West 1903 (cf *Std. triangularis*)
S. jaculiferum var. *subexcavatum* W. et G.S. West 1903 (cf *Std. cuspidicurvatus*)

Staurodesmus leptodermus (P. Lundell) Teiling   Plate 16: 10-13
Teiling 1948, p. 76
Basionym: *Staurastrum leptodermum* P. Lundell 1871, p. 58, pl. 3: 26
Cells (2-)3-radiate, about as long as broad or somewhat longer than broad, rather deeply constricted. Sinus widely open, acute-angled to subrectangled at the apex. Semicells triangular with
(slightly) convex apex and lateral sides. Apical angles upwardly produced, usually ending in a short
spine but not seldom obtuse. Semicells in apical view usually triangular with straight or convex
sides and obtusely rounded, slightly produced angles often tipped with a short spine. Zygospore
unknown. Cell length 30-60 μm, cell breadth 30-60 μm.
Ecology: in benthos and tychoplankton of acidic, oligotrophic water bodies.
Distribution: cosmopolitan. In Europe, *Std. leptodermus* is a rare species. Most of the records
originate from boreal-(sub)alpine regions.

Std. leptodermus is fairly polymorphic in both cell size and cell shape, especially regarding the formation of apical
spines. Dichotypical forms (one semicell with acute angles, the other one with obtuse angles) are regularly encountered (see, e.g., Ling & Tyler 2000). Spineless populations of *Std. leptodermus* may be confused with given *Staurastrum* species (e.g., *S. subpygmaeum* and *S. pachyrhynchum*). Infraspecific taxa characterized by a distinctly scrobiculate cell wall, such as var. *ikapoae* and var. *subcoronulatus*, are excluded from *Std. leptodermus*.

Accounted *Staurodesmus leptodermus* (var. *leptodermus*):
Staurastrum leptodermum var. *inerme* Lundberg 1931
S. leptodermum forma *minor* Eichler 1896
S. leptodermum var. *productum* Hinode 1960

Taxa excluded:
Staurodesmus leptodermus var. *americanus* (Scott et Grönblad) Teiling 1967 (cf *Std. corniculatus*)
Std. leptodermus var. *ikapoae* (Schmidle) Teiling 1967
Std. leptodermus var. *subcoronulatus* (Rich) Teiling 1967
Staurastrum leptodermum var. *alpinum* Noda et Skvortzov 1969
S. leptodermum var. *capitatum* Hirano in Yamaguchi & Hirano 1953 (cf *Std. corniculatus*)

S. leptodermum var. *compactum* Hinode 1960 (cf *Std. corniculatus*)
S. leptodermum var. *lefeburei* Laporte 1931

Staurodesmus longispinus (W. et G.S. West) Coesel et Meesters stat. et comb. nov.   Plate 5: 8-10
Basionym: *Arthrodesmus incus* var. *longispinum* W. et G.S. West 1905, p. 501, pl. 7: 22
Cells 2-radiate, exclusive of spines somewhat longer than broad with a deep, rounded, widely open sinus. Semicell body cup-shaped, abruptly widening from the slightly elongate isthmus. Spines long and strongly divergent. Semicells in apical view elliptic. Zygospore unknown. Cell length including spines 40-50 µm, cell breadh including spines 30-40 µm.
Ecology: in plankton of acidic, oligotrophic waters.
Distribution: Europe, North America. In Europe, *Std. longispinus* is a rare species with an atlantic distribution.

In the original description of *Arthrodesmus incus* var. *longispinum* in West & West (1905) not any reference is made to *Arthrodesmus incus* forma *longispina* B. Eichler et Raciborski 1893. Actually, there is but little resemblance between these two taxa. However, in the flora by West & West (1912: 96) not W. et G.S. West, but Eichler et Raciborski are presented as authors of var. *longispinus*. In our opinion, this is incorrect.

Staurodesmus megacanthus (P. Lundell) Thunmark
Cells 3(-4)-radiate, exclusive of spines about as broad as long or slightly broader than long, deeply constricted. Sinus widely open, acute-angled to subrectangled at the apex. Semicells elliptic to triangular in outline with (slightly) convex lateral sides. Lateral/apical angles ending in a stout, solid spine. Semicells in apical view tri- or quadrangular with concave sides, the angles ending in a stout spine. Zygospore unknown. Cell length including spines 30-60 µm, cell breadth including spines 50-100 µm.
Ecology: in (tycho)plankton of acidic, oligotrophic water bodies.
Distribution: Eurasia, American continents, Australia. In Europe, *Std. megacanthus* is a rare species, mainly confined to atlantic lakes.

Zygospore pictures of *Std. megacanthus* in Hegde & Bharati (1980, 1983) not only are inconsistent in themselves but likely refer to another species.

var. **megacanthus**   Plate 8: 1-6
Thunmark 1948, p. 686
Basionym: *Staurastrum megacanthum* P. Lundell 1871, p. 61, pl. 4: 1
Semicell body elliptic-rhomboid in outline, spines about parallel.

Accounted *Staurodesmus megacanthus* var. *megacanthus*:
Staurastrum megacanthum var. *tornense* Skuja 1964

var. **scoticus** (W. et G.S. West) S. Lillieroth   Plate 9: 1-3
Lillieroth 1950, p. 264
Basionym: *Staurastrum megacanthum* var. *scoticum* W. et G.S. West 1903, p. 544, pl. 16: 8
Semicell body triangular in outline, spines parallel or divergent.

Forms with divergent spines might be closer related to *Std. cuspidicurvatus* than to *Std. megacanthus.*

Questionable non-European taxa:
Staurodesmus megacanthus var. *orientalis* (A.M. Scott et Prescott) Teiling 1967 (cf var. *scoticus*)
Std. megacanthus var. *subcurvatus* (F. Rich 1935) Teiling 1967 (cf var. *scoticus*)
Staurastrum megacanthum var. *minus* Hirano 1948

Taxa excluded:
Staurodesmus megacanthus var. *kalimantanus* (A.M. Scott et Prescott) Teiling 1967
Std. megacanthus var. *triangularis* (Grönblad) Teiling 1967

Staurodesmus mucronatus (Brébisson) Croasdale
Cells (?2-)3-4-radiate, exclusive of spines somewhat broader than long to about as broad as long, deeply constricted. Sinus widely open from an acute-angled to almost rectangled apex. Semicell body elliptic, triangular or fusiform in outline with a short spine at each of the poles/apical angles. Spines about parallel.
Semicells in apical view usually triangular with concave sides. Zygospores, known from the nominate variety, globose and furnished with simple, thorn-like spines. Cell length 20-40 µm, cell breadth including spines 30-50 µm.
Ecology: in benthos and plankton of acidic, oligo-mesotrophic water bodies.
Distribution: cosmopolitan. In Europe, both the nominate variety and var. *subtriangularis* are widespread and locally common, var. *delicatulus* is more rare.
Relevant literature: Teiling 1967.

var. *mucronatus*   Plate 19: 1-5
Croasdale 1957, p. 132
Basionym: *Staurastrum mucronatum* Ralfs ex Brébisson 1856, p. 142
Original figure: Ralfs 1845, pl. 10: 5a (upper left, as *Staurastrum mucronatum*)
Semicell body elliptic in outline.

Accounted *Staurodesmus mucronatus* var. *mucronatus*:
Staurodesmus mucronatus var. *parallelus* (Nordstedt) Teiling 1967

var. *delicatulus* (G.S. West) Teiling   Plate 19: 6-8
Teiling 1967, p. 570
Basionym: *Staurastrum mucronatum* var. *delicatulum* G.S. West 1909, p. 66, pl. 5: 5
Semicell body fusiform.

Accounted *Staurodesmus mucronatus* var. *delicatulus*:
Staurastrum mucronatum var. *major* C.C. Jao 1949

var. *subtriangularis* (W. et G.S. West) Croasdale   Plate 19: 9-13
Croasdale 1957, p. 132
Basionym: *Staurastrum mucronatum* var. *subtriangulare* W. et G.S. West 1903, p. 545, pl. 17: 11
Semicell body triangular in outline.

Taxa excluded:
Staurodesmus mucronatus var. *croasdaleae* Teiling 1967 (cf *Std. megacanthus*)
Std. mucronatus var. *groenbladii* Teiling 1967
Staurastrum mucronatum forma *crassum* Grönblad 1964
S. mucronatum var. *debaryanum* (Jacobsen) Turner 1892
S. mucronatum var. *recta* Turner 1892

Staurodesmus omearae (W. Archer) Teiling   Plate 14: 5-22
Teiling 1948, p. 68
Basionym: *Staurastrum omearae* W. Archer 1858, p. 254 (as "*O'Mearii*"), pl. 21: 8-13
Cells 2-3(-4)-radiate, exclusive of spines about as long as broad or slightly longer than broad.
Sinus widely open from an obtuse-angled to about rectangled apex. Semicell body cuneate to cup-shaped in outline with (short) divergent spines at the apical angles. Semicells in apical view elliptic, tri- or quadrangular, the latter ones with about straight sides. Zygospores globose and furnished with simple spines. Cell length including spines 15-30 µm, cell breadth including spines 15-40 µm.
Ecology: in acidic, oligotrophic water bodies.
Distribution: cosmopolitan. In Europe, *Std. omearae* is widely distributed and locally very common.

In literature, forms with a relatively narrow isthmus and a less obtuse-angled sinus as reproduced on our Plate 14: 12 often are labeled *Std. spencerianus* (Maskell) Teiling (e.g., Kouwets 1987, Tomaszewicz 1988, Lenzenweger 1997). However, the original pictures of *Std. omearae* in Archer (1858) show both cells with a relatively broad and with a relatively narrow isthmus whereas the original pictures of *Std. spencerianus* (as *Staurastrum spencerianum*) in Maskell (1883, pl. 24: 12, Maskell 1889, p. 27) are rather heterogeneous as well, rendering Teiling's (1967) concept of *Std. spencerianus* a questionable one.

Accounted *Staurodesmus omearae* (var. *omearae*):
Staurodesmus omearae forma *inflatus* Nygaard 1979
Std. omearae var. *minutus* (W. West) Teiling 1967
Staurastrum omearae var. *johannensis* Irénée-Marie 1952

Taxon excluded:
Staurastrum omearae forma *parallela* Strøm 1926

Staurodesmus patens (Nordstedt) Croasdale
Cells 3(-4)-radiate, exclusive of spines about as long as broad, deeply constricted. Sinus widely open from an acute-angled apex. Semicell body bowl-shaped with a convex apex, the apical angles provided with an upwardly projected short spine. Semicells in apical view usually triangular with concave sides, the angles abruptly passing into a short spine. Zygospores, known from the nominate variety, globose and densely set with simple spines. Cell length including spines 15-40(-55?) µm, cell breadth including spines 20-55(-65?) µm.
Ecology: in (tycho)plankton of acidic, oligo-mesotrophic water bodies.
Distribution: cosmopolitan. In Europe, the nominate variety of *Std. patens* is widely distributed whereas var. *inflatus* is mainly confined to atlantic regions.

var. **patens**   Plate 18: 1-8
Croasdale 1957, p.134
Basionym: *Staurastrum dejectum* var. *patens* Nordstedt 1888a, p. 39, pl. 4: 16
Cell length up to 30 µm.

var. **inflatus** (W. West) Coesel et Meesters comb. nov.   Plate 18: 9-12
Basionym: *Staurastrum dejectum* var. *inflatum* W. West 1892, p. 170, pl. 22: 11
Cell length over 30 µm.

Taxon excluded:
Staurodesmus patens var. *maximus* Teiling 1967 (cf *Std. patens* var. *inflatus*)

Staurodesmus phimus (W.B. Turner) Thomasson   Plate 15: 9-14

Thomasson 1959, p. 75

Basionym: *Arthrodesmus phimus* W.B. Turner 1892, p. 136, pl. 12: 9

Cells 2-radiate, exclusive of spines about as long as broad with a deep, widely open, obtuse-angled sinus. Isthmus not seldom somewhat elongate. Semicell body boat-shaped, apex with a wide median indentation, Apical angles furnished with (short) divergent spines. Semicells in apical view elliptic. Zygospore unknown. Cell length including spines 14-27 µm, cell breadth including spines 18-30 µm.

Ecology: in oligotrophic, acidic water bodies.

Distribution: cosmopolitan? In Europe, *Std. phimus* is only rarely recorded, e.g., from Great Britain, France, Austria and Denmark.

The zygospore depicted in Bicudo & Ungaretti (1986) most likely does not refer to *Std. phimus* but to *Std. pterosporus*.

Accounted *Staurodesmus phimus* (var. *phimus*):
Staurodesmus. phimus var. *occidentalis* (W. et G.S. West) Teiling 1967
Std. phimus var. *robustus* Teiling 1967

Taxa excluded:
Staurodesmus phimus var. *convexus* (Prescott et A.M. Scott) Teiling 1967
Std. phimus var. *hebridarus* (W. et G.S. West) Teiling 1967 (= *Std. hebridarus*)
Std. phimus var. *menoides* (A.M. Scott et Prescott) Teiling 1967
Std. phimus var. *semilunaris* (Schmidle) Teiling 1967 (= *Std. semilunaris*)
Arthrodesmus phimus var. *koreana* Skvortsov 1932

Staurodesmus pterosporus (P. Lundell) Prescott   Plate 15: 1-8

Prescott 1966, p. 32

Basionym: *Staurastrum pterosporum* P. Lundell 1871, p. 60, pl. 3: 29

Cells (2-)3-radiate, exclusive of spines about as long as broad. Sinus widely open from an about rectangled apex. Semicell body cup- to bowl-shaped, usually with (slightly) mamillate apical angles. Angles abruptly attenuated into slightly divergent, rather short spines. Semicells in apical view usually triangular with about straight sides. Zygospores compressed-rectangular with produced angles following the outline of the enveloping, empty gametangial cells. Cell length including spines 15-25(-30?) µm, cell breadth including spines 15-25(-33?) µm.

Ecology: in acidic, oligo-mesotrophic water bodies.

Distribution: only known with certainty (zygospores!) from Europe, India, Caroline Island (Micronesia) and Brazil. In Europe records of sporulating cells are rare (Sweden, Great Britain, Brittany, Germany, Austria and the Netherlands). Presumably, however, *Std. pterosporus* is often overlooked because of possible confusion with *Std. omearae*.

Staurodesmus ralfsii (W. West) Tomaszewicz   Plate 6: 1-6

Tomaszewicz 1988, p. 62

Basionym: *Arthrodesmus ralfsii* W. West 1892a, p. 168

Original figures: Ralfs 1848, pl. 20: 4 e-h (as *Arthrodesmus incus*)

Cells 2-radiate, somewhat longer than broad exclusive of spines, with a moderately deep, obtuse-angled, widely open sinus. Isthmus often slightly elongate. Semicell body subquadrate to cuneate with straight or slightly concave apex and slightly concave to slightly convex lateral sides. Apical angles provided with parallel or convergent spines, seldom with divergent ones. Semicells in api-

cal view elliptic. Zygospores globose, furnished with simple spines. Cell length (15-)20-30 µm, cell breadth including spines (25-)30-50 µm.
Ecology: in tychoplankton of acidic, oligo-mesotrophic waters.
Distribution: Europe, North America. In Europe, *Std. ralfsii* is a rather rare species, with its main point of distribution in atlantic regions.
Relevant literature: Smith 1924.

Accounted *Staurodesmus ralfsii* (var. *ralfsii*):
Staurodesmus incus var. *ralfsii* (W. West) Teiling 1967
Arthrodesmus incus var. *ralfsii* (W. West) W. et G.S. West 1901a
A. ralfsii var. *extensus* Irénée-Marie 1852

Taxa excluded:
Arthrodesmus incus var. *ralfsii* forma *latiuscula* W. et G.S. West 1912 (cf *Std. glaber*)
A. ralfsii var. *brebissonii* (Raciborski) G. M. Smith 1924 (cf *Std. glaber*)
A. ralfsii var. *subhexagonum* (W. et G.S. West) Hirano 1968 (= *Std. subhexagonus*)

Staurodesmus recurvus (Skuja) Coesel et Meesters stat. et comb. nov.   Plate 10: 9
Basionym: *Staurastrum glabrum* var. *recurvum* Skuja 1964, p. 251, pl. 50: 9
Cells 3-radiate, about as long as broad, with a deep, widely open, undulate sinus. Semicells anvil-shaped with strongly convex apex, the basal angles furnished with vertically downwards projected, curved spines. Semicells in apical view triangular with concave sides, the broadly rounded angles as well as the somewhat laterally inserted spines clockwise bent. Zygospore unknown. Cell length 19-21 µm, cell breadth 19-21 µm.
Ecology: in benthos of acidic, oligotrophic pools.
Distribution: only known from a site in Swedish Lappland.

Staurodesmus semilunaris (Schmidle) Coesel et Meesters stat. et comb. nov.   Plate 15: 15-16
Basionym: *Arthrodesmus incus* forma *semilunaris* Schmidle 1896, p. 26, pl. 16: 9
Cells 2(-3?)-radiate, exclusive of spines about as long as broad with a deep, widely open, acute-angled sinus. Semicell body crescent. Apical angles furnished with divergent spines. Semicells in apical view elliptic. Zygospore unknown. Cell length including spines 23-30(-50) µm, cell breadth including spines 20-30(-40) µm.
Ecology: in tychoplankton of acidic, oligotrophic water bodies.
Distribution: Europe, North America, Australia. In Europe, *Std. semilunaris* is only known from Austria and the Azores.

The picture of *Staurastrum* cf *aristiferum* in Heimerl (1891) might refer to a triradiate form of *Std. semilunaris*.

Accounted *Staurodesmus semilunaris*:
Staurodesmus phimus var. *semilunaris* (Schmidle) Teiling 1967

Staurodesmus subhexagonus (W. et G.S. West) Coesel   Plate 13: 25-30
Coesel 1993, p. 110
Basionym: *Arthrodesmus incus* var. *ralfsii* forma *subhexagona* W. et G.S. West 1912, p. 96, pl. 114: 6
Cells 2-radiate, exclusive of spines about as long as broad with a deep, widely open, rounded sinus and an elongate isthmus. Semicell body (above the isthmus) hexagonal, the lateral angles furnished with short, parallel or convergent spines. Semicells in apical view elliptic. Zygospores globose, furnished with simple spines. Cell length 15-23 µm, cell breadth including spines 20-28 µm.
Ecology: in acidic, oligotrophic water bodies.

Distribution: Europe, North America. In Europe, *Std. subhexagonus* is a rare species, known from Great Britain, Sweden and the Netherlands.

Accounted *Staurodesmus subhexagonus*:
Staurodesmus triangularis var. *subhexagonus* (W. et G.S. West) Teiling 1967
Arthrodesmus ralfsii var. *subhexagonum* (W. et G.S. West) Hirano 1968

Staurodesmus subquadratus (W. et G.S. West) Coesel et Meesters stat. et comb. nov. Plate 3: 6-10
Basionym: *Arthrodesmus incus* var. *subquadratus* W. et G.S. West 1912, p. 97, pl. 114: 7
Cells 2-radiate, about one and a half times as long as broad exclusive of spines, with a shallow, widely open, obtuse-angled sinus. Semicell body subquadrate to cuneate with straight or slightly convex sides. Apical angles provided with divergent spines. Semicells in apical view elliptic. Zygospore unknown. Cell length including spines 17-26 µm, cell breadth including spines 12-22 µm.
Ecology: in acidic, oligotrophic water bodies.
Distribution: Europe, Africa? In Europe, *Std. subquadratus* is a rare species, known from England and Poland.

Contrary to Teiling (1967) we are not inclined to synonymize *Arthrodesmus incus* var. *subquadratus* W. et G.S. West 1912 with *A. incus* forma *quadrata* Schmidle 1896. Schmidle's figure (1896, pl. 16: 10) does not show the slightest indication of a sinus rendering its desmid identity somewhat doubtful.

Staurodesmus subtriangularis (Borge) Teiling
Cells 2(-3)-radiate, somewhat longer than broad exclusive of spines, with a deep, widely open sinus. Semicell body triangular-rhomboid with slightly convex lateral sides. Apical angles provided with long, about parallel spines. Apex distinctly elevated, usually with a median indentation. Semicells in apical view elliptic, rarely triangular. Zygospore unknown. Cell length including spines 25-40 µm, cell breadth including spines 55-105 µm.
Ecology: in plankton of acidic, oligotrophic lakes.
Distribution: Europe, North america. In Europe, both the nominate variety and var. *inflatus* are only known from Great Britain.

var. *subtriangularis*   Plate 6: 7-12
Teiling 1948, p. 64
Basionym: *Arthrodesmus incus* var. *subtriangularis* Borge 1897, p. 212, pl. 3: 4
Lateral sides of semicell body indented. Sinus V-shaped.

Accounted *Staurodesmus subtriangularis* var. *subtriangularis*:
Staurodesmus subtriangularis var. *robustus* (W. et G.S. West) Kurt Förster 1974

var. *inflatus* (W. et G.S. West) Teiling   Plate 6: 13-16
Teiling 1967, p. 523
Basionym: *Arthrodesmus triangularis* var. *inflatus* W. et G.S. West 1898, p. 320
Original figure: W. West 1892, pl. 24: 19 (as *Arthrodesmus triangularis* forma)
Lateral sides of semicell body not indented. Sinus rounded, giving rise to an elongate isthmus.

Staurodesmus subulatus (Kützing) Thomasson
Cells 2(-3)-radiate, exclusive of spines about as long as broad, deeply constricted. Sinus widely open from an acute-angled sinus. Semicells cuneate-ellipsoid-rhomboid with (slightly) convex apex and lateral sides, at the apical angles/poles with (rather) long, about parallel spines. Semicells in apical view ellipsoid, rarely triangular. Zygospores (only known of var. *nordstedtii*) globose

and smooth-walled. Cell length including spines 30-55 μm, cell breadth including spines 50-100 (-130) mm.
Ecology: in plankton of acidic, oligotrophic lakes.
Distribution: all continents, particularly in (sub)tropical regions. In Europe, both var. *subulatus* and var. *nordstedtii* are rare and almost exclusively known from atlantic regions.
Literature: Teiling 1967.

var. *subulatus*   Plate 2: 1-6
Thomasson 1957a, p. 17
Basionym: *Arthrodesmus subulatus* Kützing 1849, p. 176
Original figure: Bailey 1841, pl. 1: 12 (as *Euastrum*)
Semicell body elliptic-rhomboid in outline, spines laterally inserted.

Accounted *Staurodesmus subulatus* var. *subulatus*:
Staurodesmus subulatus var. *americanus* (A.M. Scott et Grönblad) Teiling 1967
Std. subulatus var. *nygaardii* Teiling 1967
Std. subulatus var. *rhomboides* (Hirano) Teiling 1967
Std. subulatus var. *subaequalis* (W. et G.S. West) Thomasson 1960
Arthrodesmus subulatus var. *subaequalis* W. et G.S. West 1912

var. *nordstedtii* (G.M. Smith) Thomasson   Plate 2: 7-11
Thomasson 1957a, p. 17
Basionym: *Arthrodesmus subulatus* var. *nordstedtii* G. M. Smith 1924, p. 127, pl. 85: 1-3
Semicell body cuneate in outline, spines apically inserted.

Accounted *Staurodesmus subulatus* var. *nordstedtii*:
Staurodesmus subulatus forma *major* (Nordstedt) Thomasson 1966
Arthrodesmus subulatus forma *incrassatus* A.M. Scott et Prescott 1958
A. subulatus forma *triquetra* F. Rich 1935

Taxa excluded:
Staurodesmus subulatus var. *depressus* Nygaard 1979 (cf *Std. triangularis*)
Std. subulatus var. *orbicularis* (Wolle) Teiling 1967
Arthrodesmus subulatus var. *americana* (W.B. Turner) W. et G.S. West 1892 (cf *Std. subtriangularis* var. *inflatus*)
A. subulatus var. *gracilis* Joshua 1885
A. subulatus var. *validus* (W. et G.S. West) F.E. Frich et F. Rich 1937 (= *Std. validus*)

Staurodesmus triangularis (Lagerheim) Teiling
Cells 2(-3)-radiate, about as long as broad or slightly longer than broad exclusive of spines, with a deep, widely open sinus. Isthmus usually (slightly) elongate. Semicell body above the isthmus triangular. Apex slightly elevated with a shallow median depression. Apical angles provided with about parallel (rarely convergent or divergent) spines. Semicells in apical view elliptic, more rarely triangular. Zygospores, known from the nominate variety, globose and furnished with simple spines. Cell length including spines (12-)20-30 μm, cell breadth including spines 25-70 μm.
Ecology: in acidic, oligo-mesotrophic water bodies.
Distribution: cosmopolitan. In Europe, the nominate variety has its main point of distribution in euplankton of atlantic, oligotrophic lakes whereas var. *brevispina* and var. *indentatus*, mainly occurring in tychoplankton of mesotrophic water bodies, are more widely distributed.

var. ***triangularis*** Plate 7: 1-6

Teiling 1948, p. 62

Basionym: *Arthrodesmus triangularis* Lagerheim 1885, p. 244, pl. 27: 22

Spines longer than ⅔ breadth of semicell body. Lateral sides of semicell body straight to slightly convex.

Accounted *Staurodesmus triangularis* var. *triangularis*:

Staurodesmus triangularis forma *longispinus* Teiling 1948

var. ***brevispina*** (V. et P. Allorge) Coesel et Meesters comb. nov. Plate 7: 7-13

Basionym: *Arthrodesmus triangularis* var. *brevispina* V. et P. Allorge 1931, p. 363, pl. 10: 42-44

Spines shorter than ⅔ breadth of semicell body. Lateral sides of semicell body straight to slightly convex.

Accounted *Staurodesmus triangularis* var. *brevispina*:

Staurodesmus triangularis var. *convergens* Thomasson 1959

Std. triangularis forma *rotundatus* (Raciborski) Teiling 1948

Arthrodesmus triangularis forma *lagerheimii* Gutwiński 1892

A. triangularis var. *minuta* Gauthier-Lièvre 1931

A. triangularis forma *triquetra* W. et G.S. West 1912

var. ***indentatus*** Coesel et Meesters var. nov. Plate 7: 14-17

Diagnosis: differs from the nominate variety by spines longer than ⅔ breadth of semicell body. Semicell body with lateral sides that, about half-way, are (slightly) indented.

Type: Coesel 1994, plate 11: 10 (as *Staurodesmus triangularis* var. *subparallelus*)

Taxa excluded:

Staurodesmus triangularis var. *acuminatus* (Hirano) Teiling 1967

Std. triangularis var. *americanus* (W.B. Turner) Teiling 1967 (cf *Std. subtriangularis* var. *inflatus*)

Std. triangularis forma *curvispina* Thomasson 1963

Std. triangularis var. *latus* (Irénée-Marie) Teiling 1967 (cf *Std. megacanthus*)

Std. triangularis var. *limneticus* Teiling 1967 (cf *Std. incus* var. *jaculiferus*)

Std. triangularis subsp. *stellatus* Teiling 1948 (cf *Std. incus* var. *jaculiferus*)

Std. triangularis subsp. *stroemii* Teiling1948 (cf *Std. incus* var. *jaculiferus*)

Std. triangularis var. *subhexagonus* (W. et G.S. West) Teiling 1967 (= *Std. subhexagonus*)

Std. triangularis var. *subparallelus* (G.M. Smith) Teiling 1948 (cf *Std. incus*)

Arthrodesmus triangularis var. *alaskanum* Hirano 1968

A. triangularis var. *hebridarum* W. et G.S. West 1903 (cf *Std. subtriangularis*)

A. triangularis var. *inflatus* W. et G.S. West 1898 (cf *Std. subtriangularis*)

A. triangularis var. *latiusculum* (W. et G.S. West) Hirano 1957

A. triangularis var. *rectangularis* A.M. Scott et Grönblad 1957

A. triangularis var. *simplex* Skuja 1934

A. triangularis var. *subtriangularis* (Borge) W. et G.S. West 1905 (= *Std. subtriangularis*)

Staurodesmus unguiferus (W.B. Turner) Thomasson Plate 15: 19-23

Thomasson 1966, p. 29

Basionym: *Staurastrum unguiferum* W.B. Turner 1892, p. 130, pl. 15: 18

Cells 3-radiate, longer than broad, moderately constricted. Sinus widely open, obtuse-angled or V-shaped. Semicell outline like a cup with convex sides and a deeply concave apex. Radial processes of the semicells lunate in shape, projected in vertical direction and smoothly passing into the

semicell body. Apical angles usually furnished with a minute spine, sometimes spineless (obtusely rounded). Semicells in apical view triangular with slightly concave to slightly convex sides and rounded angles with an intramarginal minute spine. Zygospore unknown. Cell length including spines 41-80 µm, cell breadth including spines 20-42 µm.
Ecology: in benthos and tychoplankton of acidic, oligo-mesotrophic water bodies.
Distribution: Eurasia. In Europe, *Std. unguiferus* is a rare species, almost exclusively known from Finland.

Accounted *Staurodesmus unguiferus* (var. *unguiferus*):
Staurodesmus unguiferus var. *pseudoamericanus* (Grönblad) Teiling 1967
Staurastrum unguiferum var. *extensum* Grönblad 1938
S. unguiferum var. *inerme* (W.B. Turner) W. et G.S. West 1907
S. unguiferum forma *major* W.B. Turner 1892

Taxa excluded:
Staurodesmus unguiferus var. *brasiliensis* (Grönblad) Teiling 1967
Std. unguiferus var. *corniculatus* (F.E. Fritsch et F. Rich) Thomasson 1966
Staurastrum unguiferum var. *prespanse* Petkoff 1910

Staurodesmus validus (W. et G.S. West) Thomasson

Cells 2-radiate, exclusive of spines about as long as broad or slightly longer than broad, deeply constricted. Sinus widely open from an acute-angled apex. Semicells bowl-shaped to obversely trapeziform with an almost straight apex, the apical angles furnished with long, strongly divergent spines. Semicell body in apical view elliptic. Zygospores, known from the nominate variety, globose and smooth-walled. Cell length including spines 40-110 µm, cell breadth including spines 35-75 µm.
Ecology: in (tycho)plankton of acidic, oligotrophic water bodies.
Distribution: mainly pantropical. In Europe, the nominate variety of *Std. validus* is very rare (Scotland, Ireland), var. *subincus* is somewhat less rare but also mainly confined to atlantic regions.

var. *validus*   Plate 3: 11-12
Thomasson 1960, p. 35
Basionym: *Arthrodesmus incus* var. *validus* W. et G.S. West 1898, p. 320, pl. 17: 16
Cell length including spines > 60 µm, cell breadth including spines > 55 µm.

Accounted *Staurodesmus validus* var. *validus*:
Staurodesmus validus var. *subvalidus* (Grönblad) Teiling 1967

var. *subincus* (W. et G.S. West) Coesel et Meesters comb. nov.   Plate 3: 13-17
Basionym: *Arthrodesmus bulnheimii* var. *subincus* W. et G.S. West 1912, p. 105, pl. 116: 3
Cell length including spines < 60 µm, cell breadth including spines < 55 µm.

Taxa excluded:
Staurodesmus validus var. *apertus* (Scott et Grönblad) Teiling 1967
Std. validus var. *sinuosus* (Børgesen) Teiling 1967

Staurodesmus wandae (Raciborski) Willi Krieger et Bourrelly   Plate 14: 1-4

Krieger & Bourrelly 1956, p. 163
Basionym: *Staurastrum wandae* Raciborski 1889, p. 100, pl. 7: 9
Cells (?3-)4-10-radiate, somewhat longer than broad exclusive of spines, moderately constricted.

Sinus widely open from an obtuse-angled apex. Semicells cuneate with straight to slightly convex apex and lateral sides, the apical angles furnished with a (rather) short, upwardly projected spine. Semicells in apical view usually 5-angular with straight to slightly concave sides . Zygospore elliptic-oval, smooth-walled. Cell length including spines 25-40 μm, cell breadth including spines 20-30 μm.

Ecology: in acidic, oligotrophic water bodies.

Distribution: Eurasia, South America. In Europe, *Std wandae* is a rare species, known from Lithuania, Finland and Austria.

Accounted *Staurodesmus wandae* (var. *wandae*):

Staurodesmus wandae var. *brevispinus* (Grönblad) Willi Krieger et Bourrelly 1956

Std. wandae var. *pseudopterosporus* (Förster) Teiling 1967

Taxon excluded:

Staurodesmus wandae var. *longissimus* (Borge) Teiling 1967

Genus *Staurastrum*

Usually, *Staurastrum paradoxum* Meyen 1828, being the only *Staurastrum* species dealt with by Meyen (1828) is considered the type species of this genus. However, the name of *S. paradoxum* Meyen 1828 was not validly published as it appeared before the starting point of desmid nomenclature (Ralfs 1848). Ralfs (l.c.) validated a number of species names in the genus *Staurastrum*, among them, *S. paradoxum*, but he did not explicitly designate a type species of this genus. Since we consider *Staurastrum paradoxum* as diagnosed in Ralfs (l.c.) to be relatively well defined, we herewith formally designate it the type species of the genus in question.

Type species of genus *Staurastrum* Meyen 1828 ex Ralfs 1848:
Staurastrum paradoxum Meyen 1828 ex Ralfs 1848, British Desmidieae: 138, pl. 23, fig. 8.
Type specimen: *Staurastrum paradoxum* Meyen 1828 ex Ralfs 1848, British Desmidieae: 138, pl. 23, fig. 8a.

Diagnosis of genus:
Staurastrum Meyen 1828 ex Ralfs 1848
Cells solitary, bi- to pluriradiate. Biradiate cells are provided with hollow, arm-like processes. Tri- to pluriradiate cells may lack such processes. Cell wall either smooth or furnished with granules or spines. Chloroplasts generally axial with radiating lobes or ridges, in some large-sized species parietal.

Basically, the genus *Staurastrum* is distinguished by the radial symmetry of the cells as shown in apical view. Traditionally, all desmid species characterized by solitary cells that are tri- to pluriradiate are accounted the genus *Staurastrum,* as is the case with all biradiate species marked by hollow, arm-like processes. Actually, *Staurastrum* includes species of more varied morphology than any other genus of desmids. Cell length may range from less than ten to far over a hundred micrometres and also the shape of the semicell body differs (e.g., subcircular, semicircular, elliptic, obversely triangular, campanulate, trapeziform, rectangular). In species characterized by arm-like processes, the length of those processes appears to be related to the habitat: plankton inhabitants having longer arms than benthic representatives of the same species (compare, e.g., Pl. 120, figs 6 and 1)

Valuation of diagnostic features in species identification

The general appearance of a given cell observed will give a first indication of its taxonomic classification. A reasonable number of species are so characteristic in their morphology that they are recognized immediately. In the vast majority of species, however, identification is not that easy. One of the main tools is found in the degree of radiation (number of radii, planes of symmetry through the longitudinal cell axis) but this criterion is in no way absolute. One has always to take into account, in certain species, of a radiation that deviates from the one generally observed. Also the criterion of cell dimensions is rather elastic. Both dwarf and giant forms (whether or not with increased ploidy level) are to be expected, with cell dimensions both above and below those provided in identification manuals. One of the most essential features is provided by cell wall sculpturing. A sharp boundary appears to exist between smooth-walled species and those with patterns of granules, verrucae and spines, although it should be noticed that certain smooth-walled species characterized by marked scrobiculae (pits in the cell wall) may present a seemingly granular appearance. The distribution pattern of any cell wall ornamentation holds as a relatively good taxonomic characteristic; less reliable is the expression of this ornamentation. A number of species show many kinds of transition (whether genotypically or phenotypically determined) between granules, verrucae, dentations and spines (see, e.g., Nygaard 1945, fig. 45, or Brook 1959, pl. 5: 1-6). The outline of the semicell body in combination with the shape of the sinus, on the other hand, are considered more relevant parameters. Also the diameter (cross section) of possible arm-like processes in relation to both the size of the semicell body and size of the terminal spines comes to the fore as a reliable diagnostic characteristic.

Morphological groups (sections) within *Staurastrum*

The statement in the flora by West & West (1912: 119) that 'all attempts to split up this genus on natural principles have entirely failed' unfortunately still stands. It is true, some authors tried to reduce the heterogeneity of the genus by splitting it into a number of genera, sub-genera or sections (e.g., Palamar-Mordvintseva 1976, 1976a, Hirano 1959, 1960) but, except for Teiling's (1948) genus *Staurodesmus*, none of those taxa have been accepted by the majority of desmidiologists. In practice, it appears impossible to draw strict lines of delimitation between those smaller taxonomic entities, so in the present flora they will not be referred to. In want of satisfactory insight into real (genetic) relationships, the system of (unnamed) sections handled in the present flora is purely pragmatic, focussed on ready species identification.

Morphological terminology in *Staurastrum*

One of the most relevant diagnostic features in identification of *Staurastrum* species is the shape of the semicell body. Some main types (in stylized form) are given (Fig. 5). Also the sinus form is often a distinctive feature (Fig. 4).

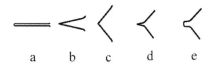

Fig. 4. Sinus forms: a = closed/linear; b = acute-angled; c = obtuse-angled; d = V-shaped; e = U-shaped.

41

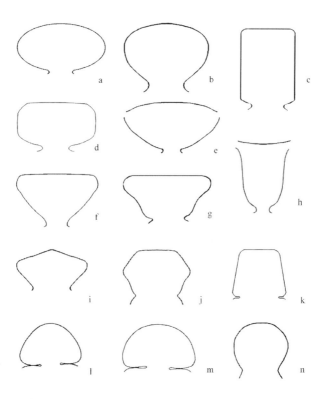

Fig. 5. Semicell body shapes *Staurastrum*: a = elliptic; b = cuneate; c = cylindric; d = rectangular; e = bowl-shaped; f = cup-shaped (triangular); g = anvil-shaped; h = vase-shaped; i = rhomboid; j = hexagonal; k = trapeziform; l = pyramidal; m = semicircular; n = subcircular.

Usually, *Staurastrum* cells show themselves in a stereotypic position. Biradiate cells almost always will be observed in frontal view (which is identical to the rear view). Triradiate cells will usually be seen in frontal view (that means resting on two radii, the third radius being directed to the observer). In rear view (resting on but one radius) a triradiate cell shows its pattern of cell wall ornamentation somewhat better, but its outline is about the same as in frontal view. Quadriradiate cells, on the contrary, have quite a different appearance depending on whether they rest on one or on two radii. When resting on one radius the cell in question will look more slender (relatively long processes) than when resting on two radii (see, e.g., Pl. 43: 7 and 8). Diagnoses of quadriradiate species given in the present flora are based on cells resting on one radius, to be called 'position 1'. In cells of penta- to pluriradiate species the exact position does not affect their appearance much.

Key to the species

Unless explicitly stated otherwise, descriptions refer to cells in frontal view (face-1 position) and to cell dimensions inclusive of processes and ornamentation.

1	Cell wall, apart from possible papillate lobes or furcate processes, quite smooth	Section 1, p. 43
-	Cell wall to a greater or lesser degree beset with granules or spines	2
2	Angles of semicells produced to form arm-like processes	Section 5, p. 51
-	No arm-like processes to be distinguished	3
3	Cell wall furnished with both granules and spines (unequal in size)	Section 4, p. 50
-	Cell wall furnished with either spines or granules which are about equal in size	4
4	Cell wall exclusively furnished with granules	Section 3, p. 48
-	Cell wall exclusively furnished with spines	Section 2, p. 46

Section 1
Cell wall quite smooth

1	Semicells with elongate, arm-like lobes/processes	2
-	Semicells without such processes	14
2	Semicell lobes bud-like swollen at the end	*S. bacillare*
-	Semicell lobes not swollen at the end	3
3	Semicell lobes entire at the end	4
-	Semicell lobes usually bifid or trifid at the end	6
4	Apex of semicell body distinctly arched	*S. subnudibranchiatum*
-	Apex of semicell body deeply concave to about straight	5
5	Isthmus elongate	*S. sublaevispinum*
-	Isthmus not elongate	*S. laevispinum*
6	Cells with one armlike process per radius (top view!)	7
-	Cells with two or more processes per radius	9
7	Apex of semicell body distinctly arched	*S. subnudibranchiatum*
-	Apex of semicell body deeply concave to about straight	8
8	Semicells about (obversely) triangular in outline	*S. laevispinum*
-	Semicells about boat-shaped	*S. brachiatum*
9	Semicells each provided with but one whorl of processes (all processes inserted at the same semicell level)	10
-	Semicells each provided with two superimposed whorls of processes: a median whorl and an apical one	11

10	Semicells in apical view with two processes symmetrically on either side of each radius	*S. laeve*
-	Semicells in apical view per radius with one process lining and one process aside	*S. clevei*
11	The median and the apical whorl with an equal number of processes: one pair at each of the angles	*S. gemelliparum*
-	Number of processes in the median whorl different from that in the apical whorl	12
12	Median whorl with three, apical whorl with two processes per semicell angle	*S. hantzschii*
-	Median whorl usually with one, apical whorl with two processes per semicell angle	13
13	Cells generally longer than 40 µm (including processes), provided with relatively stout processes which usually are tipped with three spines	*S. tohopekaligense*
-	Cells less than 40 µm in length, processes tipped with but two spines or even reduced to simple spines	*S. furcatum*
14	(1) Semicells with a papilla at each of their radii	15
-	Semicells quite smooth-lined	20
15	Cells remarkably large (length > 80 µm)	16
-	Cells not that large (length < 70 µm)	17
16	Cells longer than broad	*S. tumidum*
-	Cells broader than long	*S. conspicuum*
17	Cells small (length < 30 µm)	18
-	Cells not that small (length > 30 µm)	19
18	Cells distinctly longer than broad	*S. julicum*
-	Cells about as long as broad	*S. lanceolatum*
19	Semicells about triangular in outline	*S. aversum*
-	Semicells elliptic to bowl-shaped	*S. brevispina*
20	(14) Semicells with more or less acute lateral angles	21
-	Semicell angles broadly rounded	27
21	Cells small (length < 30 µm)	22
-	Cells not that small (length usually > 30 µm)	23
22	Semicells obversely triangular in outline	*S. sibiricum*
-	Semicells elliptic-lanceolate in outline	*S. lanceolatum*
23	Semicells triangular-rhomboid in outline	24
-	Semicells elliptic-pyramidal in outline	26
24	Cells with elongate isthmus	*S. inelegans*
-	Isthmus not distinctly elongate	25
25	Semicells (obversely) triangular in outline	*S. clepsydra*
-	Semicells rhomboid in outline	*S. angulatum*
26	Cells slightly longer than broad, cell wall at the lateral angles somewhat thickened	*S. crassangulatum*
-	Cells about as long as broad or slightly broader than long, lateral angles usually somewhat tipped up	*S. bieneanum*

27	(20) Semicell angles more or less produced or with thickened wall	28
-	Semicell angles not produced or thickened	40
28	Cells 5-8 radiate	29
-	Cells at most 4-radiate	30
29	Semicells rhomboid in outline	*S. insigne*
-	Semicells hexagonal in outline	*S. habeebense*
30	Semicells large (Br > 50 µm), about trapeziform	*S. keuruense*
-	Semicells less large, shaped otherwise	31
31	Cells generally longer than 30 µm	32
-	Cells less than 30 µm in length	36
32	Cell sinus linear and (almost) closed for a greater or lesser part	*S. crassangulatum*
-	Cell sinus open from its apex	33
33	Lateral angles truncate	*S. obscurum*
-	Lateral angles broadly rounded	34
34	Cells longer than broad, with rounded, V-shaped sinus	*S. bayernense*
-	Cells about as long as broad, with acute-angled sinus	35
35	Semicells in apical view with concave sides, angles with thickened wall	*S. pachyrhynchum*
-	Semicells in apical view with convex sides, angles not thickened	*S. subpygmaeum*
36	(31) Sinus very shallow, obtuse-angled	37
-	Sinus deeper, acute-angled or V-shaped	38
37	Cells 2-3-radiate, conspicuously twisted at the isthmus	*S. tortum*
-	Cells 3-5-radiate, not distinctly twisted	*S. minutissimum*
38	Semicells about ellipsoid with strongly produced lateral angles	*S. crassimamillatum*
-	Semicells shaped otherwise	39
39	Semicells obversely trapezoid in rough outline	*S. groenbladii*
-	Semicells campanulate in rough outline	*S. schroederi*
40	(27) Sinus shallow, breadth of isthmus > ½ cell breadth	41
-	Sinus deeper, breadth of isthmus < ½ cell breadth	44
41	Semicells cuneate in outline	42
-	Semicells subcircular or subrhomboid in outline	43
42	Cells 2-3-radiate, conspicuously twisted at the isthmus	*S. tortum*
-	Cells 3-5-radiate, not distinctly twisted	*S. minutissimum*
43	Semicells subcircular in outline	*S. subsphaericum*
-	Semicells subrhomboid in outline	*S. thomassonii*
44	Semicells about oval in outline	*S. coarctatum*
-	Semicells shaped otherwise	45
45	Semicells elliptic in outline	46
-	Semicells shaped otherwise	48

46	Cells large (> 50 μm in length)	*S. grande*
-	Cells not that large	47
47	Semicells but little broader than high (so almost subcircular)	*S. ellipticum*
-	Semicells distinctly broader than high	*S. myrdalense*
48	Sinus linear	50
-	Sinus widening from its apex	49
49	Semicells subelliptic to semicircular	*S. muticum*
-	Semicells about trapeziform	*S. retusum* var. *hians*
50	Cells about quadrangular in outline	51
-	Cells shaped otherwise	52
51	Cells more than 15 μm in length and breadth	*S. quadratulum*
-	Cells less than 15 μm in length and breadth	*S. pokljukense*
52	Semicells semi-elliptic in outline	53
-	Semicells shaped otherwise	54
53	Sinus shallow, breadth of isthmus at least ½ cell breadth	*S. cosmarioides*
-	Sinus deeper, breadth of isthmus less than ½ cell breadth	*S. extensum*
54	Semicells trapeziform	55
-	Semicells subsemicircular or pyramidal in outline	56
55	Cells large (L > 35 μm)	*S. hibernicum*
-	Cells small (L < 35 μm)	*S. retusum*
56	Semicells subsemicircular	*S. orbiculare*
-	Semicells pyramidal	57
57	Semicell apex convex	*S. ralfsii*
-	Semicell apex slightly retuse	*S. suborbiculare*

Section 2
No arm-like processes, cell wall exclusively furnished with spines

1	Spines confined to the distal part of the semicell lobes (apical view!)	2
-	Spines also occurring on the semicell body	12
2	Semicell apex provided with a series of humps	*S. gatniense*
-	Semicell apex firm-lined	3
3	As a rule, just two spines per semicell lobe	4
-	More than two spines per semicell lobe	7
4	Spines lying in the same horizontal plane	*S. bifidum*
-	Spines lying in the same vertical plane	5
5	Cells large (L > 50 μm) with stout spines	6
-	Cells relatively small (L < 45 μm) with minute spines	*S. bispiniferum*
6	Spines at each of the apical angles running about parallel	*S. longispinum*
-	Spines diverging	*S. wildemanii*

7	(3) Just three spines per semicell lobe	8
-	More than three spines per semicell lobe	9
8	Cells large (L > 50 μm)	*S. brasiliense*
-	Cells not that large (L < 30 μm)	*S. besseri*
9	Cells large (L > 55 μm)	*S. brasiliense*
-	Cells not that large (L < 55 μm)	10
10	Cells longer than broad	*S. quadrispinatum*
-	Cells about as long as broad	11
11	Semicells oblong-oval in outline	*S. hystrix*
-	Semicells trapezoid-rectangular in outline	*S. quadrangulare*
12	(1) Spines ocurring at the apex of the semicell body but not on its flanks	13
-	Spines occurring all over the semicell body	14
13	Semicells beset with a limited number (i.e., 12) of relatively long spines	*S. kanitzii*
-	Semicells beset with an indefinite number (> 50) of short spines	*S. horametrum*
14	Spines relatively long (larger ones at least 5 μm) and far apart	15
-	Spines short (≤ 5 μm) and close together	20
15	Semicells with one stout spine at each angle, for the rest with many short spines	*S. ungeri*
-	Semicell spines not partitioned in that way	16
16	Spines near the angles distinctly longer than those elsewhere on the semicell	*S. setigerum*
-	Semicell spines about equal in length	17
17	Each semicell lobe with three pairs of superimposed spines	*S. geminatum*
-	Semicell spines not arranged in such a specific pattern	18
18	Semicells about trapeziform in outline	*S. subbrebissonii*
-	Semicells elliptic-oval in outline	19
19	Cells (including spines) generally more than 60 μm long	*S. polytrichum*
-	Cells generally less than 60 μm long	*S. teliferum*
20	(14) Semicell spines arranged in three distinct, transversal series	*S. echinodermum*
-	Spines equally distributed over the semicell surface	21
21	Semicells ellipsoid-rhomboid in outline	22
-	Semicells trapezoid-semicircular in outline	24
22	Cells less than 30 μm in length	*S. erostellum*
-	Cells more than 30 μm in length	23
23	Spines short (< 3 μm) and rather tight-packed	*S. brebissonii*
-	Spines longer (3-5 μm) and further apart	*S. kouwetsii*
24	Cells at least as broad as long	*S. trapezioides*
-	Cells (slightly) longer han broad	25

25	Cells generally more than 60 μm in length, beset with thick, conical spines	*S. pyramidatum*
-	Cells generally less than 60 μm in length, beset with fine spines or acute granules	26
26	Cell wall beset with fine spines	*S. hirsutum*
-	Cell wall beset with acute granules or dentations rather than with true spines	*S. arnellii*

Section 3
No arm-like processes, cell wall exclusively furnished with (acute) granules

1	Cells but little constricted, about cylindric in shape	2
-	Cells deeply constricted, not cylindric in shape	13
2	Sinus a very shallow, V- or U-shaped incision	3
-	Sinus somewhat deeper, linear or acute-angled open	10
3	Ornamentation in the form of small, rounded granules, evenly distributed over the cell wall	*S. meriani*
-	Ornamentation in the form of variously shaped granules and verrucae, unevenly distributed over the cell wall	4
4	Semicells hexagonal in outline	*S. mutilatum*
-	Semicells rectangular to campanulate in outline	5
5	Cells subcircular in apical view	*S. rhabdophorum*
-	Cells triangular in apical view	6
6	Cell wall granules in the apical part of the semicell rather evenly distributed	7
-	Cell wall granules in the apical part of the semicell arranged in ribbon-shaped series	9
7	Semicells with retuse apex	*S. pileolatum*
-	Semicells with straight to convex apex	8
8	Granules in the apical part of the semicell simple and acute	*S. capitulum*
-	Granules in the apical part of the semicell flattened and usually merged into compound verrucae	*S. spetsbergense*
9	Semicells in apical view with straight sides, intramarginal granules/verrucae in series about parallel with the margins	*S. bifasciatum*
-	Semicells in apical view with slightly convex sides, intramarginal granules/verrucae in a hoof-shaped configuration along by each angle	*S. borgei*
10	Semicells hexagonal in outline	11
-	Semicells trapezoid or semi-elliptic in outline	12
11	Lateral angles broadly rounded	*S. polonicum*
-	Lateral angles somewhat acuminate or produced	*S. tristichum*

12	Semicells trapezoid with a U-shaped incision in the lateral sides	*S. acarides*
-	Semicells semi-elliptic with entire sides	*S. alpinum*
13	(1) Semicells rectangular with a single granule at each angle, otherwise smooth-walled	*S. kobelianum*
-	Semicells provided with (many) more granules	14
14	Semicells semicircular in outline	15
-	Semicells shaped otherwise	16
15	Sinus widely open, basal semicell angles curved downwards	*S. novae-semliae*
-	Sinus linear, no curved basal angles	*S. maamense*
16	Semicells trapeziform-pyramidal in outline	17
-	Semicells shaped otherwise	24
17	Cell wall granules arranged in marked patterns	18
-	Granules rather evenly distributed over the cell wall	19
18	Granules confined to the edges of the semicell lobes	*S. maamense*
-	Granules also occurring on the flanks of the semicells	*S. acarides*
19	Granules acute, often emarginate	20
-	Granules rounded	22
20	Cells about as long as broad	*S. scabrum*
-	Cells distinctly longer than broad	21
21	Granules all simple in shape	*S. hirsutum*
-	Granules partly emarginate or doubled	*S. arnellii*
22	Cells less than 35 μm in length	*S. donardense*
-	Cells longer than 35 μm	23
23	Cell wall coarsely granulate	*S. ricklii*
-	Cell wall finely granulate	*S. botrophilum*
24	(16) Semicells about cuneate in outline	25
-	Semicells shaped otherwise	26
25	Cell wall granules confined to the terminal parts of the semicell lobes	*S. horametrum*
-	Cell wall granules distributed over the greater part of the semicell	*S. asperum*
26	Semicells about elliptic in outline	27
-	Semicells shaped otherwise	29
27	Cells generally less than 30 μm in length	*S. alternans*
-	Cells generally longer than 30 μm	28
28	Semicells elliptic-subrhomboid	*S. lapponicum*
-	Semicells elliptic-oval	*S. turgescens*
29	Semicells about anvil-shaped	30
-	Semicells shaped otherwise	33
30	Semicell lobes more or less gradually passing into the cylindric basal part of the semicell body	31

-	Semicell lobes rather abruptly passing into the cylindric basal part of the semicell body	32
31	Cells 3-radiate, generally less than 30 μm in length	*S. alternans*
-	Cells generally 4-5-radiate and longer than 30 μm	*S. dilatatum*
32	Cell wall granules confined to the terminal parts of the semicell lobes	*S. sinense*
-	Cell wall granules distributed over the greater part of the semicell	*S. striolatum*
33	(29) Semicells ovoid-subcircular in outline	*S. punctulatoides*
-	Semicells subrhomboid in outline	34
34	Lateral angles broadly rounded	*S. punctulatum*
-	Lateral angles acute, (slightly) truncate or produced	35
35	Lateral angles acute or somewhat produced	36
-	Lateral angles (slightly) truncate	38
36	Sinus very widely open, obtuse-angled	*S. tristichum*
-	Sinus less widely open, acute-angled	37
37	Median part of the semicell apex distinctly domed	*S. trachytithophorum*
-	Semicell apex rather evenly curved	*S. acutum*
38	(35) Dorsal side of the semicell more convex than the ventral side	*S. dispar*
-	Dorsal and ventral side of the semicell about equally curved	*S. striatum*

Section 4
No arm-like processes, cell wall furnished with both granules and spines

1	Spines confined to the extreme angles	2
-	Spines also occurring elsewhere on the cell wall	3
2	At most two spines per angle	*S. avicula*
-	At least three (minute) spines per angle	*S. bohlinianum*
3	Semicells in rough outline cup-shaped, bowl-shaped or cuneate (so greatest breadth near the apex)	4
-	Semicells shaped otherwise	7
4	Semicell angles slightly arm-like extended	5
-	Semicell angles shaped otherwise	6
5	Semicell body (semi)elliptic to bowl-shaped	*S. arcuatum*
-	Semicell body cup-shaped	*S. dicroceros*
6	Apical spines remarkably long, strongly differing in size with the granules elsewhere on the cell wall	*S. pungens*
-	Apical spines not that conspicuous, often gradually fading into the cell wall granules	*S. cristatum*
7	(3) Semicells in rough outline trapeziform	8
-	Semicells in rough outline elliptic to semicircular	13

8	Semicell surface about equally set with stout furcate spines	*S. spongiosum*
-	Spines at the semicell angles distinctly larger than those on the flanks (if present at all)	9
9	Lateral sides of the semicells marked by a regular series of verrucae in combination with a huge spine at the base	*S. cornutum* var. *skujae*
-	Cell wall ornamentation not that peculiar	10
10	Semicell angles gradually attenuated into serrate spines	*S. megalonotum*
-	Semicell angles rather abruptly passing into the spines. Spines simple or furcate but not serrate	11
11	Semicell angles slightly arm-like extended	12
-	Semicell angles not extended	*S. monticulosum*
12	Each semicell lobe provided with one lateral and two apical spines. Cell wall furthermore smooth or at best with a few granules	*S. furcatum*
-	Cell wall set with a higher number of spines	*S. forficulatum*
13	(7) Outline of the semicell with a regular series of flattened verrucae and a marked spine at the base	14
-	Cell wall ornamentation not that peculiar	15
14	Basal spine very stout, often bifurcate and projected upwards	*S. cornutum*
-	Basal spine less stout, simple and projected downwards	*S. oxyrhynchum*
15	Semicell surface for the most part covered with spines and granules	16
-	Cell wall sculpturing centred round the semicell lobes	17
16	Cells relatively small, less than 30 µm in breadth	*S. simonyi*
-	Cells at least 30 µm in breadth	*S. echinatum*
17	Each semicell lobe provided with one lateral and two apical spines. Cell wall furthermore smooth or at best with a few granules	*S. furcatum* var. *aciculiferum*
-	Cell wall set with a higher number of (reduced) spines	18
18	Semicells with quite a number of stout, furcate spines	*S. forficulatum*
-	Spines for the most part reduced to acute granules	*S. podlachicum*

Section 5
Arm-like processes, cell wall furnished with granules and/or spines

1	Processes of semicells in two different planes, i.e., one apical and one median series of processes	2
-	Processes of semicells all in the same horizontal plane	5
2	Semicells with a median series of three (rarely four) processes	3
-	Semicells with a median series of at least five processes	4

3	Cells usually longer than 50 μm, with relatively short, granulate processes	*S. furcigerum*
-	Cells shorter than 50 μm, with slender processes that are smooth-walled for the greater part	*S. pseudopisciforme*
4	Apical and median processes fuse to a single series of processes before passing into the semicell body	*S. sexangulare*
-	Apical processes up to their base apart from the median processes	*S. arctiscon*
5	(1) Cells as a rule at least 5-radiate	6
-	Cells as a rule at most 4-radiate	16
6	Cells remarkably large, broader than 100 μm	7
-	Cells not that large	9
7	Processes divergent	8
-	Processes convergent to parallel	*S. ophiura*
8	Semicell body cup-shaped	*S. archeri*
-	Semicell body subcylindric	*S. verticillatum*
9	(6) Semicell body about rectangular with a basal ring of big verrucae	*S. eichleri*
-	Semicell body shaped otherwise	10
10	Cells distinctly broader than long	11
-	Cells about as long as broad or longer than broad	12
11	Processes slender, spider-like curved	*S. arachne*
-	Processes not very slender, about straight	*S. platycerum*
12	Cells stout, usually longer than 40 μm, ornamented with conical granules	*S. sexcostatum*
-	Cells less than 40 μm in length, ornamented with fine granules	13
13	Semicell body campanulate	*S. margaritaceum*
-	Semicell body elliptic to cup-shaped	14
14	Processes at their base with a pair of short accessory processes (apical view!)	*S. pertyanum*
-	Processes without accessory processes	15
15	Processes strikingly thick-set	*S. glaronense*
-	Processes more slender	*S. pentasterias*
16	(5) Cells as a rule 2-radiate	17
-	Cells as a rule 3-4-radiate	41
17	Semicell body about cylindric (whether or not inflated) at the base	18
-	Semicell body diverging from its base	26
18	Semicell body in its centre, left and right of the vertical axis, with a stout, granulate wart	*S. dimazum*
-	Semicell body without such an ornamentation	19

19	Semicell body (somewhat) longer than broad (measured at the base)	20
-	Semicell body broader than long	25
20	Processes gradually and steadily attenuating towards the end, semicell body in apical view fusiform	21
-	Processes about equal in diameter for most of their length, semicell body in apical view elliptic, circular or rhomboid	23
21	Processes (slightly) convergent, furnished with concentric series of granules	*S. bicorne*
-	Processes (slightly) divergent, furnished with (indistinctly) spiralling series of denticulations	22
22	Semicell body distinctly inflated at the base	*S. johnsonii*
-	Semicell body hardly or not inflated at the base	*S. reductum*
23	Processes slender, distinctly longer than the semicell body in cross section	24
-	Processes hardly longer than the semicell body in cross section	*S. duacense*
24	Semicell body longer than broad, with convex apex	*S. leptocladum*
-	Semicell body about as long as broad, with straight or slightly concave apex	*S. multinodulosum*
25	(19) Semicell body with a supraisthmial series of compound, dentate verrucae	*S. uhtuense*
-	Semicell body without such an ornamentation	*S. tetracerum*
26	(17) Semicel apex abruptly elevated, with a stout spine at each angle	*S. miedzyrzecense*
-	Semicell apex not distinctly elevated	27
27	Semicell centre abruptly inflated and ornamented with a circle of granules	28
-	Semicell centre not strikingly inflated	29
28	Cells more than 40 µm in length and breadth. Semicell body cup-shaped	*S. natator*
	Cells generally less than 40 µm in length and breadth. Semicell body bowl-shaped to rectangular	*S. tetracerum* var. *irregulare*
29	Semicell body longer than broad, with convex apex	*S. leptocladum*
-	Semicell body not longer than broad, with about straight or concave apex	30
30	Semicell body about as long as broad, cup-shaped	31
-	Semicell body broader than long, subrectangular to lunate	36
31	Processes at their end strikingly bifurcate	*S. bloklandiae*
-	Processes not bifurcate at their end	32
32	Processes with some 4 distinct spines at their end	33
-	Processes terminating in a number of minute spines or teeth	34

33	Face of semicell body in frontal view with 1 or 2 horizontal rows of acute granules	*S. iversenii*
-	Face of semicell body without ornamentation	*S. bullardii*
34	Semicell body ornamented with emarginate verrucae	35
-	Semicell body at best ornamented with fine granules	*S. chaetoceras*
35	Emarginate verrucae only occurring at the apex of the semicell	*S. multinodulosum*
-	Emarginate verrucae also occurring elsewhere on the semicell body	*S. levanderi*
36	(30) Sinus strikingly U-shaped	*S. lenzenwegeri*
-	Sinus shaped otherwise	37
37	Cells longer than 60 µm, processes terminating in rather stout spines	*S. nygaardii*
-	Cells usually less long than 60 µm, processes terminating in minute spines	38
38	Sinus acute-angled or V-shaped	39
-	Sinus obtuse-angled	40
39	Cells distinctly broader than long. Semicells with excavated apex	*S. subexcavatum*
	Cells about as long as broad. Semicell apex whether or not excavated	*S. tetracerum*
40	Semicells with but two processes	*S. smithii*
-	Semicells usually with four processes	*S. bibrachiatum*
41	(16) Cells strikingly elongate, L/Br > 1.5	*S. elongatum*
-	Cells less elongate	42
42	Semicell body (without processes) longer than broad, cylindric or vase-shaped	43
-	Semicell body at most as long as broad, ellipsoid, cup- or bowl-shaped	52
43	Isthmus (seemingly) elongate	44
-	Isthmus not distinctly elongate	45
44	Processes about half-way with an abrupt kink	*S. inconspicuum*
-	Processes not abruptly bent	*S. neglectum*
45	Semicell body elongate, about twice as long as broad	*S. bulbosum*
-	Semicell body not that elongate	46
46	Processes of semicells strikingly convergent (almost touching each other at their end), gradually passing into the semicell body	*S. cerastes*
-	Processes not that convergent	47
47	Processes remarkably short and thick-set, slightly convergent	*S. margaritaceum*
-	Processes longer, about parallel or divergent	48
48	Processes about parallel (sometimes slightly convergent)	*S. manfeldtii*
	Processes divergent	49

49	Processes furnished with concentric series of equally sized granules	*S. cingulum*
-	Processes furnished with series of unequally sized dents, verrucae or spines	50
50	Semicell body completely smooth-walled	*S. longipes*
-	Semicell body at least at the apex with emarginate verrucae	51
51	Semicell body at the apex with a double transversal series of emarginate verrucae	*S. johnsonii*
-	Semicell body at the apex without such a double series of verrucae	*S. pingue*
52	(42) Cell wall ornamentation, apart from possible spines at the end of the processes, in the form of equally sized granules (or minute dentations) aranged in concentric series around the processes and from there continuing onto part of the semicell body	53
-	Cell wall ornamentation either in the form of unequally sized and unequally distributed granules and/or spines, or the pattern of cell wall ornamentation (in empty cells) is not clearly visible	76
53	Semicell body ellipsoid to fusiform, processes inserted in its median part	54
-	Semicell body cup-shaped or cuneate, processes inserted near the apex	59
54	Cells distinctly broader than long	55
-	Cells hardly or not broader than long	56
55	Cells relatively large, broader than 40 µm	*S. dybowskii*
-	Cells small, at best 35 µm broad	*S. haaboeliense*
56	Cells rather stout, longer than 35 µm, with a supraisthmial whorl of granules	*S. proboscideum*
-	Cells small, less than 35 µm, without such a whorl	57
57	Processes, if occurring at all, extremely short	*S. bohlinianum*
-	Processes, although sometimes being very short, well to be distinguished	58
58	Processes tipped with minute dentations	*S. hexacerum*
-	Processes tipped with distinct spines	*S. polymorphum*
59	(53) Processes distinctly divergent	60
-	Processes about parallel or convergent	67
60	Processes distinctly bi- or trifurcate at their end	61
	Processes tipped with a number of (rather) small spines	63
61	Cells relatively small, less than 35 µm in length	*S. subcruciatum*
-	Cells rather large, over 35 µm in length	62
62	Processes very short and thick-set	*S. pelagicum*
	Processes relatively slender	*S. pseudopelagicum*

63	Processes very slender, length > 10x cross section	*S. chaetoceras*
-	Processes not that slender	64
64	Semicell body scarcely granulate	65
-	Semicell body densely granulate, at the base with a supraisthmial whorl of granules	66
65	Processes tipped with minute dents	*S. pseudotetracerum*
-	Processes tipped with spines	*S. paradoxum*
66	Cells relatively large, over 50 µm in breadth	*S. cingulum* var. *obesum*
-	Cells smaller, less than 50 µm in breadth	*S. boreale*
67	(59) Cells remarkably stout (L > 60 µm), furnished with coarse, conical granules	*S. petsamoense*
-	Cells not that stout (L < 60 µm)	68
68	Processes distinctly convergent	*S. cyrtocerum*
-	Processes at best slightly convergent	69
69	Cells longer than broad	*S. margaritaceum*
-	Cells at best about as long as broad	70
70	Cells broader than long	71
-	Cells about as long as broad	72
71	Cells relatively large, L > 30 µm	*S. gracile*
-	Cells smaller	*S. boreale*
72	Processes, if to be distinguished at all, very short and thick-set	73
-	Processes not that short; slightly convergent	75
73	Angles of the semicells with short dents	*S. bohlinianum*
-	Angles of the semicells with distinct spines	74
74	Semicell body with scarse, distant granules	*S. paradoxoides*
-	Semicell body with rather densely set granules	*S. polymorphum*
75	Semicell body relatively gradually passing into the (thick-set) processes	*S. borgeanum*
-	Semicell body rather abruptly passing into the (more slender) processes	*S. subnivale*
76	(52) Semicell body elliptic-fusiform, processes inserted in the median part	77
-	Semicell body sub-elliptic, cup-shaped, cuneate or about rectangular, processes generally inserted near the apex	80
77	Cells relatively large, B > 60 µm	*S. anatinum*
-	Cells small, B < 45 µm	78
78	Cell wall ornamentation in the form of granules or (at the apex) emarginate verrucae	*S. crenulatum*
-	Cell wall at least partly furnished with spines	79
79	Semicells in apical view with two short, bifurcate, diverging spines at the base of each process	*S. suchlandtianum*
-	Semicells without such bifurcate spines at the base of their processes	*S. heimerlianum*

80	(76) Sinus U-shaped, semicell body about rectangular	81
-	Sinus V-shaped or acute-angled, semicell body shaped otherwise	83
81	Cell wall hardly or not ornamented	*S. inconspicuum*
-	Cell wall furnished with acute granules and/or spines	82
82	Width of isthmus more than 1/4 of total cell width	*S. chavesii*
-	Width of isthmus less than 1/4 of total cell width	*S. dentatum*
83	Semicell angles produced to form pairs of processes lying in the same horizontal plane	*S. barbaricum*
-	Semicell angles produced to form singular processes	84
84	Cells slender, processes distinctly longer than the width of the semicell body	85
-	Cells not that slender	90
85	Cells strikingly small, at most 20 μm in length inclusive of processes	*S. minimum*
-	Cells larger	86
86	Processes more or less parallel or slightly divergent	87
-	Processes distinctly divergent	88
87	Semicell apex elevated and smooth-walled	*S. platycerum*
-	Semicell apex flattened and ornamented with emarginate verrucae	*S. anatinum*
88	Processes tipped with small dents or minute spines	*S. chaetoceras*
-	Processes tipped with distinct, divergent spines	89
89	Apex of the semicell body with two emarginate verrucae	*S. bullardii*
-	Apex of the semicell body smooth-walled	*S. longipes*
90	(84) Cells small, both length and breadth ≤ 30 μm	91
-	Cells larger	94
91	Processes somewhat dilated at their end and tipped with distinct spines	92
-	Processes not dilated, tipped with minute dents	93
92	Semicell body bowl-shaped	*S. micron*
-	Semicell body cup-shaped	*S. micronoides*
93	Cells very small, at best 20 μm in length	*S. minimum*
-	Cells not that small	*S. pseudotetracerum*
94	(90) Processes only ornamented at their top and their base, otherwise smooth-walled	95
-	Processes ornamented over most part of their length	96
95	Processes slender, furcate at the top	*S. subboergesenii*
-	Processes thick-set, tipped with a whorl of small granules	*S. alandicum*
96	Processes curved upwards or recurved downwards	97
-	Processes not curved in that way	102
97	Processes evenly curved upwards	*S. informe*
-	Processes recurved	98

98	Processes, where abruptly bent, with a couple of stout spines	*S. diacanthum*
-	Processes not as such	99
99	Processes from a broad base gradually attenuating to the tip	*S. magdalenae*
-	Processes over most of their length hardly changing in thickness	100
100	Processes furnished with fine, acute granules	*S. saltator*
-	Processes furnished with spines	101
101	Cells large, more than 60 μm broad	*S. anatinum*
-	Cells smaller, less than 60 μm broad	*S. floriferum*
102	(96) Processes on their dorsal side, about halfway, furnished with a couple of extra large spines	*S. diacanthum*
-	Processes not as such	103
103	Processes convergent	104
-	Processes divergent or about straight	109
104	Cell wall only furnished with granules and verrucae	*S. traunsteineri*
-	Cell wall (also) furnished with (emarginate) spines	105
105	Semicells in apical view with processes that are all bent in one direction (clockwise) and twisted in their length as well (seemingly asymmetric ornamentation)	*S. controversum*
-	Processes not bent and twisted as such	106
106	Cells stout, L > 60 μm	*S. sebaldi*
-	Cells moderate in size, L < 60 μm	107
107	Processes thick-set, usually masked by stout spines	*S. aculeatum*
-	Processes more slender, well to be distinguished	108
108	Semicell body furnished with a transverse series of verrucae/spines in its median part	*S. vestitum*
-	Semicell body without such an ornamentation	*S. oxyacanthum*
109	(103) Semicells in apical view with processes that are all bent in one direction (clockwise) and twisted in their length as well (seemingly asymmetric ornamentation)	*S. controversum* var. *semivestitum*
-	Processes not bent and twisted as such	110
110	Processes thick-set, usually masked by stout spines	*S. aculeatum*
-	Processes more slender, well to be distinguished	111
111	Cells more than 60 μm broad	112
-	Cells less than 60 μm broad	113
112	Cells more than 75 μm broad	*S. anatinum*
-	Cells less than 75 μm broad	*S. subosceolense*
113	Semicell body in apical view with a pair of stout, intramarginal spines projecting on each side	*S. oxyacanthum* var. *sibiricum*
-	Semicell body without such apical spines	*S. floriferum*

Staurastrum acarides Nordstedt

Plate 59: 5-8

Nordstedt 1872, p. 40, pl. 7: 26

Cells longer than broad, slightly to moderately constricted. Sinus linear, more or less dilate at its apex. Semicells trapezoid in outline, with a slightly retuse apex and slightly convex lateral sides, basal angles almost rectangled, apical angles often more rounded. Lateral sides about half-way with a U-shaped incision so that a definite apical region is delimited. Cell wall provided with granulate or denticulate verrucae, arranged in irregular, more or less concentric series around the semicell lobes. Semicells in apical view 3-angular with almost straight sides and bluntly rounded angles. Zygospore unknown. Cell length 37-50 µm, cell breadth 26-36 µm.

Ecology: subatmophytic on moist rocks and other wet substrates.

Distribution: arctic-alpine species, only known from Eurasia, North America and Antarctica. In Europe it is locally not rare in arctic regions, incidentally it has also been reported from the Alps and montane areas in Great Britain.

Accounted *Staurastrum acarides* (var. *acarides*):
Staurastrum acarides var. *eboracense* W. West 1889a

Questionable non-European taxon:
Staurastrum acarides var. *caucasicum* Woronichin 1926

Taxa excluded:
Staurastrum acarides var. *dyscritum* Scott et Grönblad 1957
S. acarides var. *hexagonum* W. West 1889

Staurastrum aculeatum (Ehrenberg) Ralfs

Plate 83: 1-7

Meneghini ex Ralfs 1848, p. 142, pl. 23: 2
Basionym: *Desmidium aculeatum* Ehrenberg 1838, p. 143, pl. 10: 12

Cells slightly broader than long, deeply constricted. Sinus widely open and acute-angled. Semicells subelliptical or subfusiform, the lateral angles produced to form (very) short, stout, slightly converging to almost horizontal processes. Processes tipped with 3-4 strong spines and also towards the semicell body furnished with (bifurcate) denticulations and strong spines, arranged in indistinct concentric series. Semicell body furnished with two transverse series of stout (bifurcate) spines: one series on the apex and one series in the median part. Semicells in apical view 3-4(-5)-radiate with about straight sides, the angles produced to form (very) short processes. Along the sides a marginal and an intramarginal series of stout (bifurcate) spines. Zygospore reported to be globose, furnished with long spines which are bifurcate at the apex. Cell length (33-)40-60 µm, cell breadth (48-)55-75(-85) µm.

Ecology: in benthos and tychoplankton of acidic, oligotrophic water bodies.

Distribution: Eurasia, North America. In Europe, *S. aculeatum* is widely distributed but usually not common.

The only zygospore description is by Lundell (1871) but without any illustration.

Accounted *Staurastrum aculeatum* (var. *aculeatum*):
Staurastrum aculeatum var. *aquitanicum* Capdevielle 1978
S. aculeatum var. *braunii* Reinsch 1867
S. aculeatum var. *japonicum* Hirano 1950

Taxa excluded:
Staurastrum aculeatum var. *bifidum* Schmidle 1898 (cf *S. forficulatum*)

S. aculeatum subsp. *cosmospinosum* Børgesen 1889 (= *S. echinatum*)
S. aculeatum var. *depauperatum* Wille 1879
S. aculeatum var. *intermedium* Wille 1881 (cf *S. controversum*)
S. aculeatum var. *ornatum* Nordstedt 1872 [= *S. sexcostatum* var. *ornatum* (Nordstedt) Kurt Förster 1963]
S. aculeatum var. *tibeticum* J.Y. Chen, in Wei 1984
S. aculeatum forma *torta* Børgesen 1894

Staurastrum acutum Brébisson

Cells about as long as broad or a little longer than broad, deeply constricted. Sinus widely open, acute-angled. Semicells subrhomboid (dorsal and ventral margins about equally convex) with narrowly rounded or acuminate lateral angles. Cell wall granulate, the granules around the angles arranged in more or less distinct concentric series. Semicells in apical view (2-)3(-4)-angular with slightly concave to slightly convex sides and acuminate angles. Zygospore unknown. Cell length 28-42 µm, cell breadth 27-44 µm.
Ecology: in benthos and plankton of (slightly) acidic, oligo-mesotrophic water bodies.
Distribution: unclear because of easy confusion with other species (see below). In Europe, both var. *acutum* and var. *varians* are rather widely distributed.

S. acutum Brébisson was made a variety of *S. granulosum* Ehrenberg ex Ralfs 1848 by West & West (1902). However, the original illustration of *S. granulosum* in Ehrenberg (1839) only shows an apical, little-informative view of the cell. The pictures in West & West (1912) suggest *S. granulosum* to be a reduction form of *S. avicula* var. *lunatum* Ralfs rather than to represent a separate species. *S. acutum*, on the contrary, is pretty well to be distinguished from *S. avicula* var. *lunatum* by its subrhomboid semicells (versus cup-shaped/lunate in *S. avicula* var. *lunatum*).

var. *acutum*   Plate 55: 11-15
Brébisson 1856, p. 143, pl. 1: 26
Semicells about evenly covered with series of cell wall granules.

Accounted *Staurastrum acutum* var. *acutum*:
Staurastrum acutum var. *paxilliferum* (G.S. West) Coesel 1996

var. *varians* (Raciborski) Coesel et Meesters stat. et comb. nov.   Plate 55: 16-20
Basionym: *Staurastrum varians* Raciborski 1885, p. 86, pl. 3 (= 12): 1
Semicells with a distinct unsculptured zone in between the semicell lobes. Lateral angles often slightly produced.

Accounted *Staurastrum acutum* var. *varians* (forma *varians*):
Staurastrum varians var. *badense* Schmidle 1894
S. varians forma *truncata* Gutwiński 1909
S. acutum var. *badense* (Schmidle) Coesel 1996

Staurastrum alandicum Cedercreutz et Grönblad   Plate 93: 1
Cedercreutz et Grönblad 1936, p. 4, pl. 2: 40
Cells about as long as broad, deeply constricted. Sinus V-shaped. Semicells campanulate, the apical angles produced to form short, thick-set, diverging processes. Processes tipped with some 5 small spines and around their base furnished with a circle of acute granules. Semicell body at the apex with two (emarginate) verrucae, otherwise smooth-walled. Semicells in apical view 3-radiate with about straight sides, the angles produced to form short processes. Within each margin of the semicell body two median (emarginate) verrucae. Zygospore unknown. Cell length 49, cell breadth 53 µm.

Ecology: in oligo-mesotrophic water.
Distribution: only known from a site in Åland (Finland).

S. alandicum is a questionable species as it might refer to a reduction form of some other *Staurastrum* species.

Staurastrum alpinum Raciborski   Plate 52: 11-15
Raciborski 1888, p. 108
Raciborski 1889, p. 99, pl. 3 (=7): 6
Cells distinctly longer than broad, only little constricted. Sinus slightly open to closed. Semicells semi-elliptic to semi-oval in outline with paw-shaped basal angles (best to be seen in oblique position). Cell wall densely granulate. Semicells in apical view 3-angular with slightly convex sides, the angles broadly rounded to a bit mamillate. Semicells in isthmial view with produced angles. Zygospore unknown. Cell length 28-40 µm, cell breadth 20-25 µm.
Ecology: on wet, calcareous substrates.
Distribution: Rare, only kown from Poland, Latvia, Austria and Slovakia.

Taxon excluded:
Staurastrum alpinum var. *tropicum* Lagerheim 1888

Staurastrum alternans Ralfs   Plate 54: 1-7
Brébisson ex Ralfs 1848, p. 132, pl. 21: 7
Cells about as long as broad, deeply constricted. Sinus open and acute-angled. Semicells subelliptic-subtriangular with broadly rounded angles. Cells usually twisted at the istmus. Cell wall finely and densely granulate, granules arranged in concentric rings around the angles; sometimes also a distinct supraisthmial ring of granules may be distinguished. Semicells in apical view 3-angular with concave sides and broadly rounded angles. Zygospore smooth-walled, compressed spherical, in top view with undulate margins. Cell length (15-)20-30(-35) µm, cell breadth (15-)20-30 (-37) µm.
Ecology: in benthos and tychoplankton of slightly acidic to slightly alkaline, meso-oligotrophic water bodies.
Distribution: probably cosmopolitan. In Europe, *S. alternans* is widely distributed and, particularly in mesotrophic habitats, rather common.

As Ralfs' (1848) description of *S. alternans* is fairly poor, the concept of this species as given in West & West (1912) is adopted. *S. alternans* is closely related to *S. dilatatum*, *S. striolatum* and *S. sinense*, a group of species that make up a taxonomically problematic complex.

Accounted *Staurastrum alternans* (var. *alternans*):
Staurastrum alternans var. *minus* Turner 1892

Taxa excluded:
Staurastrum alternans var. *basichondrum* Schmidle 1898 (cf *S. dispar*)
S. alternans var. *coronatum* Schmidle 1895
S. alternans var. *divergens* W. et G.S. West 1902
S. alternans forma *minimum* Beck-Managetta 1931
S. alternans var. *pulchrum* Wille 1879
S. alternans var. *spinulosum* Irénée-Marie et Hilliard 1963
S. alternans var. *subalternans* Maskell 1889

Staurastrum anatinum Cooke et Wills

Cells broader than long, deeply constricted. Sinus widely open and acute-angled. Semicells sub-fusiform to cup-shaped, the apical angles produced to form (rather) long, stout, (slightly) diverging processes. Processes tipped with 2-4 stout spines and towards the semicell body furnished with acute granules or denticulations being arranged in (indistinct) concentric series. Semicell body furnished with transverse series of acute granules, verrucae, denticulations or spines, but almost exclusively on and just below the apex. Semicells in apical view (2-)3-4-radiate with about straight sides, the angles produced to form processes. Along the sides a marginal and an intramarginal series of emarginate verrucae or spines. Zygospore unknown. Cell length (35-)45-80(-110) µm, cell breadth (60-)75-100(-150) µm.

Ecology: in benthos and plankton of acidic (to slightly alkaline?), oligotrophic (to eutrophic?) water bodies.

Distribution: Europe, North America. In Europe, var. *anatinum* and var. *subanatinum* are widely distributed and, particularly in the atlantic region, rather common; var. *longibrachiatum* is only known from Scotland, var. *denticulatum* only from Scandinavia and var. *armatum* only from a single lake in Les Landes (France).

S. anatinum is highly polymorphic with respect to cell dimensions, relative length of processes, size of the terminal spines and overall cell wall ornamentation (e.g., Brook 1959). Presumably, several different species are at issue. The species (or species complex) is closely allied to S. vestitum and (in a somewhat less degree) S. aculeatum. According to Brook (l.c.) last-mentioned two species could be considered mere forms of S. anatinum. However, in case of synonymy on species level the name of S. vestitum Ralfs 1848 or S. aculeatum Ralfs 1848 would have priority.

var. ***anatinum***   Plate 84: 1-4
Cooke et Wills, in Cooke 1881, p. 92
First illustration: Cooke 1880, pl. 139: 6
Semicell body in apical view with marginal and intramarginal series of emarginate verrucae (sometimes much reduced). Processes about straight, not much longer than the semicell body in cross-section, and furnished with acute granules or denticulations.

Accounted *Staurastrum anatinum* var. *anatinum*:
Staurastrum anatinum subsp. *biradiatum* W. West 1892a
S. anatinum var. *curtum* G.M. Smith 1922
S. anatinum var. *grande* W. et G.S. West 1902
S. anatinum var. *lagerheimii* (Schmidle) W. et G.S. West 1909
S. anatinum var. *pelagicum* W. et G.S. West 1902
S. anatinum var. *robustum* Capdevielle 1978
S. anatinum var. *truncatum* W. West 1892a

var. ***armatum*** Coesel et Meesters var. nov.   Plate 86: 1-3
Diagnosis: differs from the nominate variety by semicell body in apical view with marginal and intramarginal series of dentations, the median marginal ones sometimes lengthened to stout spines. Processes about horizontally projected but curved downwards in the terminal part. Processes longer than the breadth of the semicell body, tipped with stout spines and towards the semicell body furnished with marked dentations, incidentally lengthened to stout spines.
Type: Capdevielle 1978, pl. 17: 1-4, pl. 18: 1 (as *Staurastrum anatinum* var. *subfloriferum* Thomasson 1963).

var. ***denticulatum*** G.M. Smith   Plate 87: 1-7
Smith 1924, p. 95, pl. 75: 21-25
Semicell body in apical view with marginal and intramarginal series of emarginate verrucae or
spines (sometimes much reduced). Processes about straight, not much longer than the breadth of
the semicell body, and furnished with marked spines.

Most records of *S. anatinum* var. *denticulatum* originate from pH-neutral to slightly alkaline, rather eutrophic waters (Florin 1957). This suggests it to represent a separate (eco)species.

var. ***longibrachiatum*** W. et G.S. West   Plate 85: 4
W. & G.S. West 1905, p. 504, pl. 7: 8-9
Semicell body in apical view with an intramarginal series of emarginate verrucae. Processes about
straight, distinctly longer than the breadth of the semicell body, and furnished with small dentations.

var. ***subanatinum*** (W. et G.S. West) Coesel et Meesters comb. nov.   Plate 85: 1-3
Basionym: *Staurastrum vestitum* var. *subanatinum* W. et G.S. West 1902, p. 54, pl. 1: 28
Semicell body in apical view with two prominent (bifurcate) spines projecting from the middle of
each margin. Processes about straight, not much longer than the semicell body in cross-section,
and furnished with acute granules or denticulations.

Taxa excluded:
Staurastrum anatinum var. *aculeatum* (Ralfs) Brook 1959 (= *S. aculeatum*)
S. anatinum var. *controversum* (Ralfs) Brook 1959 (= *S. controversum*)
S. anatinum var. *convergens* Ostenfeld et Nygaard 1925
S. anatinum forma *glabrum* Brook 1959
S. anatinum forma *hirsutum* Brook 1959
S. anatinum var. *nodulosum* (W. West) Prescott, in Prescott, Bicudo & Vinyard 1982
S. anatinum forma *paradoxum* Brook 1959
S. anatinum forma *parvum* (W. West) Prescott, in Prescott, Bicudo & Vinyard 1982
S. anatinum forma *semivestitum* (W. West 1892) Brook 1959 (= *S. controversum* var. *semivestitum*)
S. anatinum var. *simplicius* Croasdale et Grönblad 1964 (cf *S. paradoxum*)
S. anatinum forma *tortum* Brook 1959
S. anatinum var. *subfloriferum* Thomasson 1963
S. anatinum var. *subglabrum* G.S. West 1907
S. anatinum forma *vestitum* (Ralfs) Brook 1959 (= *S. vestitum*)

Staurastrum angulatum W. West
Cells deeply constricted. Sinus widely open and acute-angled. Semicells rhomboid in outline
with an obtusely angled to rounded apex and narrowly rounded to acute lateral angles. Cell wall
smooth. Semicells in apical view 3-angular with slightly concave sides and subacute angles. Zygospore unknown. Cell length 47-79 μm, cell breadth 44-62 μm.
Ecology: in acidic, oligotrophic water bodies.
Distribution: North America (var. *angulatum*) and Europe (var. *planctonicum*). In Europe, var.
planctonicum is only known from a single lake in Scotland.

Descriptions of both the nominate variety and var. *planctonicum* are based on only a few cells. Apart from the original illustrations, no other reliable finds are known. Especially the figures of the nominate variety somewhat suggest
to refer to collapsed cells, so *S. angulatum* could be considered a questionable species.

var. *angulatum*   Plate 26: 16
W. West 1889, p. 20, pl. 3: 20

Synonym: *Staurodesmus angulatus* (W. West) Teiling 1948

Cells a little longer than broad. Lateral angles narrowly rounded. Cell length 76-79 µm, cell breadth 60-62 µm.

Accounted *Staurastrum angulatum* var. *angulatum*:
Staurastrum angulatum var. *subangulatum* W. West 1889a

var. *planctonicum* W. et G.S. West   Plate 26: 17
W. et G.S. West 1903, p. 551, pl. 16: 10

Synonym: *Staurodesmus angulatus* var. *planctonicum* (W. et G.S. West) Teiling 1967

Cells about as long as broad. Lateral angles subacute. Cell length 47 µm, cell breadth 44-46 µm.

Staurastrum arachne Ralfs
Cells broader than long, deeply constricted. Sinus V- to U-shaped. Semicell body cup-shaped, the apical angles produced to form long, remarkably slender, usually converging processes. Processes tipped with 3-4 minute spines and towards the semicell body furnished with (indistinct) series of granules or denticulations. Semicell body not seldom at the apex provided with a number of (emarginate) granules and/or at the base with a transversal series of acute granules, each granule corresponding with the insertion site of a process. Semicell body otherwise unsculptured. Semicells in apical view (4-)5-radiate with about straight sides, the angles abruptly produced to form long, slender, usually slightly bent processes. On each of the sides of the semicell body usually an (emarginate) intramarginal granule/denticulation. Zygospore unknown. Cell length 20-40 µm, cell breadth 40-70 µm.

Ecology: in (tycho)plankton of oligotrophic, acidic water bodies.

Distribution: Eurasia, North America. In Europe, var. *arachne* is widely distributed but far from common; var. *gyrans* and var. *curvatum* are rare.

var. *arachne*   Plate 79: 1-5
Ralfs ex Ralfs 1848, p. 136, pl. 23: 6

Original figure: Ralfs 1845, pl. 11: 6

Semicells with convergent processes. Processes only furnished with small granules/denticulations.

Accounted *Staurastrum arachne* var. *arachne*:
Staurastrum arachne var. *arachnoides* W. West 1892a
S. arachne var. *basiornatum* Capdevielle et Couté 1980

var. *curvatum* W. et G.S. West   Plate 79: 6
W. & G.S. West 1903, p. 549, pl. 18: 9

Semicells with divergent processes. Processes only furnished with small granules/denticulations.

var. *gyrans* (L.N. Johnson) A.M. Scott et Grönblad   Plate 79: 7-9
Scott & Grönblad 1957, p. 32

Basionym: *Staurastrum gyrans* L.N. Johnson 1894, p. 290, pl. 211: 4

Semicells with convergent processes. Processes, in addition to small granules/denticulations, with a remarkably stout, dorsal spine near their base.

Relevant literature: Scott & Grönblad (1957)

Accounted *Stauurastrum arachne* var. *gyrans*:
Staurastrum arachne var. *basiornatum* forma *pseudogyrans* Capdevielle et Couté 1980
S. arachne var. *incurvatum* Messikommer 1942

Taxon excluded:
Staurastrum arachne var. *sumatranum* A.M. Scott et Prescott 1961

Staurastrum archeri W. West   Plate 116: 1-2
West, W. 1892a, p. 183, pl. 23: 15
Cells broader than long, deeply constricted. Sinus widely open with obtuse apex. Semicells cup-shaped with convex apex, the apical angles produced to form slender, slightly diverging arm-like processes. Cell wall of the processes ornamented with many (at least 10) concentric series of denticulations and some 3 teeth at the tip. Cell wall of the semicell body smooth. Semicells in apical view circular with a whorl of 9-10 radiating processes. Zygospore unknown. Cell length 80-100 μm, cell breadth 130-140 μm.
Ecology: in plankton of acidic, oligotrophic water bodies.
Distribution: Only known from a few sites in western Ireland.

S. archeri differs from *S. ophiura* only by its unsculptured semicell body and its slightly diverging processes. Maybe, *S. archeri* has to be considered a (less ornamented) variety of *S. ophiura*, or even a mere reduction form of that latter species.

Staurastrum arctiscon (Ralfs) P. Lundell
Cells about as long as broad, deeply constricted. Sinus widely open with subacute apex. Semicells broadly elliptic to oval in outline, the lateral sides produced to form arm-like processes which are almost as long as the body of the semicell is broad. Apex of semicell with a series of similar processes, ascending obliquely. Cell wall of the processes with concentric series of granulations/denticulations, cell wall of the semicell body smooth. Semicells in apical view almost circular with two whorls of radiating processes: a marginal series of 9 processes and an intramarginal series of 6 processes. Zygospore unknown. Cell length 90-155 μm, cell breadth 85-160 μm.
Ecology: in plankton of acidic, oligotrophic water bodies.
Distribution: Eurasia and the American continents. In Europe, most of the records originate from Great Britain, Ireland and Scandinavia. Outside the atlantic region it is very rare.

S. arctiscon var. *glabrum* as originally described by West & West (1896) has to be considered a mere reduction form of the nominate variety. However, most of the later records of *S. arctiscon* var. *glabrum* refer to other species, allied to *S. tohopekaligense*.

var. *arctiscon*   Plate 119: 1-3
Lundell 1871, p. 70
Basionym: *Xanthidium arctiscon* Ehrenberg ex Ralfs 1848, p. 212
First illustration: Bailey 1841, pl. 1: 15 (as *Xanthidium* no. 2)
Arm-like processes tipped with 2-3 spines.

Accounted *Staurastrum arctiscon* var. *arctiscon*:
Staurastrum arctiscon var. *glabrum* W. et G.S. West 1896

Non-European variety:
var. *truncatum* Irénée-Marie
Irénée-Marie 1938, p. 335, pl. 57: 1
Arm-like processes bifurcate-truncate and tipped with blunt denticulations.

Taxa excluded:
Staurastrum arctiscon var. *brevibrachiatum* Borge 1903 (cf *S. tohopekaligense* and allied species)
S. arctiscon var. *crenulatum* Harvey 1892

Staurastrum arcuatum Nordstedt
Cells about as broad as long or a little broader than long, deeply constricted. Sinus open, acute-angled. Semicells (sub)elliptic to bowl-shaped, the lateral/apical angles produced to form bifurcate processes. A couple of similar, or more simple processes is also present on the apex in between each pair of angles. Cell wall of the semicell lobes provided with concentric series of granules. Semicells in apical view 3(-4)-radiate with concave to straight sides, the angles ending in a stout spine. On either side of each angle a (bifurcate) intramarginal spine. Zygospore unknown. Cell length 25-40 µm, cell breadth 25-50 µm.
Ecology: in (tycho)plankton of both oligo and meso-eutrophic water bodies.
Distribution: Eurasia, American continents. In Europe, both var. *arcuatum* and var. *subavicula* are widely distributed and rather common.

var. ***arcuatum***   Plate 66: 15-18
Nordstedt 1873, p. 36, fig. 18
Semicells in frontal view bowl-shaped, in apical view with concave sides, the angles ending into relatively long, bifurcate processes.

Accounted *Staurastrum arcuatum* var. *arcuatum* (forma *arcuatum*):
Staurastrum arcuatum forma *aciculifera* W. et G.S. West 1896
S. arcuatum var. *guitanense* W. West 1892a

var. ***subavicula*** (W. West) Coesel et Meesters stat. nov.   Plate 67: 1-6
Basionym: *Staurastrum arcuatum* subsp. *subavicula* W. West 1892, p. 732, fig. 25
Synonym: *S. subavicula* (W. West) W. et G.S. West 1894
Semicells in frontal view ellipsoid, in apical view with about straight sides, the angles ending into relatively short, bifurcate sprocesses.

Accounted *Staurastrum arcuatum* var. *subavicula*:
Staurastrum arcuatum var. *lapponicum* Schmidle 1898
S. arcuatum var. *vastum* Schmidle 1894a
S. subavicula var. *tyrolense* (Schmidle) Messikommer 1935

Taxa excluded:
Staurastrum arcuatum var. *pseudopisciforme* (B. Eichler et Gutwiński) W. et G.S. West 1895 (= *S. pseudopisciforme*)
S. arcuatum var. *senarium* Grönblad 1942 (cf *S. forficulatum*)
S. subavicula var. *nigrae-silvae* (Schmidle) Grönblad 1942 (cf. *S. forficulatum*)

Staurastrum arnellii Boldt   Plate 50: 13-15
Boldt 1885, p. 112, pl. 5: 21
Cells somewhat longer than broad, (rather) deeply constricted. Sinus linear, closed for the greater part. Semicells in outline trapeziform with rounded angles and slightly convex, more or less crenate lateral margins. Cell wall covered with acute, for the most part emarginate granules/dentations arranged in (obscure) circles around the angles. Semicells in apical view 3-angular with almost straight sides and obtuse angles. Zygospore unknown. Cell length 35-55 µm, cell breadth 30-40 µm.

Ecology: in benthos of acidic to neutral, oligo-mesotrophic water bodies.
Distribution: arctic-alpine. Rare, only known with certainty from Siberia and some sites in Austria.

S. arnellii is often confused with some other species, in particular *S. hirsutum* and *S. simonyi* (cf West, West & Carter 1923, Coesel & Meesters 2007).

Taxa excluded:
Staurastrum arnellii var. *inornatum* J. Roy 1890 (nomen dubium)
S. arnellii var. *spiniferum* W. et G.S. West 1902 (cf *S. simonyi*)

Staurastrum asperum Ralfs   Plate 51: 2-5
Ralfs 1848, p. 139, pl. 22: 6, pl. 23: 12
Cells longer than broad, deeply constricted. Sinus widely open, almost rectangled. Semicells sub-oval to sub-elliptic, widest above the midline. Lobes with 3 or 4 concentric series of acute granules, granule series in the median part of the semicell flank usually incomplete (only present near the apex), at the apex emarginate ot doubled. Semicells in apical view 3-angular with straight sides and broadly rounded angles, a curved series of (emarginate) granules within each lateral margin. Zygospore globose, provided with stout, bifurcate spines. Cell length 42-54 µm, cell breadth 34-47 µm.
Ecology: in acidic, oligotrophic water bodies.
Distribution: Europe, *S. asperum* is a rare species, known from France and Great Britain, with a few questionable records from other countries.

Because of the low number of reliable records, morphological variability of *S. asperum* is poorly known.

Taxon excluded:
Staurastrum asperum var. *proboscideum* Ralfs 1848 (= *S. proboscideum*)

Staurastrum aversum P. Lundell   Plate 22: 1-4
Lundell 1871, p. 59, pl. 3: 27
Synonym: *Staurodesmus aversus* (P. Lundell) Lillieroth 1950, p. 264
Cells somewhat longer than broad, deeply constricted. Sinus widely open from its acute-angled apex. Semicells cup-shaped with broadly rounded apical angles which are furnished with a papilla. Semicells in apical view 3-angular with slightly concave sides and broadly rounded, papillate angles. Zygospore unknown. Cell length 35-60 µm, cell breadth 30-45 µm.
Ecology: in plankton and benthos of acidic, oligotrophic water bodies.
Distribution: Europe and North America. In Europe, *S. aversum* is particularly known from the western lake areas of the British Islands, anywhere else it is very rare.

Taxa excluded:
Staurastrum aversum var. *bipapillosum* Skuja 1949
S. aversum var. *inflatum* Kossinskaja 1949

Staurastrum avicula Ralfs
Cells about as broad as long, or a little broader than long, deeply constricted. Sinus open and acute-angled. Semicells in rough outline bowl- to cup-shaped with one or two spines (rarely more) on each apical angle. Cell wall finely granulate, the granules around the angles arranged in concentric series, the granules on the flanks of the semicell body may be much reduced (light-microscopically even invisible) or, on the contrary, fused to emarginate verrucae. Semicells in apical view (2-)3(-4)-angular with concave to almost straight sides, angles terminating in one or two spines.

Zygospore globose and furnished with a number of conical spines, 2-3-fid at the apex. Cell length 22-35(-55) μm, cell breadth 24-50(-75) μm.

Ecology: in benthos and plankton of slightly acidic to slightly alkaline, oligo-mesotrophic water bodies.

Distribution: possibly cosmopolitan, but with the main point of distribution in temperate climatic regions. In Europe, all four below-described varieties are widely distributed and fairly common.

S. avicula is considered a highly polymorphic species. Ralfs (1848) described *S. avicula* and *S. lunatum* as separate species, characterized by forked and single spines at the apical angles, respectively. However, since then in literature many transitional forms have been figured, in particular dichotypical cells combining monospinous and bispinous semicells (see our Pl. 67:20). The zygospore depicted on Pl. 67 is copied from an unlabeled illustration by G.S. West that almost certainly is to be attributed to *S. avicula*, see West, West & Carter (1923: 39). The zygospore of *S. avicula* forma depicted in Roy & Bisset (1894) might refer to *S. subcruciatum* rather than to *S. avicula*. Relevant literature: Brook 1967.

var. *avicula*   Plate 67: 8-21; Plate 68: 1-5
Brébisson ex Ralfs 1848, p. 140, pl. 23: 11
Cells including spines less than 50 μm in breadth. As a rule, two spines at each apical angle.

var. *lunatum* (Ralfs) Coesel et Meesters stat. et comb. nov.   Plate 68: 6-14
Basionym: *Staurastrum lunatum* Ralfs 1848, p. 124, pl. 34: 12
Cells including spines less than 50 μm in breadth. As a rule, one upwardedly projected spine at each apical angle.

Accounted *Staurastrum avicula* var. *lunatum*:
Staurastrum avicula var. *turficola* Tarnavschi et Radulescu 1956
S. lunatum forma *alpestris* Schmidle 1896
S. lunatum forma *groenlandica* Børgesen 1894
S. lunatum var. *hinganicum* Noda et Skvortzov 1969
S. lunatum var. *messikommeri* Grönblad 1948
S. lunatum var. *ovale* Grönblad 1942
S. lunatum var. *subarmatum* W. et G.S. West 1894
S. lunatum var. *tvaerminneense* Grönblad 1942

var. *planctonicum* (W. et G.S. West) Coesel et Meesters comb. nov.   Plate 69: 1-7
Basionym: *Staurastrum lunatum* var. *planctonicum* W. et G.S. West 1903, p. 546, pl. 16: 11-12
Cells relatively large, more than 50 μm in breadth. Semicells with one or two stout spines (> 5 μm) at each angle. Planktonic form.

Accounted *Staurastrum avicula* var. *planctonicum*:
Staurastrum lunatum var. *borgei* (Strøm) Skuja 1964

Taxa excluded:
Staurastrum avicula var. *aciculiferum* W. West 1889 (= *S. furcatum* var. *aciculiferum*)
S. avicula var. *bifidum* (A.M. Scott et Prescott) Coesel 2004
S. avicula var. *rotundatum* W. et G.S. West 1907
S. avicula var. *trispinosa* Huber-Pestalozzi 1936
S. lunatum forma *luxurians* Lütkemüller 1910 (cf *S. arcuatum* Nordstedt 1873)
S. lunatum var. *triangularis* Børgesen 1894

Staurastrum bacillare Ralfs

Cells about as broad as long or somewhat broader than long, deeply constricted. Sinus widely open, more or less rectangled. Semicells obversely triangular with almost straight to slightly convex sides. Semicell body at the apical angles gradually passing into armlike-processes which are capitate at the ends. Cell wall smooth. Semicells in apical view 3-5-angular with strongly concave sides and capitate angles. Zygospore unknown. Cell length 17-34 μm, cell breadth 21-38 μm.
Ecology: in benthos of acidic, oligotrophic water bodies, also sub-atmophytic on wet substrates.
Distribution: rare species, only known with certainty from North America and Eurasia. In Europe, var. *bacillare* is extremely rare; var. *obesum* is somewhat less scarce, particularly in arctic-alpine regions.

Kouwets (1987) makes mention, in var. *obesum*, of the presence of distinct cell wall pores (presumably scrobicles) in the swollen ends of the processes. It is not clear in how far this is a species-specific characteristic.

var. *bacillare*   Plate 35: 1
Brébisson ex Ralfs 1848, p. 214, pl. 35: 21
Arm-like processes of the semicells relatively slender, diverging.

var. *obesum* P. Lundell   Plate 35: 2-4
Lundell 1871, p. 57, pl. 3: 24
Arm-like processes of the semicells stout, more or less parallel.

Taxon excluded:
Staurastrum bacillare var. *undulatum* W. et G.S.West in West, West & Carter 1923 (presumably an anomalous form of some other species)

Staurastrum barbaricum W. et G.S. West 1902   Plate 75: 8
W. & G.S. West 1902, p. 53, pl. 1: 23
Cells slightly broader than long, deeply constricted. Sinus widely open from an acute-angled apex. Semicells broadly cup-shaped with undulate margins and convex apex. Apical angles produced to form pairs of short, slightly divergent processes which are all arranged in the same horizontal plane. Processes tipped with some three tiny spines and towards the semicell body furnished with a few concentric series of denticulations. Semicells in apical view 3(-4?)-radiate, each radius being bilobed thus giving rise to a 6-angular semicell with concave sides. Zygospore unknown. Cell length 30 μm, cell breadth 38-41 μm.
Ecology: in benthos of acidic, oligotrophic bogs.
Distribution: Europe, North America? In Europe, *S. barbaricum* is only known from the type locality, a bog in North Ireland.

S. barbaricum is a most questionable species. Croasdale & Grönblad (1964) figure an asymmetrical 4-radiate cell, labeled '*Staurastrum ? barbaricum*' suggesting that *S. barbaricum* might be considered a mere anomalous form of some other *Staurastrum* species (with singular processes).

Staurastrum bayernense Coesel et Meesters spec. nov.   Plate 28: 17
Diagnosis: cells somewhat longer than broad, rather deeply constricted. Sinus open, about U-shaped. Semicells cuneate with slightly convex apices and produced, broadly rounded, apical angles. Cell wall without ornamentation, coarsely punctate except for the apical angles. Semicells in apical view 3-angular with slightly convex sides and broadly rounded, produced angles. Cell length 33-37 μm, cell breadth 29-34 μm.
Type: Dick 1919, plate 16: 6 (as *Staurastrum* spec.)

Ecology: in acidic, oligotrophic water.
Distribution: only known from a site in southern Germany.

Staurastrum besseri Woloszynska   Plate 40: 3
Woloszynska 1921, p. 140, fig. 18, 19
Cells slightly broader than long, deeply constricted. Sinus widely open from its acute-angled apex. Semicells cuneate-rectangular. Apical angles provided with two short, inflated processes, projected upwards and abruptly passing into an acute spine. Basal angles provided with one similar process, projected downwards. Cell wall smooth. Semicells in apical view 3-angular with concave sides and broadly rounded, trispinate angles. Zygospore unknown. Cell length 25 μm, cell breadth 40 μm.
Ecology: in acidic, oligotrophic water.
Distribution: only known from a site in Poland.

Staurastrum bibrachiatum Reinsch   Plate 113: 1-6
Reinsch 1875, p. 85, pl. 16: 2
Cells about as broad as long, deeply constricted and usually twisted at the isthmus. Sinus widely open, rectangled or obtuse-angled. Semicell body about bowl-shaped with convex apex, the apical angles usually extended into an upper and a lower process, the apical semicell processes being turned upwards, the basal ones almost horizontal. Not seldom, however, semicells occur in which the basal, horizontal processes are wanting. Processes tipped with some 2-4 minute spines and towards the semicell body furnished with a spiralling series of undulations/denticulations. Semicell body smooth-walled. Semicells in apical view 2-radiate with straight or slightly convex sides, the poles attenuated into processes. Zygospore unknown. Cell length 25-60 μm, cell breadth 28-64 μm,
Ecology: in plankton of meso-eutrophic water bodies.
Distribution: cosmopolitan. In Europe, *S. bibrachiatum* is a rare species, known from Italy, France, Austria and Great Britain.

Cell forms of *S. bibrachiatum* with only two instead of four processes per semicell are hardly or not to be distinguished from *S. smithii*. Possibly, *S. bibrachiatum* and *S. smithii* refer to one and the same species.
Relevant literature: Grönblad & Scott (1955), Brook (1982).

Accounted *Staurastrum bibrachiatum* (var. *bibrachiatum*):
S. bibrachiatum var. *cymatium* W. et G.S. West 1895

Taxa excluded:
Staurastrum bibrachiatum var. *elegans* (W. et G.S. West) Prescott, in Prescott, Bicudo & Vinyard 1982
S. bibrachiatum forma *excavatum* Brook 1982
S. bibrachiatum forma *smithii* Brook 1982

Staurastrum bicorne Hauptfleisch   Plate 100: 4-7
Hauptfleisch 1888, p. 95, pl. 3: 21, 24, 27
Cells slightly broader than long, deeply constricted. Sinus V-shaped, getting very wide towards the exterior. Semicell body vase- to cup-shaped, the apical angles produced to form fairly long, almost horizontal processes. Processes tipped with 2-3 short spines and towards the semicell body furnished with concentric series of acute granules or denticulations. Semicell body at the apex with 2(-3) transversal series of emarginate verrucae, just above the isthmus with one or two concentric whorls of granules. Semicells in apical view biradiate with a fusiform semicell body and gradually

attenuating processes. Semicell body with two parallel series of (intra)marginal verrucae along each of the long sides. Zygospore unknown. Cell length 50-70 μm, cell breadth 70-105 μm.
Ecology: in tychoplankton of acidic, oligo-mesotrophic water bodies.
Distribution: Europe, North America. In Europe, *S. bicorne* is of occasional occurrence in arctic/alpine regions, elsewhere very rare.

In view of cell size, cell shape and cell wall ornamentation pattern *S. bicorne* might be a mere biradiate form of *S. manfeldtii*. Compare, e.g., bi- and triradiate forms of *S. johnsonii* W. et G.S. West, in Scott & Grönblad 1957.

Accounted *Staurastrum bicorne* (var. *bicorne*):
Staurastrum bicorne var. *danicum* De Toni 1889
S. bicorne forma *reducta* Messikommer 1927

Taxa excluded:
Staurastrum bicorne var. *australis* Raciborski 1892
S. bicorne var. *boreale* Schmidle 1898 (cf *S. duacense* W. et G.S. West 1909a)
S. bicorne var. *longebrachiatum* Borge 1896 [= *S. longebrachiatum* (Borge) Gutwiński 1902]
S. bicorne var. *quadrifidum* Grönblad 1921

Staurastrum bieneanum Rabenhorst   Plate 27: 1-9
Rabenhorst 1862, no. 1410
First illustration: W. & G.S. West 1896a, pl. 3: 27
Cells about as long as broad, deeply constricted. Sinus widely open, at its apex, however, often narrow and sublinear. Semicells subelliptic in outline, the dorsal margin less convex than the ventral margin, the lateral sides more or less acuminate. Cell wall smooth. Semicells in apical view 3-(4-) angular with concave to almost straight sides, the angles generally somewhat acuminate. Zygospore globose, furnished with numerous simple, slightly curved, acute spines. Cell length (24-)30-38(-45) μm, cell breadth (27-)30-43 μm.
Ecology: in benthos and plankton of acidic or circumneutral, oligo-mesotrophic water bodies.
Distribution: only known with certainty from Europe, Asia, and the American continents. In Europe, *S. bieneanum* is rather widely distributed.

The original concept of *S. bieneanum* is rather unclear, see, e.g., Teiling (1967: 497). Our present concept is based on that in West & West (1912) exclusive of var. *ellipticum* Wille.

Accounted *Staurastrum bieneanum* (var. *bieneanum* forma *bieneanum*):
Staurastrum bieneanum var. *angulatum* Nygaard 1949
S. bieneanum var. *depressum* Messikommer 1960
S. bieneanum forma *spetsbergensis* Nordstedt 1875
S. bieneanum var. *subellipticum* Messikommer 1966

Taxa excluded:
Staurastrum bieneanum var. *brasiliense* Grönblad 1945
S. bieneanum var. *connectens* Boldt 1885
S. bieneanum var. *ellipticum* Wille 1879 (cf *S. lapponicum*)
S. bieneanum forma *groenlandica* Larsen 1904 (cf *S. punctulatum*)
S. bieneanum var. *myrdalense* Strøm 1926 (= *S. myrdalense*)
S. bieneanum var. *orientale* W. et G.S. West 1901

Staurastrum bifasciatum Lütkemüller

Cells longer than broad, slightly constricted. Sinus a V- or U-shaped incision. Semicells campanulate in outline, with concave lateral margins and a (slightly) convex apex. Angles (broadly) rounded, the apical ones often a little produced. Cell wall across the apex with a series of elongate verrucae and a similar series of verrucae across the median part of the flanks of the semicell. Semicells in apical view 3-angular with broadly rounded angles and straight sides. Zygospore unknown. Cell length 33-42 μm, cell breadth 26-32 μm.
Ecology: in benthos of acidic to neutral, oligo-mesotrophic water bodies.
Distribution: rare species, only known from alpine regions in central Europe.

var. *bifasciatum*   Plate 58: 15-18
Lütkemüller 1900, p. 77, pl. 1: 43-47
At the base of the semicell a supraisthmial ring of vertically elongated verrucae.

var. *subkaiseri* (Messikommer) Coesel et Meesters stat. et comb. nov.   Plate 58: 19-20
Basionym: *Staurastrum subkaiseri* Messikommer 1956, p. 138, pl. 2: 31
At the base of the semicell small groups of verrucae in disposition corresponding with the apical angles.

Staurastrum bifidum Ralfs

Cells somewhat broader than long, deeply constricted. Sinus widely open from its acute-angled to subrectangled apex. Semicells bowl- to cup-shaped. Apical angles provided with two stout spines which lie in about the same horizontal plane and project obliquely downwards. Cell wall smooth. Semicells in apical view 3(-4)-angular with concave to almost straight sides and broad, bifid angles. Zygospore unknown. Cell length 28-35 μm, cell breadth 40-56 μm.
Ecology: in benthos and tychoplankton of slightly acidic, oligo-mesotrophic water bodies.
Distribution: widely distributed in the Indo-Malaysian/N. Australian region. Also known from the American continents and Africa. In Europe, the nominate variety of *S. bifidum* has been recorded from a series of countries (e.g., UK, Sweden, France, Poland) but obviously it is quite rare all over the continent; var. *hexagonum* is only known from a site in Hungary.

Hegde & Bharati (1983) figure zygospores attributed to *S. bifidum*. However, in view of the position of the two spines at the apical angles of the adhering semicells, i.e., positioned in the same vertical plane, instead of the same horizontal plane, the spores in question are considered to refer to another species.

var. *bifidum*   Plate 39: 1-2
(Ehrenberg) Brébisson ex Ralfs 1848, p. 215
Basionym: *Desmidium bifidum* Ehrenberg 1838, p. 141, pl. 10: 11
Semicell body more or less bowl-shaped. Spines at each of the apical angles relatively close together (top view!).

Accounted *Staurastrum bifidum* var. *bifidum*:
Staurastrum bifidum forma *tetragona* W.B. Turner 1892
S. bifidum var. *tortum* W.B. Turner 1892

var. *hexagonum* Schaarschmidt   Plate 39: 3
Schaarschmidt 1883, p. 273, fig. 19
Semicell body cup-shaped, almost triangular. Spines at each of the apical angles far apart (top view!).

Staurastrum bispiniferum Coesel et Meesters spec. nov.   Plate 39: 4
Diagnosis: cells about as long as broad, deeply constricted. Sinus open, acute-angled. Semicells cuneate-subrhomboid with convex apices and slightly convex lateral sides, the apical angles produced and furnished with two minute spines. Cell wall smooth. Semicells in apical view 4-angular with convex sides and produced angles. Zygospore unknown. Cell length 39-40 µm, cell breadth 36 µm.
Type: Capdevielle 1978, plate 14: 9 (as *Staurodesmus subpygmaeus* var. *spiniferus*)
Ecology: in slightly acidic, oligotrophic water.
Distribution: only known from a site in S.W. France.

S. bispiniferum is a somewhat questionable species. In our opinion it is different from *Staurastrum subpygmaeum* var. *spiniferum* described by Scott & Grönblad (1957) from the southeastern United States. Obviously, Capdevielle (1978) encountered but one or a few cells. Maybe, those cells refer to an anomalous form of *Staurodesmus leptodermus*.

Staurastrum bloklandiae Coesel et Joosten   Plate 112: 1-3
Coesel & Joosten 1996, p. 9, fig. 1-6
Cells about as long as broad, deeply constricted. Sinus V-shaped. Semicell body cup-shaped, the apical angles produced to form rather long, divergent processes that usually are faintly bent upwards. Processes tipped with two stout spines and towards the semicell body furnished with a spiralling series of robust dentations. Semicell body at the apex with two marginal dents, otherwise smooth. Semicells in apical view 2-radiate, the semicell body rectangular-elliptic with an intramarginal granule near each of the angles. Zygospore unknown. Cell length 25-45 µm, cell breadth 29-48 µm.
Ecology: in plankton of eutrophic, pH-neutral to alkaline waters.
Distribution: up to then only known from a number of European countries (the Netherlands, Germany, France, Great Britain, Austria, Poland, Serbia) but presumably wider distributed.

Staurastrum bohlinianum Schmidle 1898
Cells about as long as broad, deeply constricted. Sinus open. Semicells elliptic to cup-shaped, the lateral/apical poles sometimes a little produced. Cell wall provided with tiny denticulations arranged in concentric series around the angles. Denticulations at the angles more pronounced than elsewhere on the cell wall. Semicells in apical view 3(-4)-angular with (slightly) concave sides and broadly rounded to truncate angles. Zygospore unknown. Cell length 19-30 µm, cell breadth 20-31 µm.
Ecology: in benthos of mesotrophic water bodies.
Distribution: Europe, North America. *S. bohlinianum* may be readily confused with other small-sized, denticulate *Staurastrum* species, so its distribution is poorly known. Most of the (few) records are from arctic-alpine regions.

var. ***bohlinianum***   Plate 71: 1-7
Schmidle 1898, p. 53, pl. 3: 3
Sinus acute-angled.

var. ***subpygmaeum*** Růžička   Plate 71: 8-9
Růžička 1972, p. 474, pl. 63: 26-27
Sinus about rectangular.

Taxon excluded:
Staurastrum bohlinianum var. *capense* Hodgetts 1926

Staurastrum boreale W. et G.S. West   Plate 73: 8-16

W. & G.S. West 1905a, p. 27, pl. 2: 25

Cells (slightly) broader than long, deeply constricted. Sinus widely open and acute-angled (the apex often V-shaped). Semicells cup-shaped, the apical angles produced to form rather short to moderately long, parallel or slightly divergent processes tipped with short spines. Cell wall furnished with (acute) granules or denticulations arranged in concentric series around the processes and continuing onto part of the semicell body, those at the apex being emarginate. Semicells at the base with a supraisthmial whorl of granules. Semicells in apical view 3-4(-5)-radiate with about straight sides, the angles produced into processes. The centralmost series of intramarginal verrucae/denticulations on the semicell body emarginate, often also at each of the sides a number of emarginate verrucae. Zygospore unknown. Cell length 22-32 µm, cell breadth 25-50 µm.

Ecology: in benthos and tychoplankton of (slightly) acidic, oligo-mesotrophic water bodies.

Distribution: Europe, North America. In Europe, *S. boreale* is a widespread and locally rather common species.

S. boreale sometimes is hardly to be distinguished from *S. gracile*. Mutual differences are such relative that the identity of *S. boreale* as a separate species may be questioned.

Accounted *Staurastrum boreale* (var. *boreale*):
Staurastrum boreale var. *quadriradiatum* Korshikov 1941
S. boreale var. *robustum* Messikommer 1951

Taxon excluded:
Staurastrum boreale var. *planctonicum* Brook 1959 (cf *S. cingulum*)

Staurastrum borgeanum Schmidle   Plate 76: 7-10

Schmidle 1898, p. 60, pl. 3: 7

Cells about as long as broad, deeply constricted. Sinus V-shaped. Semicell body cup-shaped or subcuneate, the apical angles produced to form slightly incurved processes tipped with acute granules or short spines. Cell wall furnished with coarse, conical granules, arranged in concentric series around the angles, those at the apex usually emarginate. Flanks of the semicells just below the apical series of granules usually with an unsculptured zone. Semicells at the base with a supraistmial whorl of (simple) granules. Semicells in apical view 3-4 angular with straight to concave sides and truncate angles; along the sides with intramarginal series of emarginate granules. Zygospore unknown. Cell length 30-50 µm, cell breadth 30-55 µm.

Ecology: in benthos and tychoplankton of acidic, oligotrophic water bodies.

Distribution: Eurasia, North America. In Europe, *S. borgeanum* is widespread and rather common.

West, West & Carter (1923) consider *S. borgeanum* Schmidle a synonym of *S. proboscideum* Ralfs. As well as Grönblad (1926), however, we are of opinion that we have to do with a separate species.

Accounted *Staurastrum borgeanum* (var. *borgeanum*):
Staurastrum borgeanum var. *compactum* Grönblad 1926

Taxa excluded:
Staurastrum borgeanum forma *minor* Schmidle 1898 (cf *S. subnivale*)
S. borgeanum var. *parvum* Messikommer1949 (cf *S. pentasterias*)
S. borgeanum var. *tatricum* Gutwiński 1909 (cf *S. subnivale*)

Staurastrum borgei (Kaiser) Coesel et Meesters stat. nov.   Plate 59: 1-2
Basionym: *Staurastrum capitulum* (Ralfs) forma *borgei* Kaiser 1916, p. 37
Original figure: Borge 1911, fig. 16
Cells longer than broad, slightly constricted. Sinus a V- or U-shaped incision. Semicells subquad-rate in outline, with concave lateral margins and truncate to retuse apices. Apical and basal angles broadly rounded. Cell wall at the base of the semicell with a broad ring of longitudinally directed, granulated ridges. Just below the apex a similar series of granulate-denticulate verrucae. Also the apical margin itself provided with variably pronounced, emarginate verrucae. Semicells in apical view 3-angular with slightly convex sides and very broadly rounded angles, each angle associated with a hoof-shaped series of intramarginal verrucae. Zygospore unknown. Cell length 45-65 µm, cell breadth 30-40 µm.
Ecology: subatmophytic on moist rocks, also in benthos of small, pH-circumneutral alpine pools.
Distribution: rare arctic-alpine species, only known from Europe and North America.

S. rhabdophorum forma *trigona*? in Bourrelly (1987, taxon not formally described) may be accounted this species.

Staurastrum botrophilum Wolle   Plate 52: 6-9
Wolle 1881, p. 2, pl. 6: 13a-c
Cells longer than broad, deeply constricted. Sinus linear with a dilate extremity. Semicells pyram-idate-truncate with broadly rounded angles. Cell wall granulate, granules arranged in concentric series around the angles. Semicells in apical view 3-angular with slightly retuse sides and rounded angles. Zygospore unknown. Cell length 41-54 µm, cell breadth 31-43 µm.
Ecology: in benthos of acidic, oligotrophic water bodies.
Distribution: North America, Europe. In Europe, *S. botrophilum* is very rare; more or less reliable records are only known from Great Britain.

S. botrophilum is a questionable species. The original description of *S. botrophilum* by Wolle (1881) from the United States is most confusing as, according to the accompanying illustrations, obviously two different species are mixed up. The illustration by West & West (1912) was made after Wolle's original material sent by that author, so refers to North American material. Since then the species in question has hardly been reported again. Possibly, some figures published under the name of *S. botrophilum* refer to triradiate *Cosmarium* species, e.g., *C. botrytis* or *C. vexatum* (Brook & Williamson 1983).

Staurastrum brachiatum Ralfs
Cells about as broad as long or broader than long, deeply constricted. Sinus widely open, more or less rectangled. Cells about X-shaped. Semicells largely consisting of robust, arm-like processes which are 2-3(-5)-fid at the ends, resulting in shorter or longer, blunt, diverging terminal teeth. Cell wall smooth. Semicells in apical view (2-)3-4-angular with strongly concave sides, not seldom with a slight median inflation. Zygospore, as described by Ralfs (1848), more or less quadrangular, the angles topped with a couple of spines. Cell length (16-)25-45(-75) µm, cell breadth (20-)30-45(-75)µm.
Ecology: in benthos and plankton of acidic, oligotrophic water bodies.
Distribution: cosmopolitan. In Europe, *S. brachiatum* is widely distributed and rather common in its appropriate habitat.

S. brachiatum is highly polymorphic with respect to cell size, slenderness and length of the terminal furcations. Most likely it may be split into a number of different species, see, e.g., our Pl. 36: 19 showing a form with a rounded, spineless zygospore. Markedly slender forms usually are labeled var. *gracilius*. The original illustration of this variety by Maskell (1889) from New Zealand, however, shows a cup-shaped semicell body and remarkably acute terminal furcations, rendering its affiliation to *S. brachiatum* doubtful. Cells of *S. brachiatum* may be united to form short filaments.

var. **brachiatum**   Plate 36: 1-18
Ralfs 1848, p. 131, pl. 23: 9
Cells about as broad as long with divergent processes.

Accounted *Staurastrum brachiatum* var. *brachiatum* (forma *brachiatum*):
Staurastrum brachiatum var. *major* Wade 1957
S. brachiatum forma *minor* Lütkemüller 1900
S. brachiatum var. *notarisii* Rabenhorst 1868
S. brachiatum forma *robustum* Scott et Grönblad 1957

Non-European variety:
var. **parallelum** (Coesel) Coesel et Meesters stat. nov.
Basionym: *Staurastrum brachiatum* forma *parallelum* Coesel 1987, p. 133, pl. 3: 5, 6
Cells distinctly broader than long with parallel processes.

Questionable non-European taxon:
S. brachiatum var. *gracilius* Maskell 1889

Taxa excluded:
Staurastrum brachiatum var. *bicorne* Nygaard 1991 (cf *S. subnudibranchiatum*)
S. brachiatum var. *compactum* Grönblad 1942 (= *S. laevispinum* var. *compactum*)
S. brachiatum var. *elongatum* Thérézien 1985
S. brachiatum var. *longipedum* Raciborski 1895
S. brachiatum var. *tenerrimum* Skuja 1976

Staurastrum brasiliense Nordstedt   Plate 40: 4-5; Plate 41: 1-2
Nordstedt 1870, p. 227, pl. 4: 39
Cells about as broad as long or somewhat broader than long, deeply constricted. Sinus widely
open, excavated at its apex. Semicells cup- or bowlshaped with almost straight to slightly convex
sides. Apical angles truncate and provided with 3(-5) long, stout, diverging spines, usually placed
in two different horizontal planes (in case of three spines: two in one plane, the third superimposed). Cell wall smooth. Semicells in apical view (3-)4-5(-6)-radiate with truncate angles (provided with diverging spines) and deeply concave sides. Zygospore unknown. Cell length (55-)100-
150(-200) µm, cell breadth (65-)100-160(-210) µm.
Ecology: in plankton of acidic, oligotrophic water bodies.
Distribution: known from the American continents, Europe and Africa. In Europe, most of the
records originate from Great Britain, Ireland and Scandinavia. Outside the atlantic region it is
very rare.

From the Amazon region Förster (1969) and Thomasson (1971) reported, under the name of *S. brasiliense* var. *porrectum*, a form characterized by a strikingly swollen base of its angular spines. Possibly this form may be distinguished as a separate variety, but not as var. *porrectum* since the original diagnosis of the latter taxon (Borge 1925)
does not mention that feature.

Accounted *Staurastrum brasiliense* (var. *brasiliense* forma *brasiliense*):
Staurastrum brasiliense var. *lundellii* W. et G.S.West 1896
S. brasiliense var. *lundellii* forma *major* Irénée-Marie 1959
S. brasiliense forma *maior* Ryppowa 1927
S. brasiliense var. *porrectum* Borge 1925
S. brasiliense var. *triangulare* Thomasson 1960

Taxa excluded:
Staurastrum brasiliense forma *pentagona* Lagerheim 1885 (*nomen nudum*)
S. brasiliense var. *triquetrum* Wolle 1887 (cf *S. subtrifurcatum* W. et G.S.West 1896)

Staurastrum brebissonii W. Archer
Cells about as broad as long or a little broader than long, deeply constricted. Sinus open from an acute-angled apex. Semicells subelliptic-rhomboid with broadly rounded angles, equably and closely covered with short, fine spines arranged in concentric series around the angles (the spines near the angles often slightly longer than the other ones). Semicells in apical view 3(-4)-angular with concave sides and broadly rounded angles. Zygospore (only known of the nominate variety) globose and furnished with stout, three- or fourfold furcate spines. Cell length (30-)35-55µm, cell breadth (30-)40-65µm.
Ecology: in benthos and tychoplankton of mesotrophic, slightly acidic water bodies.
Distribution: Eurasia, American continents, Africa. In Europe, both var. *brebissonii* and var. *ordinatum* are widely distributed.

Because of a poor original description, the taxonomic concept of *S. brebissonii* is not very clear (compare West, West & Carter 1923: 62, and Grönblad 1942: 41).

var. *brebissonii*   Plate 48: 6-8
Archer in Pritchard 1861, p. 739
Original illustration: Brébisson 1856, pl. 2: 49 (as *Staurastrum pilosum*)
Semicells subelliptic-rhomboid in outline with greatest width in the median part of the semicell.

Accounted *Staurastrum brebissonii* var. *brebissonii* (forma *brebissonii*):
Staurastrum brebissonii var. *maximum* Cedercreutz 1932
S. brebissonii forma *minor* Boldt 1888

var. *ordinatum* Schmidle 1898   Plate 48: 9-12
Schmidle 1898, p. 53, pl. 3: 1
Semicells rhomboid to bowl-shaped in outline, greatest width in the apical part of the semicell.

In our opinion, *S. brebissonii* var. *ordinatum* is identical to *S. erasum* Brebisson 1856.

Accounted *Staurastrum brebissonii* var. *ordinatum*:
S. brebissonii var. *laticeps* Grönblad 1942

Taxa excluded:
Staurastrum brebissonii var. *brasiliense* Grönblad 1945
S. brebissonii var. *brevispinum* W. West 1892 (probably another species but hard to identify)
S. brebissonii var. *curvispinum* Grönblad 1945
S. brebissonii var. *heteracanthum* W. et G.S. West 1896
S. brebissonii var. *paucispina* G.M. Smith 1921
S. brebissonii var. *truncatum* Grönblad 1926 (= *S. trapezioides*)

Staurastrum brevispina Ralfs
Cells deeply constricted. Sinus acute-angled and widely open, near its apex often sublinear. Semicells generally broadly elliptic-oval, less often almost semicircular or bowl-shaped, the lateral sides furnished with a papilla. Semicells in apical view 3-angular with concave sides and broadly rounded, papillate angles. Zygospores, known from the nominate variety, spherical to broadly elliptic, smooth-walled. Cell length (?25-)30-50(-65) µm, cell breadth (?25)30-50(-55) µm.

Ecology: in benthos and tychoplankton of acidic to slightlly alkaline, oligo-mesotrophic water bodies.

Distribution: only known with certainty from Europe, the American continents and Japan. In Europe, both var. *brevispina* and var. *boldtii* are widespread and rather common, var. *obversum* is rare.

Relevant literature: Teiling (1967).

Small forms of *S. brevispina*, usually labeled *Staurastrum brevispina* forma *minima* Lütkemüller, may be confused with *Staurodesmus mucronatus*. In general, the latter species is smaller sized and somewhat broader than long, whereas the lateral sides are furnished with a short spine or acute mucro.

var. *brevispina*   Plate 22: 5-8
Brébisson ex Ralfs 1848, p. 124, pl. 34: 7a, b
Synonym: *Staurodesmus brevispina* (Ralfs) Croasdale 1957, p. 122
Cells about as long as broad; semicells elliptic-oval in outline. Papillae laterally located.

Accounted *Staurastrum brevispina* var. *brevispina* (forma *brevispina*):
Staurastrum brevispina forma *hexagona* Eichler et Gutwiński 1894
S. brevispina forma *major* W. et G.S. West 1909
S. brevispina var. *prespanse* Petkoff 1910
S. brevispina var. *retusum* Borge forma *galiciensis* Gutwiński 1896

var. *boldtii* Lagerheim   Plate 23: 1-4
Lagerheim 1893, p. 163
Original figure: Boldt 1885, pl. 5: 30 (as *Staurastrum brevispinum* forma)
Synonym: *Staurodesmus brevispina* var. *boldtii* (Lagerheim) Croasdale 1957, p. 124
Cells up to 1.4 times as long as broad; semicells almost semicircular in outline. Papillae laterally located.

Accounted *Staurastrum brevispina* var. *boldtii*:
Staurastrum brevispina var. *altum* W. et G.S. West 1905
S. brevispina var. *retusum* Borge 1894

var. *obversum* W. et G.S. West   Plate 22: 9-11
W. & G.S. West 1905, p. 502, pl. 7: 15
Synonym: *Staurodesmus brevispina* var *obversus* (W. et G.S. West) Croasdale 1962, p. 32
Cells about as long as broad or slightly broader than long; semicells about bowl-shaped. Papillae located at the apical angles.

Taxa excluded:
Staurastrum brevispina var. *basidentatum* Cushman 1905
S. brevispina forma *boldtii* Turner 1892
S. brevispina var. *canadense* Taft 1945
S. brevispina var. *inerme* Wille 1879
S. brevispina var. *masogamum* Claassen 1961
S. brevispina forma *minima* Lütkemüller 1900
S. brevispina var. *reversum* Virieux 1913
S. brevispina var. *tumidum* G.M. Smith 1924
Staurodesmus brevispina var. *kossinskajae* Teiling 1967 (cf *Std. patens* var. *inflatus*)

Staurastrum bulbosum (W. West) Coesel

Cells (slightly) broader than long, deeply constricted. Sinus V-shaped. Semicell body shaped as an elongate vase with inflated base, the apical angles produced to form long processes. Processes tipped with 2-3 rather short spines and towards the semicell body furnished with concentric series of minute denticulations. Semicell body at the apex with a series of (emarginate) granules/denticulations, just above the isthmus sometimes with two concentric whorls of granules. Semicells in apical view 3-radiate with straight to slightly concave sides gradually passing into the processes. Semicell body with a series of intramarginal (emarginate) granules along each of the sides. Zygospore unknown. Cell length 50-70 μm, cell breadth 70-100 μm.

Ecology: in (tycho)plankton of acidic, oligo-mesotrophic water bodies.

Distribution: Europe, North America, Australia. In Europe, *S. bulbosum* is a rare species, particularly known from atlantic regions (UK, France, Netherlands).

S. bulbosum unmistakably is affiliated to the *S. manfeldtii / S. pingue-planctonicum* species complex rather than to *S. gracile* (Teiling 1947, Thomasson & Tyler 1971, Coesel 1996a). It might be considered a variety of *S. manfeldtii* but in our opinion it is not identic to *S. planctonicum* var. *bullosum* Teiling 1946 as stated in Croasdale, Flint et Racine (1994).

var. *bulbosum*   Plate 102: 1

Coesel 1996a, p. 102

Basionym: *Staurastrum gracile* ssp. *bulbosum* W. West 1892a, p. 182, pl. 23: 11

Synonym: *Staurastrum gracile* var. *bulbosum* (W. West) W. et G.S. West 1902, p. 54

Processes tipped with two slender, curved teeth. Apex of the semicel body beset with series of single granules.

var. *cyathiforme* (W. et G.S. West) Coesel et Meesters comb. nov.   Plate 102: 2-5

Basionym: *Staurastrum gracile* var. *cyathiforme* W. et G.S. West 1895, p. 77, pl. 9: 2

Processes tipped with three short, straight spines. Apex of the semicell body set with series of emarginate verrucae.

Staurastrum bullardii G.M. Smith   Plate 109: 4-8

Smith 1924, p. 91, pl. 74: 19-23

Cells about as broad as long or somewhat broader than long, deeply constricted. Sinus V-shaped. Semicells cup-shaped, the apical angles rather abruptly produced to form long, slender, diverging processes. Processes tipped with some four stout spines and towards the semicell body furnished with spiralling series of marked dentations. Semicell body at the apex with two emarginate verrucae, otherwise smooth-walled. Semicells in apical view (2-)3-radiate, semicell body on either side with two intramarginal verrucae. Zygospore unknown. Cell length 50-85 μm, cell breadth 80-100 μm.

Ecology: in euplankton of meso-eutrophic lakes.

Distribution: North America, Europe, Asia. In Europe, *S. bullardii* is only known from Sweden and Denmark.

Accounted *Staurastrum bullardii* (var. *bullardii*):
Staurastrum bullardii var. *alandicum* Teiling 1942

Taxa excluded:
Staurastrum bullardii var. *brasiliense* Kurt Förster 1969
S. bullardii var. *glabrum* Kurt Förster 1969
S. bullardii var. *suecica* Borge 1939 (cf *S. subosceolense*)

Staurastrum capitulum Ralfs   Plate 58: 8-10
Brébisson ex Ralfs 1848, p. 214, pl. 35: 25
Cells distinctly longer than broad, slightly constricted. Sinus a V- or U-shaped incision. Semicells campanulate in outline, with concave lateral margins and a slightly convex to slightly retuse apex. Angles broadly rounded, the apical ones usually a little produced. Cell wall in the apical region provided with acute granules or denticulations, near the angles arranged in concentric series. At the base of the semicell a ring of 2-4-denticulate verrucae, separated from the apical ornamentation by an unsculptured zone. Semicells in apical view 3-angular with narrowly rounded to subacute angles and almost straight sides. Zygospore unknown. Cell length (18-)30-45 µm, cell breadth (13-)23-35 µm.
Ecology: in benthos of acidic to neutral, oligo-mesotrophic water bodies, also sub-atmophytic on wet substrates.
Distribution: typical upland species, only known with certainty from Eurasia, North America and New Zealand. In Europe, *S. capitulum* is not rare in arctic-alpine areas and submontane regions.
Relevant literature: West & West (1912).

S. capitulum differs from *S. pileolatum* in having its apical angles horizontally spread (not upwardly directed). Illustrations of *S. capitulum* in which apical angles are extended into distinct processes (see, e.g., Hirano 1968, Förster 1964) are considered to relate to other species.

Accounted *Staurastrum capitulum* (var. *capitulum*):
Staurastrum capitulum var. *acanthophorum* Nordstedt, in Nordstedt et Wittrock 1876
S. capitulum var. *amoenum* (Hilse) Rabenhorst 1868
S. capitulum var. *dimidio-minus* Croasdale et Grönblad 1964
S. capitulum var. *intermedium* Gutwiński 1909

Questionable non-European taxa:
Staurastrum capitulum var. *tumidiusculum* (Nordstedt) W. et G.S.West 1912 (4-radiate form of *S. spetsbergense?*)
S. capitulum var. *australe* Skuja 1976

Taxa excluded:
Staurastrum capitulum forma *borgei* Kaiser 1916 (= *S. borgei*)
Staurastrum capitulum var. *italicum* (Nordstedt) W. et G.S. West 1912 (cf *S. bifasciatum* var. *subkaiseri*)
S. capitulum var. *magnum* Förster 1964
S. capitulum forma *quadrata* Schmidle 1898

Staurastrum cerastes P. Lundell   Plate 100: 1-3
Lundell 1871, p. 69, pl. 4: 6
Cells about as broad as long or a little broader than long, deeply constricted. Sinus V-shaped. Semicell body vase-shaped to almost cylindric with convex apex. Semicell body at the apical angles smoothly passing into rather long, gradually attenuating and gracefully incurved processes. Processes tipped with some 2-3 minute spines and along the apical and the lateral sides furnished with longitudinal series of emarginate verrucae. Both the apical and the lateral series continue along the semicell body so that they stretch from tip to tip of adjacent processes. Semicells in apical view 3-4-radiate, the angles produced into gradually tapering processes, along each side with both a marginal and an intramarginal series of emarginate verrucae. Zygospore unknown. Cell length (45-)50-65(-80) µm, cell breadth (55-)60-70(-80) µm.
Ecology: in tychoplankton of acidic, oligotrophic water bodies.
Distribution: Eurasia, North America. In Europe, *S. cerastes* is a rare, atlantic species, confined to the north-western part of the continent.

Accounted *Staurastrum cerastes* (var. *cerastes*):
Staurastrum cerastes var. *triradiatum* G.M. Smith 1922

Questionable non-European taxon:
Staurastrum cerastes var. *ceylanicum* W. et G.S. West 1902a

Taxa excluded:
Staurastrum cerastes var. *coronatum* Willi Krieger 1932
S. cerastes var. *delicatissimum* Hinode 1966
S. cerastes var. *pulchrum* A.M. Scott et Grönblad 1957
S. cerastes var. *simplicius* Grönblad 1948 (erroneously labeled 'var. *gracilius*' in the plate caption)

Staurastrum chaetoceras (Schröder) G.M. Smith   Plate 110: 1-9
Smith 1924, p. 139
Basionym: *Staurastrum polymorphum* var. *chaetoceras* Schröder in Zacharias 1898, p. 131, text-fig. a-c
Cells somewhat longer than broad to slightly broader than long, deeply constricted. Sinus V-shaped. Semicell body triangular in outline, the apical angles produced to form long, slender, divergent processes. Processes tipped with some four minute spines and towards the semicell body furnished with concentric series of acute granules/denticulations, advancing on part of the semicell body. An additional transverse supraisthmial series of similar granules may be present (on the other hand, granules on the semicell body sometimes are so much reduced that they are invisible under the light microscope). Semicells in apical view 2-3-radiate with about straight sides, the poles attenuated into processes. Zygospore globular, furnished with stout, bi- or trifurcate spines. Cell length (22-)35-80(-105) μm, cell breadth (30-)45-85(-100) μm.
Ecology: in plankton of both oligotrophic and eutrophic, both acidic and alkaline water bodies, may cause a bloom in eutrophic waters.
Distribution: Europe, North America. In Europe, *S. chaetoceras* is widely distributed and locally very common in eutrophic waters.

The only zygospore record is by Watanabe, Watanabe & Saitow (1980) obtained by inducing conjugation in a Japanese clonal culture.

Accounted *Staurastrum chaetoceras* (var. *chaetoceras*):
Staurastrum chaetoceras var. *convexum* Grönblad 1960

Taxon excluded:
Staurastrum chaetoceras var. *tricrenatum* Skuja 1956 (cf *S. multinodulosum*)

Staurastrum chavesii Bohlin
Cells slightly broader than long, rather deeply constricted. Sinus U-shaped. Semicell body inversely trapeziform, the apical angles produced to form stout, short to fairly long, diverging processes. Processes tipped with 3-4 minute spines/dentations, towards the semicell body furnished with 2-3 concentric series of dentations. Sinus at its extremity bordered by a pronounced (emarginate) dentation. Semicells in apical view 4-radiate with concave to straight sides, the angles produced into processes. Zygospore unknown. Cell length 10-23 μm, cell breadth 15-32 μm.
Ecology: in benthos and tychoplankton of acidic, oligo-mesotrophic water bodies.
Distribution: *S. chavesii* is but rarely recorded; var. *chavesii* only from the Azores, Ireland and the Netherlands, var. *latiusculum* only from Ireland.

var. *chavesii*   Plate 114: 3-4
Bohlin 1901, p. 56, fig. 15
Processes furnished with dentations only.

var. *latiusculum* (W. et G.S. West) Coesel et Meesters stat. et comb. nov.   Plate 114 : 5-6
Basionym: *Staurastrum latiusculum* W. et G.S. West 1902, p. 53, pl. 1: 20
Processes furnished, in addition to denticulations, with three small, but distinct spines at the end
and two at the upper side of their base (best to be seen in apical view).

Staurastrum cingulum (W. et G.S. West) G.M. Smith
Cells usually slightly broader than long, deeply constricted. Sinus V- or U-shaped. Semicells about
cup-shaped, the apical angles produced to form (rather) long, diverging processes, tipped with
short spines. Cell wall furnished with (acute) granules, arranged in concentric series around the
processes and continuing onto part of the semicell body, those at the apex often emarginate. Semicells at the base with a supraisthmial whorl of granules. Semicells in apical view (2-)3-radiate with
about straight sides, the angles produced into (rather) long processes; the inner series of granules
on the semicell body arranged in pairs (actually emarginate). Zygospore unknown. Cell length
(25-)35-70(-100) µm, cell breadth (40-)50-80(-105) µm.
Ecology: in plankton of acidic to slightly alkaline, oligo-mesotrophic water bodies.
Distribution: Eurasia, North America. In Europe, var. *cingulum* is rather rare and possibly confined to atlantic regions, var. *obesum* is widely distributed and much more common.
Relevant literature: Brook 1959.

var. *cingulum*   Plate 93: 2-5
Smith 1922, p. 353
Basionym: *Staurastrum paradoxum* var. *cingulum* W. et G.S. West 1903, p. 548, pl. 18: 6-7
Semicell body campanulate, subcylindrical in its basal portion. Processes relatively long.

Accounted *Staurastrum cingulum* var. *cingulum*:
S. cingulum forma *annulatum* Brook 1959
S. cingulum var. *floridense* A.M. Scott et Grönblad 1957
S. cingulum var. *tortum* G.M. Smith 1922

var. *obesum* G.M. Smith   Plate 94: 1-9
Smith 1922, p. 353, pl. 12: 3-5
Semicell body more or less triangular. Processes relatively short.

Accounted *Staurastrum cingulum* var. *obesum*:
Staurastrum cingulum var. *affine* (W. et G.S. West) Brook 1959
S. cingulum var. *latum* Korshikov 1941

Taxa excluded:
Staurastrum cingulum var. *inflatum* Nygaard 1949 (cf *S. chaetoceras*)
S. cingulum var. *ornatum* Irénée-Marie 1949

Staurastrum clepsydra Nordstedt   Plate 26: 1-4
Nordstedt 1870, p. 224, pl. 4: 47, 48
Synonym: *Staurodesmus clepsydra* (Nordstedt) Teiling 1948
Cells about as broad as long or slightly broader than long, deeply constricted. Sinus open, acute-
angled to almost rectangled. Semicells bowl-shaped with slightly produced, more or less acumi-

nate apical angles which are often somewhat downturned. Cell wall smooth. Semicells in apical view 3-angular with concave sides and narrowly rounded to acuminate/mucronate angles. Zygospore unknown. Cell length 28-44 µm, cell breadth 31-54 µm.
Ecology: in acidic, oligotrophic water bodies.
Distribution: South America, Australia, Europe. In Europe, *S. clepsydra* is only rarely reported and most of the records are questionable.

S. clepsydra is mainly known from tropical regions (see, e.g., Förster 1969, Ling & Tyler 2000). Records from Europe, on the contrary, for the majority originate from arctic-alpine habitats. According to the algal illustrations provided most of those refer to other species. Also records from other continents (particularly those of var. *obtusum*) are often highly questionable.

Accounted *Staurastrum clepsydra* (var. *clepsydra*):
Staurastrum clepsydra var. *acuminatum* Nordstedt 1870
S. clepsydra var. *obtusum* Nordstedt 1870

Taxa excluded:
S. clepsydra forma *biradiata* Eichler 1896
S. clepsydra var. *minimum* Scott et Prescott 1958
S. clepsydra var. *sibiricum* (Borge) W. et G.S. West 1912 (= *S. sibiricum*)

Staurastrum clevei (Wittrock) J.Roy

Cells about as broad as long or somewhat broader than long, deeply constricted. Sinus widely open with acute-angled apex. Semicells bowl-shaped to subelliptic, with convex sides. Angles produced to form arm-like processes which are bifurcate (rarely trifurcate) at the ends. Near each of the radial processes (and just above it) a similar, additional process is inserted. Cell wall smooth. Semicells in apical view 3-(4)-radiate, each radius giving rise to an arm-like process. Within the circle of radial processes an additional circle of arm-like processes, each of them being obliquely oriented to the nearby radial axis. Zygospore unknown. Cell length (25-)40-70 µm, cell breadth (35-)40-70 µm.
Ecology: in benthos and tychoplankton of acidic, oligo-mesotrophic water bodies.
Distribution: rather rare species, known from Europe, North America, northern Africa, Japan and Australia. In Europe, var. *clevei* is widely distributed (but scarce), var. *inflatum* is only known from Swedish Lapland.

Occasionally, next to the regular processes a variable number of additional processes may be developed (Teiling 1957, Messikommer 1960, see our Pl. 37: 4). Being an inconsistent feature, distinction of a separate taxonomic status for such forms is not justified.

var. *clevei*   Plate 37: 1-4
Roy 1893, p. 179
Basionym: *Staurastrum laeve* var. *clevei* Wittrock 1869, p. 18, pl. 1: 9
Semicell body bowl-shaped, lateral processes divergent to parallel.

Accounted *Staurastrum clevei* var. *clevei*:
Staurastrum clevei var. *africanum* Gauthier-Lièvre 1958
S. clevei var. *octocornis* Ryppowa 1927
S. clevei var. *variabile* Messikommer 1960

var. *inflatum* (Schmidle) Coesel et Meesters comb. nov.   Plate 37: 5
Basionym: *Staurastrum kitchelii* Wolle var. *inflatum* Schmidle 1898, p. 52, pl. 2: 41
Semicell body subelliptic, lateral processes slightly convergent.

In addition to the above-mentioned differentiating features, var. *inflatum* in its original description and illustration is also characterized by a small cell size (25 x 36 µm) and a narrow isthmus (ca 7 µm). Despite those differences var. *inflatum* is presented here with some reserve since the alga in question was found at but one single site and in very small cell numbers, only one cell being illustrated (Schmidle l.c.).

Staurastrum coarctatum Brébisson
Cells about as long as broad, deeply constricted. Sinus open. Semicells transversely oblong, with broadly rounded angles and almost straight or retuse apex. Cell wall smooth. Semicells in apical view 3-angular with concave sides and broadly rounded angles. Zygospore unknown. Cell length 20-33 µm, cell breadth 16-40 µm.
Ecology: in acidic, oligotrophic water bodies.
Distribution: Europe, S.E. Asia, New Zealand. In Europe, *S. coarctatum* has been only very incidentally recorded, notably from France and Great Britain.

At first glance the nominate variety as described by Brébisson and var. *subcurtum* as described by Nordstedt seem to belong to different species. However pictures of both varieties given by Hirano (1959) represent somewhat intermediate forms, see our Pl. 31: 17. Nevertheless, *S. coarctatum* is a poorly known, problematic species. Particularly var. *subcurtum* may be readily confused with other species, such as *S. ellipticum* and *S. muticum*.

var. *coarctatum*   Plate 31: 16-17
Brébisson 1856, p. 144, pl. 1: 29
Cells slightly broader than long. Sinus acute-angled.

var. *subcurtum* Nordstedt   Plate 31: 18-19
Nordstedt 1887, p. 158
Original illustration: Nordstedt 1888a, pl. 4: 20
Cells slightly longer than broad. Sinus broadly rounded at its apex, resulting in a somewhat elongate isthmus.

Questionable non-European variety:
Staurastrum coarctatum var. *curtum* Nordstedt 1870

Taxa excluded:
Staurastrum coarctatum var. *horii* Förster 1972
S. coarctatum var. *solitarium* Flint, in Croasdale, Flint & Racine 1994
S. coarctatum var. *subcurtum* Nordstedt forma *maius* Grönblad, in Grönblad, Scott & Croasdale 1964

Staurastrum conspicuum W. et G.S. West   Plate: 25: 1-2
W. & G.S. West 1903, p. 547, pl. 14: 4
Synonym: *Staurodesmus conspicuus* (W. et G.S. West) Teiling 1967
Cells distinctly broader than long, deeply constricted. Sinus narrowly open and more or less linear in its median part, widely opening outward. Semicells elliptic-fusiform in outline, the lateral sides acuminate/mamillate. Cell wall smooth. Semicells in apical view 3-angular with concave sides and mamillate angles. Zygospore unknown. Cell length 83-103 µm, cell breadth 111-134 µm.
Ecology: in benthos and tychoplankton of acidic, oligotrophic water bodies.
Distribution: only known with certainty from some sites in Scotland.

After West & West (1912) no reliable pictures of *S. conspicuum* have been published. Illustrations of *S. conspicuum* by Irenée-Marie (1938) and Skuja (1964) should be accounted *S. grande*.

Taxa excluded:
Staurastrum conspicuum forma *montrealense* Prescott, in Prescott, Bicudo & Vinyard 1982 (= *S. grande*)
S. conspicuum forma *minor* Woronichin 1930

Staurastrum controversum Ralfs

Cells slightly broader than long to about as broad as long, deeply constricted. Sinus widely open and acute-angled. Semicells subelliptical or bowl-shaped, the lateral/apical angles produced to form (rather) short, stout, converging to about parallel processes. Processes tipped with 3-4 short spines and also towards the semicell body furnished with smaller or larger denticulations arranged in indistinct concentric series. Denticulations/spines on the dorsal side of the processes larger than those on the other sides, often being bifurcate and continuing onto the apex of the semicell body; usually with a pair of extra large spines near the base of each process. Semicells in apical view 3-4(-5)-radiate with concave or straight sides, the angles produced into processes that are bent in a particular way (usually clockwise). Semicell body with an intramarginal series of emarginate verrucae or (bifurcate) spines in series with the spines on the processes which are usuallly displaced by twisting of the processes. Zygospore (of the nominate variety) irregularly globose, provided with a number of broad protuberances each of which is furnished with several rather short, much branched appendages. Cell length (20-)25-40(-65?) μm, cell breadth (25-)30-55(-75?) μm.
Ecology: in benthos and tychoplankton of acidic, oligotrophic water bodies.
Distribution: Eurasia, North America, New Zealand. In Europe, both the nominate variety and var. *semivestitum* are widely distributed and locally common, particularly in mountainous areas.

var. ***controversum***   Plate 81: 1-9
Brébisson ex Ralfs 1848, p. 141, pl. 23: 3
Semicells subelliptical, the lateral angles produced to form converging processes.

Accounted *S. controversum* var. *controversum* (forma *controversum*):
Staurastrum controversum forma *elegantior* Strøm 1920

var. ***semivestitum*** (W. West) Coesel   Plate 81: 10-13
Coesel 1996, p. 20, fig. 11
Basionym: *Staurastrum vestitum* var. *semivestitum* W. West 1892, p. 732, pl. 9: 38
Semicells cup- or bowl-shaped, processes parallel to slightly divergent.

Staurastrum cornutum W. Archer

Cells including spines (slightly) broader than long, deeply constricted. Sinus closed or narrowly open. Semicells semicircular to trapezoid. Cell wall provided with series of crenel-like verrucae, running from apex to semicell base along each of the semicell lobes. Near the base of each semicell lobe a stout, often furcate spine. Semicells in apical view 3(-4)-angular with straight to concave sides and broad angles furnished with a stout spine. Zygospore unknown. Cell length 37-50 μm, cell breadth 40-57 μm.
Ecology: in benthos and tychoplankton of acidic to neutral, oligo-mesotrophic water bodies.
Distribution: rare; only known from North America and Europe. Within Europe, the nominate variety is known from Scotland, Ireland, Sweden and Finland, var. *skujae* only from a few sites in Uppland (Sweden).

var. *cornutum*   Plate 64: 2-4
Archer 1881, p. 232
First illustration: Roy & Bisset 1894, pl. 4: 5
Semicells about semicircular in outline.

var. *skujae* Coesel et Meesters var. nov.   Plate 64: 5
Diagnosis: Differs from the nominate variety by semicells trapezoid in outline.
Type: Skuja 1948, pl. 18: 6 [as *Staurastrum maamense* var. *atypicum* (Magnotta) Skuja]

Staurastrum cosmarioides Nordstedt   Plate 33: 20-21
Nordstedt 1870, p. 223, pl. 4: 43
Cells distinctly longer than broad, rather deeply constricted. Sinus linear. Semicells semi-elliptic
to subpyramidate in outline with broadly rounded apical angles and narrowly rounded, almost
rectangled basal angles. Cell wall smooth. Semicells in apical view 3(-4)-angular with straight
to slightly concave sides and broadly rounded angles. Zygospore unknown. Cell length (?64-)80-
100(-142?) μm, cell breadth 35-50(-62?) μm.
Ecology: in benthos of oligo-mesotrophic water bodies.
Distribution: South America, particularly Brazil. Records from other continents, including Europe, are doubtful. In Europe, *S. cosmarioides* has been recorded from Great Britain.

According to West & West (1912) it is possible that all the records of *S. cosmarioides* refer to trigonal and tetragonal
forms of *Cosmarium* species. Anyhow, there is a lot of morphological variation among the algal forms labeled *S.
cosmarioides*. As Nordstedt (1870) explicitly states the cell wall to be delicately punctate, the exclusion of all forms
described to be coarsely punctate/scrobiculate seems advisable.

Accounted *Staurastrum cosmarioides* (var. *cosmarioides*):
Staurastrum cosmarioides var. *tropicum* Borge 1918

Taxa excluded:
Staurastrum cosmarioides var. *callosum* Kurt Förster 1969
S. cosmarioides forma *elevatum* Prescott et A.M. Scott 1942
S. cosmarioides var. *minor* Irénée-Marie 1957
S. cosmarioides forma *minutum* Kurt Förster 1963
S. cosmarioides forma *maxima* Kurt Förster 1964
S. cosmarioides forma *procerum* Kurt Förster 1963

Staurastrum crassangulatum Coesel   Plate 27: 10-14
Coesel 2007, p. 10.
Synonym: *Staurastrum orbiculare* var. *angulatum* Kaiser 1919, p. 228, fig. 32 (original description)
Synonym: *Staurastrum kaiseri* Růžička 1972, p. 477 (invalid homonym of *S. kaiseri* Pevalek 1925)
Cells slightly longer than broad to almost as broad as long, deeply constricted. Sinus closed at
its apex, then opening widely. Semicells in outline pentagonal with (broadly) rounded angles.
Cell wall smooth, usually thickened at the lateral angles. Semicells in apical view 3-angular with
slightly concave sides and broadly rounded angles. Zygospore unknown. Cell length 33-44 μm, cell
breadth 30-41μm.
Ecology: in benthos and tychoplankton of slightly acidic or circumneutral, mesotrophic water
bodies.
Distribution: Europe; known from Germany, Austria, Czech Republic and the Netherlands, but
most likely wider distributed.

Staurastrum crassimamillatum Coesel et Meesters stat. et nom. nov.   Plate 29: 6

Synonym: *Staurastrum ecorne* var. *podlachicum* B. Eichler et Gutwiński 1895, p. 174, pl. 5: 47

Synonym: *Staurodesmus subpygmaeus* var. *podlachicus* (B. Eichler et Gutwiński) Teiling 1967

Cells about as long as broad, rather deeply constricted. Sinus at its apex more or less linear over a short distance, then opening widely. Semicells campanulate with slightly convex apex. Lateral sides subapically with a mamilla-like outbulging. Cell wall smooth. Semicells in apical view 3-angular with slightly convex sides and broadly rounded, distinctly produced angles. Zygospore unknown. Cell length 23 µm, cell breadth 23 µm.

Ecology: in benthos/tychoplankton of oligo-mesotrophic water body.

Distribution: only known from a site in Poland.

S. crassimamillatum is a questionable species, being only described from a single site. Originally described as a variety of *Staurastrum ecorne* W.B. Turner, Teiling (1967: 502) transferred it (as a variety) to *Staurodesmus subpygmaeus*. In our opinion, however, neither of those options is satisfactory.

Staurastrum crenulatum (Nägeli) Delponte   Plate 74: 12-21

Delponte 1878, p. 68, pl. 12: 1-11

Basionym: *Phycastrum crenulatum* Nägeli 1849, p. 129, pl. 8: B

Cells somewhat broader than long to about as broad as long, deeply constricted. Sinus a V-shaped incision. Semicell body subfusiform with a short-cylindric base, the lateral sides produced to form processes of variable length. Processes about horizontal, distinctly articulate, tipped with minute spines and towards the semicell body furnished with acute granules or denticulations being arranged in concentric series but on the dorsal and ventral side usually much more prominent than on the lateral sides. Semicell body also furnished with denticulations or emarginate verrucae but almost exclusively on the apex. Semicells in apical view 3-5-radiate with straight to slightly concave sides, the angles produced to form processes, semicell body with a number of intramarginal, emarginate verrucae on each side. Zygospore globose, furnished with long spines that are 1-2 times furcate at the apex. Cell length (17-)20-30 µm, cell breadth (20-)25-35(-45) µm.

Ecology: in benthos and tychoplankton of slightly acidic to slightly alkaline, mesotrophic water bodies.

Distribution: possibly cosmopolitan. In Europe, *S. crenulatum* is widely distributed and presumably rather common (although most of the records appear to refer to other species).

The taxonomic concept of *S. crenulatum* is rather obscure. Nägeli (1849) in his original illustrations represents hardly any cell wall sculpturing. West (1899a), on the contrary, depicts prominent emarginate verrucae, in particular on the dorsal side of the semicell. Most (if not all) figures provided by Delponte refer to other species. As Nägeli's original concept does not provide enough characteristics to enable distinction from a number of superficially resembling species West's 1899 concept is adopted.

Accounted *Staurastrum crenulatum* (var. *crenulatum*):

Staurastrum crenulatum var. *britannicum* Messikommer1927

S. crenulatum var. *nepalense* Hirano 1955

Taxon excluded:

Staurastrum crenulatum var. *continentale* Messikommer 1927 (cf *S. polymorphum*).

Staurastrum cristatum (Nägeli) W. Archer

Cells about as broad as long or somewhat broader than long, (rather) deeply constricted. Sinus widely open from an acute-angled apex, or narrowly open and more or less linear in its apical part and then widely opening outward. Semicells bowl-shaped to almost hexagonal, with straight or convex sides. Semicell lobes on the dorsal side furnished with a series of short spines extending from the lateral angles to the apex, on the other sides with a variable number of (acute) granules arranged in semiconcentric series. Median part of the semicell flanks without ornamentation. Semicells in apical view 3(-4)-angular with straight to slightly concave sides and rounded angles ending in a spine, each lobe with two series of short spines diverging from the angle to the (unsculptured) semicell centre. Zygospore unknown. Cell length (30-)35-50(-55) μm, cell breadth (30-)40-65(-70) μm.

Ecology: in benthos of acidic, oligo-mesotrophic water bodies.

Distribution: Eurasia, North America. In Europe, both var. *cristatum* and var. *oligacanthum* are rather widely distributed but not particularly common.

Although Archer (1866) stated *S. cristatum* and *S. oligacanthum* to be distinctly different species, in practice they appear to be interconnected by transitional forms.

var. *cristatum*   Plate 61: 1-3
Archer 1861, p. 738
Basionym: *Phycastrum cristatum* Nägeli 1849, p. 127, pl. 8, C: 1
Semicells bowl-shaped. Apical angles of semicells not produced.

Accounted *Staurastrum cristatum* var. *cristatum*:
Staurastrum cristatum var. *japonicum* Hirano 1951

var. *cuneatum* Hinode   Plate 61: 4-5
Hinode 1967, p. 77, fig. 3: 6
Semicells cuneate. Apical angles of semicells hardly or not produced.
Relevant literature: Coesel 1996.

Accounted *Staurastrum cristatum* var. *cuneatum*:
Staurastrum cristatum var. *navigiolum* (Grönblad) Coesel 1996

var. *oligacanthum* (W. Archer) Coesel et Meesters stat. et comb. nov.   Plate 61: 6-8
Basionym: *Staurastrum oligacanthum* Brébisson ex W. Archer 1866, p. 189
First illustration: Nordstedt 1875, pl. 8: 39
Semicells cuneate-hexagonal. Lateral angles of semicells distinctly produced.

Accounted *Staurastrum cristatum* var. *oligacanthum*:
Staurastrum oligacanthum var. *incisum* W. West 1892a

Taxa excluded:
Staurastrum oligacanthum forma *evoluta* Laporte 1931
S. oligacanthum var. *podlachicum* (B. Eichler et Gutwiński) W. et G.S. West 1895a (= *S. podlachicum*)

Staurastrum cyrtocerum Ralfs

Cells slightly broader than long to about as broad as long, deeply constricted. Sinus widely open with a V-shaped apex. Semicell body cup-shaped, the apical angles produced to form inflexed processes tipped with minute spines. Cell wall furnished with (acute) granules or (emarginate) denticulations, arranged in concentric series around the processes. Semicells in apical view 3(-4)-angular with (slightly) concave sides; along the sides with intramarginal series of (paired) granules. Zygospore globose, furnished with slender, furcate spines. Cell length 19-40 μm, cell breadth 20-45(-60?) μm.

Ecology: in benthos and tychoplankton of slightly acidic to slightly alkaline, mesotrophic water bodies.
Distribution: probably cosmopolitan.In Europe, var. *inflexum* is widely distributed and common whereas the nominate variety and var. *brachycerum* are rather rare.

The differences between the nominate variety, var. *brachycerum* and var. *inflexum* are only in degree, so rather arbitrary; see also comment on Lütkemüller's drawing of a zygospore in West, West et Carter (1923: 109). The distribution pattern of granules in apical cell view as represented in literature is varying. In the original illustrations of var. *brachycerum* and var. *inflexum* by Brébisson (1856) no distinct pattern is visible. However, on account of Grönblad's (1934) picture after an exsiccate originating from Brébisson's collection (see our Plate 78: 14) var. *brachycerum* in apical cell view most likely is characterized by intramarginal series of granules along the sides. The taxonomic relevance of an isthmial whorl of granules as figured in a number of papers is unclear.

var. *cyrtocerum*   Plate 78: 1-2
Ralfs 1848, p. 139, pl. 22: 10
Cells stout: cell length > 30 μm.

Accounted *Staurastrum cyrtocerum* var. *cyrtocerum* (forma *cyrtocerum*):
Staurastrum cyrtocerum var. *compactum* W. et G.S. West 1905a
S. cyrtocerum forma *tetragona* W. West 1890

Taxa excluded:
Staurastrum cyrtocerum var. *major* Wolle 1884
S. cyrtocerum var. *pentacladum* Wolle 1884

var. *brachycerum* (Brébisson) Coesel et Meesters stat. et comb. nov.   Plate 78: 13-15
Basionym: *Staurastrum brachycerum* Brébisson 1856, p. 139, pl. 1: 24
Synonym: *Staurastrum inflexum* var. *brachycerum* (Brébisson) Coesel 1996
Cells rather small: cell length ≤ 30 μm. Processes relatively short and strongly incurved.

Taxon excluded:
Staurastrum brachycerum var. *destitutum* Messikommer 1942

var. *inflexum* Coesel et Meesters stat. et comb. nov.   Plate 78: 3-12
Basionym: *Staurastrum inflexum* Brébisson 1856, p. 140, pl. 1: 25
Cells rather small: cell length ≤ 30 μm. Processes relatively long and only slightly incurved.

Staurastrum dentatum Willi Krieger   Plate 114: 7-8
Krieger 1932, p. 197, pl. 16: 20
Cells a little broader than long, deeply constricted. Sinus U- or V-shaped. Semicell body rectangular to lunate, the apical angles produced to form stout, diverging processes. Processes tipped with 3 (rather) stout spines, towards the semicell body furnished with some series of short spines/denticulations. At the base of each process, both dorsal and ventral, two distinct spines, the ventral ones being disposed quite close to the sinus. Semicells in apical view 3-radiate with deeply concave sides (including the processes) and with two median spines projecting from each margin. Zygospore unknown. Cell length 22-28 μm, cell breadth 28-31 μm.
Ecology: in tychoplankton of oligotrophic water bodies.
Distribution: Eurasia. In Europe, *S. dentatum* is only known from a mountain lake in Austria.

In view of the different dimensions of the spines at the end of the processes and the somewhat differently shaped sinus it is not quite sure that the Austrian alga depicted by Lenzenweger (1997) really refers to the same species as described by Krieger (1932) from Sumatra.

Accounted *Staurastrum dentatum* (var. *dentatum*):
Staurastrum dentatum var. *gracilis* Hirano 1959a

Staurastrum diacanthum A. Lemaire   Plate 91: 8-15
Lemaire 1890, p. 4, fig. 2

Cells broader than long to about as broad as long, deeply constricted. Sinus widely open and acute-angled to subrectangular. Semicells campanulate or cup-shaped, the apical angles produced to form rather long, stout, parallel or slightly diverging processes often abruptly bent downwards at some distance from the semicell body. Processes tipped with 3-5 spines and towards the semicell body furnished with sparse denticulations arranged in indistinct concentric series; about halfway, on the dorsal side, usually with two extra large, spine-like dentations. Semicell body also furnished with sparse granules but almost exclusively on the apex. Semicells in apical view 3-4-radiate with about straight sides, the angles produced to form processes. Within each margin of the semicell body a single series of scattered granules. Zygospore unknown. Cell length 30-50 μm, cell breadth 30-75 μm.

Ecology: in benthos and tychoplankton of acidic, oligotrophic water bodies.
Distribution: Eurasia, North America. In Europe, *S. diacanthum* is rather rare.

S. diacanthum is a problematic taxon. The original figure by Lemaire (1890) much resembles *S. acestrophorum* described by W. & G.S. West (1902a) from Ceylon. Homfeld (1929), on the contrary, considers *S. diacanthum* a variety of *S. paradoxum*. Forms of *S. diacanthum* in which the dorsal spines on the processes are much reduced sometimes can hardly be distinguished from *S. paradoxum* indeed. Possibly, under the name of *S. diacanthum* two different species are figuring, one affiliated to the (sub)tropical species *S. acestrophorum*, the other one to *S. paradoxum*.

Accounted Staurastrum *diacanthum* (var. *diacanthum*):
Staurastrum diacanthum var. *glabrius* (Grönblad) Prescott, in Prescott, Bicudo & Vinyard 1982

Taxa excluded:
Staurastrum diacanthum var. *americanum* A.M. Scott et Grönblad 1957 (cf *S. dentatum*)
S. diacanthum var. *evolutum* A.M. Scott et Grönblad 1957 (cf *S. acestrophorum* W. et G.S. West 1902)

Staurastrum dicroceros Růžička   Plate 67: 7
Růžička 1963, p. 73, fig. 1

Cells about as long as broad, deeply constricted. Sinus widely open, acute-angled. Semicells cup-shaped, the apical angles produced to form short, thick-set, slightly divergent processes. Processes tipped with (3-)4 blunt spines and at their dorsal side, at the transition to the semicell body, provided with two short, bifurcate processes that are disposed obliquely upwards. Semicell apex with two emarginate verrucae, semicell lobes with 2 concentric series of distant, acute granules. Semicells in apical view 3-radiate with about straight sides and truncate, spiny angles. On either side of each angle a stout, bifurcate, intramarginal spine. Within each margin of the semicell body a couple of median verrucae. Zygospore unknown. Cell length 29-35 μm, cell breadth 24-30 μm.
Ecology: in benthos of oligotrophic, acidic water bodies.
Distribution: only known from a site in the Slovakian Tatra Mountains.

Maybe, *S. dicroceros* is just a form of *S. arcuatum* var. *subavicula*.

Staurastrum dilatatum Ralfs
Plate 53: 9-15

Ehrenberg ex Ralfs 1848, p. 133, pl. 21: 8

Cells about as long as broad, deeply constricted. Sinus widely open, apex acute-angled to about rectangled. Semicells more or less rhomboid with broadly rounded angles; the basis of the semicell short-columnar and consequently the lateral angles often a little produced. Cell wall finely and densely granulate, granules arranged in concentric rings around the angles; usually also a distinct supraisthmial ring of granules may be distinguished. Semicells in apical view (3-)4-5-angular with concave sides and broadly rounded angles. Zygospore smooth-walled, compressed spherical, in top view with undulate margins. Cell length (21-)30-40(-46) μm, cell breadth (21-)30-40(-46) μm. Ecology: in benthos and tychoplankton of acidic, meso-oligotrophic water bodies. Distribution: possibly cosmopolitan. In Europe, *S. dilatatum* is widely distributed and, particularly in mesotrophic habitats, rather common.

Like the original illustration of *S. alternans*, Ralfs' (1848) figure of *S. dilatatum* is poor in quality, reason to adopt the concept of this species as given in West & West (1912). As compared to *S. alternans*, cells of *S. dilatatum* in average are larger, usually 4-5 radiate (versus 3-radiate) and more rhomboid in shape.

Accounted *Staurastrum dilatatum* (var. *dilatatum*):
Staurastrum dilatatum var. *hibernicum* W. et G.S. West 1912
S. dilatatum var. *obtusilobum* De Notaris 1867

Taxa excluded:
Staurastrum dilatatum forma *australica* Schmidle 1896
S. dilatatum var. *extensum* Borge 1906
S. dilatatum var. *indicum* Turner 1892 [= *S. striolatum* (Nägeli) W. Archer 1861]
S. dilatatum var. *insigne* Raciborski 1892 [= *S. sinense* var. *insigne* (Raciborski) Compère 1983]
S. dilatatum forma *productum* A.M. Scott et Prescott 1957
S. dilatatum var. *thomassonii* Ricci 1990
S. dilatatum forma *trigranulatum* G.W.F. Carlson 1913

Staurastrum dimazum (Lütkemüller) Grönblad
Plate 106: 1-3

Grönblad 1920, p. 62

Basionym: *Staurastrum natator* (W. West 1892a) ssp. *dimazum* Lütkemüller 1910, p. 498, pl. 3: 16-18

Cells about as long as broad, deeply constricted. Sinus V- or U-shaped. Semicell body cup-shaped, the apical angles produced to form moderately long, divergent processes. Processes tipped with some three short spines and towards the semicell body furnished with denticulations arranged in (indistinct) concentric series. Denticulations increasing in size towards the base of the processes, often getting emarginate and proceeding on the lateral sides of the semicell body. Semicell body at the apex with two marked, emarginate verrucae; just above the isthmus near the lateral sides an additional (emarginate) granule. Face of the semicell body with two marked protuberances, left and right of the median vertical axis, each of them ornamented with a circle of granules. Semicells in apical view 2-radiate with an ellipsoid semicell body passing into the processes. Semicell body on either side with two pronounced dentate protuberances and two intramarginal emarginate verrucae. Zygospore unknown. Cell length 50-65(-76) μm, cell breadth 50-70(-95) μm. Ecology: in tychoplankton of acidic, oligo-mesotrophic water bodies. Distribution: Eurasia. In Europe, *S. dimazum* is a rare species; only a few records are known, i.e. from Scandinavia, Switzerland, Austria and the Netherlands.

Accounted *Staurastrum dimazum* (var. *dimazum*):
Staurastrum dimazum var. *elegantius* Grönblad 1920

Taxon excluded:
Staurastrum dimazum var. *reductum* Messikommer 1927 (= *S. reductum*)

Staurastrum dispar Brébisson   Plate 55: 7-10
Brébisson 1856, p. 144, pl. 1: 27

Cells about as long as broad, deeply constricted. Sinus open and acute-angled. Semicells fusiform-subrhomboid, dorsal side more convex than the ventral side, lateral angles narrowly rounded-truncate. Cells usually twisted at the isthmus. Cell wall finely granulate, granules arranged in concentric series around the angles. Semicells in apical view 3-angular with concave sides and narrowly rounded-truncate angles. Zygospore unknown. Cell length 24-37 µm, cell breadth 23-38 µm.

Ecology: in benthos and tychoplankton of slightly acidic-slightly alkaline, meso-oligotrophic water bodies.

Distribution: unclear because of problematic identification (see below). In Europe, *S. dispar* is locally not rare, particularly in mesotrophic habitats.

The original illustration of *S. dispar* by Brébisson (1856) is somewhat confusing as cell angles in frontal view are much more acute than in apical view. According to Grönblad & Růžička (1959) and Růžička (1972) shape of the cell angles may vary from broadly rounded to produced-truncate. It is questionable, however, whether all those algal forms have been identified correctly (compare, e.g., *S. hexacerum*).

Accounted *Staurastrum dispar* (var. *dispar*):
Staurastrum dispar var. *semicirculare* (Wittrock) Coesel 1996

Staurastrum donardense W. et G.S. West   Plate 52: 10
W. & G.S. West 1902, p. 50, pl. 2: 33

Cells a little longer than broad, deeply constricted. Sinus linear with a dilated extremity. Semicells subtrapezoid with broadly rounded angles. Cell wall very minutely granulate, granules arranged in indistinct concentric series around the angles. Semicells in apical view 3-angular with almost straight sides and broadly rounded angles. Zygospore unknown. Cell length 25-31 µm, cell breadth 21-25 µm.

Ecology: in benthos of acidic, oligotrophic water bodies and on wet rocks.

Distribution: only known for certain from Great Britain.

After the original decription by West & West (1902) only a very few, often dubious finds have been recorded. Maybe, *S. donardense* refers to a triradiate form of some *Cosmarium* species (*C. punctulatum?*).

Taxon excluded:
Staurastrum donardense var. *major* Borge 1918

Staurastrum duacense (W. West) W. et G.S. West   Plate 101: 1-3
W. & G.S. West 1909a, p. 202

Basionym: *Staurastrum pseudosebaldi* Wille subsp. *duacense* W. West 1892a, p. 184, pl. 24: 1

Cells somewhat broader than long, deeply constricted. Sinus V-shaped. Semicell body more or less cylindric, the apical angles produced to form long, horizontal processes. Processes tipped with 2-4 short spines and towards the semicell body furnished with concentric series of granules or denticulations, the dorsal and ventral ones being much better developed than the lateral ones. Dorsal series of process denticulations continuing on the apex of the semicell body as emarginate verrucae. Semicell body just above the isthmus usually with two concentric whorls of granules. Semicells in apical view biradiate with a circular, ellipsoid or rhomboid semicell body, rather

abruptly passing into the processes. Semicell body with two series of verrucae more or less in an even line with those on the processes. Zygospore unknown. Cell length 32-42 µm, cell breadth 55-73 µm.
Ecology: in acidic, oligo-mesotrophic water bodies.
Distribution: Eurasia. In Europe, *S. duacense* has been encountered only a few times (Ireland, Scandinavia).

Staurastrum duacense is an incidentally recorded, most questionable species. Probably it is a mere biradiate form of *S. manfeldtii* but as yet mixed populations or dichotypical forms are unknown.

Staurastrum dybowskii Woloszynska   Plate 72: 12-15
Woloszynska 1919, p. 38, pl. 3: 53-54
Cells about twice as broad as long, deeply constricted. Sinus open and acute-angled. Semicells fusiform, the poles produced to form rather long, parallel processes, tipped with some four spines. Cell wall furnished with (acute) granules, arranged in concentric series around the processes and continuing onto part of the semicell body, those at the apex in the form of emarginate verrucae. Semicells in apical view 3-radiate with concave sides, the angles produced into rather long processes; the inner series of granules on the semicell body arranged in pairs or threesomes (actually emarginate verrucae). Zygospore unknown. Cell length 21-27 µm, cell breadth 40-60 µm.
Ecology: in plankton of oligo-mesotrophic, slightly acidic water bodies.
Distribution: Europe, North America? From Europe, only a few records are known (Poland, the Netherlands, Austria, Sweden).

Staurastrum echinatum Ralfs   Plate 62: 1-6
Ralfs 1848, p. 215, pl. 35: 24
Cells about as long as broad, deeply constricted. Sinus open from an acute-angled apex. Semicells elliptic, covered with rather short, thorn-like spines and acute granules arranged in concentric series around the angles. Spines at the apex and the angles larger than those at the faces and then usually partly furcate, those at the base of the lateral sides usually projected downwards. Semicells in apical view 3-4-angular with almost straight sides and acuminate angles. Zygospore unknown. Cell length (25-)30-40(-45) µm, cell breadth (30-)35-45(-50) µm.
Ecology: in benthos of oligotrophic, acidic water bodies.
Distribution: only known for sure from Europe where it is widely distributed (including the Azores) but rather rare.
Relevant literature: Heimans 1926.

Accounted *Staurastrum echinatum* (var. *echinatum*):
Staurastrum echinatum var. *alpinum* Messikommer 1942
S. echinatum var. *pecten* (Perty) Rabenhorst 1868

Taxa excluded:
Staurastrum echinatum var. *sicaeferum* Irénée-Marie 1952
S. echinatum forma *spinulosum* Patel et Kumar 1980

Staurastrum echinodermum W. et G.S. West   Plate 43: 11
W. & G.S. West 1903a, p. 76, pl. 446: 13
Cells about as long as broad, deeply constricted. Sinus open from an acute-angled apex. Semicells sub-elliptical, covered with short spines which are arranged in three distinct, transversal series: two series running over the semicell flanks and one series along the apex, the spines in the apical series arranged in pairs. Semicells in apical view 4-5-angular with slightly concave sides and

rounded angles, series of spines along the margins, 2 pairs of intramarginal spines on each of the sides. Zygospore unknown. Cell length 33 µm, cell breadth 31 µm.
Ecology: in oligotrophic, acidic water.
Distribution: only known from a site in North Wales.

Staurastrum eichleri Raciborski
Plate 115: 3-5

Raciborski in Eichler & Raciborski 1893, p. 123, pl. 3: 25

Cells about as long as broad, deeply constricted. Sinus widely open. Semicells obversely trapeziform to almost rectangular, the apical angles produced to form relatively short, divergent processes. Processes biarticulate with a circle of granules at the base of each articulation and furcate at the top. Semicell body at the apex with a series of emarginate verrucae. At the base of the semicell a ring of big, about 6-dentate verrucae, separated from the apical ornamentation by an unsculptured zone. Semicells in apical view 5-radiate, the semicell body abruptly passing into the 5 processes. Processes at their base on either side with a broad, dentate verruca. Midregion of the semicell body with a ring of emarginate verrucae. Zygospore unknown. Cell length 28-48 µm, cell breadth 28-48 µm.
Ecology: in slightly acidic, oligo-mesotrophic water bodies.
Distribution: Europe. Very rare, only known from Poland and SW Finland.

S. eichleri has not been recorded outside of Europe. However, it is closely allied (if not identic) to *S. fuellebornei* Schmidle, a species widely distributed on the African continent.

Staurastrum ellipticum W. West
Plate 34: 1-3

W. West 1892, p. 731, pl. 9: 28

Cells ca $1^1/_2$ times longer than broad, rather deeply constricted. Sinus open and acute-angled. Semicells broadly elliptic in outline. Cell wall smooth. Semicells in apical view 3(-4)-angular with almost straight to slightly concave sides and broadly rounded angles. Zygospore unknown. Cell length (20-)40–46 µm, cell breadth (15-)29-32 µm.
Ecology: in benthos of acidic, oligotrophic water bodies, also among mosses on dripping rocks.
Distribution: Europe, North America, New Zealand. In Europe, *S. ellipticum* is only known with certainty from England, Swedish Lapland and Carelia (N.W. Russia).

As only a few incidental finds have been recorded, *S. ellipticum* is a poorly-known desmid species. Possibly it concerns a 3-4-radiate form of *Cosmarium contractum* Kirchner 1878.

Accounted *Staurastrum ellipticum* (var. *ellipticum*):
Staurastrum ellipticum var. *minor* Skuja 1964

Taxa excluded:
Staurastrum ellipticum forma *minus* Bicudo 1969
S. ellipticum forma *scrobiculata* Bourrelly et Couté 1991

Staurastrum elongatum J. Barker
Cells distinctly longer than broad, deeply constricted. Sinus a V-shaped incision. Semicells in outline vase-shaped, with a swollen base; the apical angles extended into relatively short, stout processes, usually slightly curving upwards. Processes ornamented with a number of concentric series of acute denticulations, and ending in 3-5 teeth. Semicell body just above the isthmus with 3-4 concentric series of conical granules or verrucae. Semicells in apical view 3-5-angular with concave sides, angles extending into processes; each side with 2 emarginate and 2 intramarginal median verrucae. Zygospore unknown. Cell length 50-80 µm, cell breadth 35-50 µm.

Ecology: in benthos and tychoplankton of acidic to neutral, oligo-mesotrophic water bodies. Distribution: rare; only known from Europe and the American continents. Within Europe, distribution is confined to atlantic regions.

var. *elongatum*   Plate 115: 1-2
Barker 1869, p. 424
First illustration: Wolle 1884, p. 130, pl. 46: 11-12
Usually 3-radiate. Isthmus breadth $^1/_4 - ^1/_5$ of total cell breadth. Cell wall ornamentation relatively modest, in the form of granules or small teeth.

Accounted *Staurastrum elongatum* var. *elongatum* (forma *elongatum*):
Staurastrum elongatum forma *pentagona* Irenée-Marie 1952
S. elongatum var. *quadratum* Irenée-Marie 1939
S. elongatum forma *quadratum* (Irenée-Marie) Prescott, in Prescott, Bicudo & Vinyard 1982
S. elongatum var. *tetragonum* Wolle 1884
S. elongatum forma *tetragona* (Wolle) Irenée-Marie 1952

Non-European variety:
var. *amazonense* A.M. Scott et Croasdale
Scott, Grönblad & Croasdale 1965, p. 53, fig. 206
Usually 5-radiate. Isthmus breadth ca $^1/_3$ of total cell breadth. Cell wall ornamentation relatively elaborate, with a long subterminal spine on the ventral side of each of the processes.

Staurastrum erostellum W. et G.S. West   Plate 49: 6
W. & G.S. West 1900, p. 296
Basionym: *Staurastrum rostellum* var. *erostellum* W. et G.S. West 1897, p. 493, pl. 6: 18
Cells about as long as broad, deeply constricted. Sinus open from an acute-angled apex. Semicells subelliptic-oval, covered with rather short spines arranged in concentric series around the angles; the spines near the angles somewhat longer than elsewhere. Semicells in apical view 3-angular with slightly concave sides and broadly rounded angles. Zygospore unknown. Cell length ca 20 µm, cell breadth ca 20 µm.
Ecology: in oligotrophic, acidic water bodies.
Distribution: Europe, North America? In Europe, *S. erostellum* is known from but one single site in the south of England.

S. erostellum is a poorly-known species. Maybe, it refers to a mere minor form of *S. brebissonii*.

Staurastrum extensum (Nordstedt) Coesel et Meesters stat. nov.   Plate 33: 17-19
Basionym: *Staurastrum orbiculare* var. *extensum* Nordstedt 1873, p. 26, pl. 1: 10
Cells distinctly longer than broad, deeply constricted. Sinus linear with a dilate extremity. Semicells semi-elliptic in outline, with broadly rounded angles. Cell wall smooth. Semicells in apical view 3-angular with slightly concave sides and broadly rounded angles. Zygospores globose and furnished with simple, acute spines. Cell length 38-50 µm, cell breadth 25-36 µm.
Ecology: in oligo-mesotrophic water bodies.
Distribution: unclear because of easy confusion with other Staurastra and triradiate Cosmaria. In Europe, records of *S. extensum* are mainly known from Great Britain, only incidentally from other countries (Czech Republic, Sweden).

Staurastrum floriferum W. et G.S. West   Plate 87: 8-11
W. & G.S. West 1896, p. 267, pl. 18: 1

Cells (slightly) broader than long, deeply constricted. Sinus acute-angled to almost rectangular. Semicells cup-shaped (triangular), often with a somewhat elevated apex, the apical angles produced to form rather long, parallel or diverging processes. Processes tipped with 3-4 spines, towards the semicell body furnished with a spiralling series of short spines/dentations. Semicells in apical view 3-radiate with about straight sides, the angles produced to form processes, the semicell body on each side with two intramarginal emarginate verrucae together forming a hexagonal ring. Zygospore unknown. Cell length 23-40 μm, cell breadth 45-55(-85?) μm.

Ecology: in plankton of meso-eutrophic, neutral to alkaline lakes. Also reported from slightly brackish water.

Distribution: USA, Europe. In Europe only recorded from Scandinavia.

Pictures of *S. floriferum* encountered in Scandinavian samples differ from the original figure of this species by W. & G.S. West (1896, from northern America) in having diverging processes instead of parallel ones. Smith (1924) however, also from N. America, provides pictures both resembling the forms represented by Florin (1957) and that by W. & G.S. West (1896). Both Smith (l.c.) and Florin (l.c.) suggest a possible relationship with *S. anatinum* (var. *denticulatum*).

Staurastrum forficulatum P. Lundell

Cells slightly broader than long to somewhat longer than broad, deeply constricted. Sinus acute-angled, open or (almost) closed near the isthmus and then widely opening. Semicells subtrapeziform to subhexagonal, the lateral/basal angles slightly produced and ending in two (rarely more) stout, diverging spines lying in the same vertical plane, the apical angles provided with a prominent, usually bifurcate spine. Similar spines on the semicell flanks near the base of the semicell lobes. Semicell lobes, particularly on their dorsal and ventral side, furnished with emarginate verrucae, small spines or acute granules arranged in semiconcentric series. Semicells in apical view (2-)3-4-radiate, the angles ending in a stout (bifurcate) spine, the sides furnished with two (furcate) spines projecting from each margin and with two others just within the margin. Zygospore unknown . Cell length 30-60(-75) μm, cell breadth 30-70(-95) μm.

Ecology: in benthos and tychoplankton of acidic, oligo-mesotrophic water bodies.

Distribution: Eurasia, North America, Australia. In Europe, all three varieties are widely distributed and locally rather common, particularly in boreal and (sub)alpine regions.

S. forficulatum is a most polymorphic species and often confused with other species, such as *S. arcuatum* and *S. monticulosum*. In literature, *S. forficulatum* var. *verrucosum* can also be found under the name of *S. senarium* Ralfs (e.g., Kouwets 1987, Coesel & Meesters 2007) a species, however, that was described by Ralfs (1848) very incompletely so gives rise to confusion.

Relevant literature: Péterfi (1972).

var. *forficulatum*   Plate 64: 6-9
Lundell 1871, p. 66, pl. 4: 5

Spines on the lateral angles much stouter than those elsewhere on the semicell; cell wall sculpturing on the ventral side of the semicell lobes in the form of emarginate verrucae.

Accounted *Staurastrum forficulatum* var. *forficulatum*:
Staurastrum forficulatum var. *cornutiforme* Wade 1957
S. forficulatum var. *exacutum* Cedergren 1932
S. forficulatum var. *eximium* A.M. Scott et Grönblad 1957
S. forficulatum var. *heteracanthum* Grönblad 1920

S. forficulatum var. *longicornis* Schmidle1898
S. forficulatum var. *subheteroplophorum* Grönblad 1920

var. *verrucosum* Grönblad

Plate 65: 4-8

Grönblad 1920, p. 64, pl. 3: 47, 50-51

Spines on the lateral angles hardly or not stouter than those elsewhere on the semicell; cell wall sculpturing on the ventral side of the semicell lobes in the form of marked, acute granules.

Accounted Staurastrum *forficulatum* var. *verrucosum*:
Staurastrum forficulatum var. *evolutum* Prescott, in Prescott, Bicudo & Vinyard 1982
S. forficulatum var. *subspongiosum* Grönblad 1920

var. *subsenarium* (W. et G.S. West) Coesel et Meesters comb. nov.

Plate 65: 1-3

Basionym: *Staurastrum furcatum* var. *subsenarium* W. et G.S. West 1894, p. 10, fig. 53

Spines on the lateral angles hardly or not stouter than those elsewhere on the semicell; cell wall sculpturing on the ventral side of the semicell lobes in the form of minute, often even wanting granules.

Questionable non-European taxa:
Staurastrum forficulatum var. *americanum* W. West 1891
S. forficulatum var. *ellipticum* C.C. Jao 1949
S. forficulatum var. *enoplon* W. West 1891
S. forficulatum var. *simplicius* C.C Jao 1949

Taxa excluded:
Staurastrum forficulatum var. *africanum* Bourrelly 1957
S. forficulatum var. *granulato-furcigerum* Huber-Pestalozzi 1928 (cf *S. arcuatum*)
S. forficulatum var. *kerrii* W.R. Taylor
S. forficulatum var. *minus* (F.E. Fritsch et F. Rich) Grönblad et A.M. Scott 1958

Staurastrum furcatum (Ralfs) Brébisson

Cells slightly longer than broad to somewhat broader than long, deeply constricted. Sinus open and acute-angled. Semicells subelliptical to trapezoid in outline, the angles produced to form short, simple or bifurcate processes. Near the base of the processes some distant granules may be present, cell wall otherwise smooth. Semicells in apical view 3(-4?)-radiate with straight to slightly concave sides, the angles ending in a stout (bifurcate) spine. On either side of each of the angular spines a (bifurcate) intramarginal spine. Zygospore globose, furnished with long spines that are bifid at the apex. Cell length (20-)30-40 μm, cell breadth (20-)30-40 μm.

Ecology: in benthos and tychoplankton of acidic, oligotrophic water bodies.

Distribution: cosmopolitan. In Europe, both var. *furcatum* and var. *aciculiferum* are widely distributed and common.

Relevant literature: Compère (1976), Péterfi (1973).

var. *furcatum*

Plate 66: 1-7

Brébisson 1856, p. 136

Basionym: *Xanthidium furcatum* Ehrenberg ex Ralfs 1848, p. 213

Original figure: Ehrenberg 1838, pl. 10: 25a,b (as *Xanthidium furcatum*)

Semicells trapeziform in outline. Apical and lateral projections about equal in size.

Accounted *Staurastrum furcatum* var. *furcatum* (forma *furcatum*):
Staurastrum furcatum var. *aculeatum* Schmidle 1895
S. furcatum var. *asymmetricum* Grönblad et A.M. Scott, in Grönblad, Prowse & Scott 1958
S. furcatum forma *elegantior* Irénée-Marie 1938
S. furcatum forma *indica* W.B. Turner 1892
S. furcatum forma *spinosa* Wittrock et Nordstedt, in Grönblad 1920

var. *aciculiferum* (W. West) Coesel   Plate 66: 8-14
Coesel 1996, p. 21
Basionym: *Staurastrum avicula* var. *aciculiferum* W. West 1889a, p. 293, pl. 291: 12
Synonym: *Staurastrum aciculiferum* (W. West) Andersson 1890, p. 11
Semicells elliptic in outline. Apical projections less pronounced than the lateral ones.

Accounted *Staurastrum furcatum* var. *aciculiferum*:
Staurastrum aciculiferum var. *burkartii* Tell 1980
S. aciculiferum var. *pulchrum* (W. et G.S. West) Kurt Förster 1970

Questionable non-European taxa:
Staurastrum furcatum var. *aristeron* A.M. Scott et Prescott 1961
S. furcatum var. *spinatum* Irénée-Marie 1952

Taxa excluded:
Staurastrum furcatum var. *candianum* (Delponte) Cooke 1887 (cf *S. arcuatum?*)
S. furcatum var. *cercophorum* Skuja 1949
S. furcatum forma *elliptica* Irénée-Marie 1952 (cf *S.arcuatum*)
S. furcatum var. *lyaense* Hinode 1962 (cf *S. forficulatum* var. *subsenarium*)
S. furcatum forma *minor* F.E. Fritsch et F. Rich 1937
S. furcatum var. *pisciforme* Irénée-Marie 1938 (cf *S. pseudopisciforme*)
S. furcatum forma *richae* Croasdale, in Grönblad & Croasdale 1971
S. furcatum var. *scaevum* A.M. Scott et Grönblad 1957
S. furcatum var. *senarium* [(Ehrenberg) ex Ralfs] Joshua 1886
S. furcatum var. *subsenarium* W. et G.S. West 1894 (= *S. forficulatum* var. *subsenarium*)
S. furcatum var. *taylorii* W.E. Wade 1957
S. furcatum var. *trifurcatum* Kurt Förster 1964
S. furcatum var. *vanoisii* Baïer 1978 (cf *S. forficulatum* var. *subsenarium*)

Staurastrum furcigerum (Ralfs) W. Archer   Plate 119: 4; Plate 120: 1-6
Archer in Pritchard 1861, p. 743, pl. 3: 32, 33
Basionym: *Staurastrum furcigerum* Brebisson 1840, p. 226
Synonym: *Didymocladon furcigerus* (Brébisson) ex Ralfs 1848, p. 144, pl. 33: 12
Cells about as broad as long, deeply constricted. Sinus (almost) closed at its apex, then opening widely. Semicells elliptic in outline, the lateral sides produced into relatively short, arm-like processes, tipped with 2-3(-5) spines. Apex of semicell with a series of similar processes, projecting obliquely upwards. Cell wall of the processes with concentric series of granulations/denticulations, cell wall of the semicell body for the greater part smooth. Semicells in apical view 3(-4)-radiate with straight to concave margins, each angle tapering into a short process; associated with each of the angular processes one or two similarly shaped, intramarginal processes. Zygospore globose, provided with numerous long spines that are double-bifurcate at the apex. Cell length (35-)50-80 μm, cell breadth (45-)55-75 μm.

Ecology: in benthos and plankton of slightly acidic to slightly alkaline, oligo-mesotrophic water bodies.
Distribution: Eurasia and the American continents. In Europe, *S. furcigerum* is widely spread.

S. furcigerum is highly polymorphic with respect to both length and number of cell processes as well as markedness of cell wall sculpture. Relating to those characteristics quite a number of infraspecific taxa have been described. Because of the frequent finds of intermediate forms (e.g., dichotypical cells) distinction of those taxa does not seem to be justified.

Accounted *Staurastrum furcigerum* (var. *furcigerum* forma *furcigerum*):
Staurastrum furcigerum forma *armigera* (Brébisson) Nordstedt 1888
S. furcigerum var. *armigerum* (Brébisson) G.M. Smith 1924
S. furcigerum var. *crassum* B. Schröder 1897
S. furcigerum forma *eustephana* (Ralfs) Nordstedt 1888
S. furcigerum forma *longicornis* Schmidle 1898
S. furcigerum var. *montanum* (Raciborski) W. et G.S. West 1898a
S. furcigerum var. *pseudofurcigerum* (Reinsch) Nordstedt 1880
S. furcigerum forma *pseudosenarium-tetragonum* Dick 1926
S. furcigerum var. *reductum* W. et G.S. West 1906
S. furcigerum var. *simplicissimum* Brook 1958

Taxa excluded:
Staurastrum furcigerum var. *egestosum* Beck-Mannagetta 1926
S. furcigerum forma *inequale* Beck-Mannagetta 1926

Staurastrum gatniense W. et G.S. West   Plate 43: 10
W. & G.S. West 1902, p. 48, pl. 2: 35
Cells slightly broader than long, deeply constricted. Sinus (almost) closed near its apex, then opening widely outward. Semicells subelliptic-trapeziform, the broad apex provided with a pair of swellings. Basal angles each furnished with a stout, downwardly projected spine. Lateral margins with some distant, shorter spines. Semicells in apical view 3-radiate with slightly concave sides, very broadly rounded angles and a series of 3 x 3 intramarginal humps. Angles encircled by a number of short spines/denticulations. Zygospore unknown. Cell length 27.5 µm, cell breadth 33.5 µm.
Ecology: acidic, oligotrophic.
Distribution: only known from Lough Gatny, in the north of Ireland.

Of *S. gatniense* only one reliable record is known. Although West & West (1902) considered it a peculiar *Staurastrum*, not very closely allied to any other British species, it was never found again. Presumably, judging from the wanting range in cell dimensions in the original description, but a single specimen was encountered. Possibly it refers to an anomalous form of some other species.

Staurastrum gemelliparum Nordstedt   Plate 37: 15-17
Nordstedt 1870, p. 230, pl. 4: 54
Cells about as long as broad, deeply constricted. Sinus open and acute-angled. Semicells about hexagonal in outline, the lateral and apical angles produced to form a pair of processes disposed in the same horizontal plane. Processes bifid at the apex, the two teeth lying in the same vertical plane. Cell wall smooth. Semicells in apical view 3(-4)-radiate, each radius being bilobed thus giving rise to a 6 (or 8)-angular semicell, at each angle a lower pair visible under an upper pair. Zygospore unknown. Cell length 25-35 µm, cell breadth 20-40 µm.
Ecology: in benthos and tychoplankton of acidic, oligotrophic water bodies.

Distribution: American continents, Eurasia, Africa. In Europe, *S. gemelliparum* is a rare species, known from Great Britain (?), Poland, Switzerland and Portugal.

Accounted *Staurastrum gemelliparum* (var. *gemelliparum* forma *gemelliparum*):
Staurastrum gemelliparum var. *africanum* Claassen 1961
S. gemelliparum forma *simplex* Gutwiński 1896

Taxon excluded:
Staurastrum gemelliparum var. *fabrisii* Tell 1980

Staurastrum geminatum Nordstedt   Plate 45: 1-3
Nordstedt 1873, p. 30, fig. 13
Cells about as long as broad, deeply constricted. Sinus rather widely open, acute-angled. Semicells elliptic in outline, each of the lobes provided with two superimposed stout, often curved terminal spines and two similar superimposed spines on both sides just proximal of it. Cell wall otherwise smooth. Semicells in apical view 3-angular with almost straight sides, each side (including the angles) furnished with 4 marginal and 4 intramarginal spines. Zygospore unknown. Cell length 40-45 µm, cell breadth 40-50 µm.
Ecology: in oligo-mesotrophic water bodies.
Distribution: poorly known; reliable records are only known from some sites in Norway.

Accounted *Staurastrum geminatum* (var. *geminatum*):
Staurastrum geminatum var. *longispinum* Printz 1915

Taxa excluded:
Staurastrum geminatum var. *heteracanthum* Grönblad 1921
S. geminatum var. *minus* S. Peterfi 1943
S. geminatum var. *numeraria* Istvánffi 1887
S. geminatum var. *rotundatum* Boldt 1885

Staurastrum glaronense Messikommer   Plate 74: 1-2
Messikommer 1951, p. 65, pl. 2: 25
Cells about as long as broad, deeply constricted. Sinus open, (sub)obtuse-angled. Semicells elliptic, the lateral sides produced to form short, thick-set, slightly convergent processes. Processes tipped with a series of minute spines and also towards the semicell body furnished with some concentric series of dentations. Semicell body at the apex with some scattered minute spines and at the base with a supraisthmial series of granules, otherwise smooth-walled. Semicells in apical view (4-)5-radiate, the short, thick-set processes rounded-truncate at their apex. Semicell body with a cluster of granules near the base of each process. Zygospore unknown. Cell length 19-22 µm, cell breadth 22-24 µm.
Ecology: in oligo-mesotrophic water bodies.
Distribution: only known from a site in the Swiss Alps.

Staurastrum gracile Ralfs   Plate 73: 1-7
Ralfs 1848, p. 136, pl. 22: 12
Cells broader than long, deeply constricted. Sinus widely open and acute-angled. Semicells cuneate to cup-shaped, the apical angles produced to form rather long, parallel processes, tipped with short spines. Cell wall furnished with (acute) granules, arranged in concentric series around the processes and continuing onto the semicell body, those at the apex usually emarginate. Flanks of the semicells just below the apical series of granules sometimes with an unsculptured zone. Semi-

cells at the base usually with a supraisthmial whorl of granules. Semicells in apical view 3(-5)-radiate with about straight sides, the angles produced to form rather long processes; the inner series of granules on the semicell body arranged in pairs (actually emarginate). Zygospore globose, furnished with slender spines that are once or twice divided at the apex. Cell length 30-40(-50) μm, cell breadth 40-60(-70) μm.

Ecology: in benthos and tychoplankton of (slightly) acidic, oligo-mesotrophic water bodies.

Distribution: unclear because of diverse taxonomic conceptions. In Europe, *S. gracile* (as conceived above) is widespread but of rather occasional occurrence.

S. gracile is known as one of the most notorious catch-all species. No doubt this has to do with the way in which this species is treated in the flora of West, West & Carter (1923) where, under the name of *S. gracile*, apparently different species are united. The original description (including illustration) in Ralfs (1848) is not very informative. Brook (1959) examined type material collected by Ralfs. Taking that Brook in his decription of that material was dealing with the same species as Ralfs (which is a bit questionable in view of differences in both cell size and cell shape) the diagnosis of *S. gracile* as provided in the present flora is primarily based on Brook (l.c.). We herewith exclude all known varieties of *S. gracile*, either as they are inconsistent with our present conception of the species, or because their original description is too little informative to enable any classification.

Taxa excluded:
Staurastrum gracile var. *bicorne* Bulnheim 1861
S. gracile var. *bulbosum* W. West 1892 (= *S. bulbosum*)
S. gracile var. *convergens* W. et G.S. West 1895
S. gracile var. *coronulatum* Boldt 1885
S. gracile var. *curtum* Nordstedt 1870
S. gracile var. *cyathiforme* W. et G.S. West 1895
S. gracile var. *elegantulum* W. et G.S. West 1902
S. gracile var. *elongatum* A.M. Scott et Prescott 1958
S. gracile var. *granulosum* Schmidle 1898
S. gracile forma *kriegeri* A.M. Scott et Prescott 1961
S. gracile forma *maior* E.M. Lind et Pearsall 1945
S. gracile forma *minimum* Prescott et A.M. Scott 1952
S. gracile var. *nanum* Wille 1881
S. gracile var. *naravashae* H. Bachmann 1938
S. gracile var. *nyansae* G.S. West 1907
S. gracile var. *ornatum* Willi Krieger 1932
S. gracile var. *planctonicum* E.M. Lind et Pearsall 1945
S. gracile var. *protractum* G.S. West 1907
S. gracile var. *pusillum* W. et G.S. West 1895
S. gracile var. *splendidum* Messikommer 1928 (= *S. manfeldtii* var. *splendidum*)
S. gracile var. *subornatum* Schmidle 1898
S. gracile var. *subtenuissimum* Woronichin 1926
S. gracile var. *subventricosum* Børgesen 1890
S. gracile var. *tenuissima* Boldt 1885
S. gracile var. *uniseriatum* W. et G.S. West 1895
S. gracile var. *verrucosum* W. et G.S. West 1895

Staurastrum grande Bulnheim

Plate 30: 1-6

Bulnheim 1861, p. 51, pl. 9: 14

Synonym: *Staurodesmus grandis* (Bulnheim) Teiling 1967

Cells about as long as broad, deeply constricted. Sinus widely opening from its acute apex. Semi-cells elliptic in outline, the lateral sides not seldom acuminated/ mamillate. Cell wall smooth. Semicells in apical view 3-angular with slightly concave to almost straight sides and rounded or acuminate/mamillate angles. Zygospore described as angular-globose, furnished with scattered, stout, often slightly curved spines. Cell length (50-)60-100 μm, cell breadth (45-)60-110 μm.

Ecology: in benthos and tychoplankton of acidic, oligotrophic water bodies.

Distribution: only known with certainty from Europe and North America. In Europe, *S. grande* is a rare species, mainly confined to atlantic regions.

Zygospores are only described by Cushman (1905) from New England (USA). The zygospores depicted are not attended with vegetative cells so that species identification cannot be verified.

Accounted *Staurastrum grande* (var. *grande* forma *grande*):

Staurastrum grande forma *intermedia* Schmidle 1898

S. grande forma *major* Lundell 1871

S. grande var. *parvum* W. et G.S. West 1894

Questionable non-European taxon:

Staurastrum grande var. *glabrum* Cushman 1905

Taxa excluded:

Staurastrum grande var. *angulosum* Grönblad 1920 (= *S. keuruense*)

S. grande var. *rotundatum* W. et G.S. West 1896

Staurastrum groenbladii Skuja

Plate 29: 1-4

Skuja 1931, p. 17, pl. 1: 16, 17

Cells about as long as broad, rather deeply constricted. Sinus narrowly open and more or less lin-ear in its apical part, widely opening outward. Semicells obversely trapezoid with broadly rounded angles, concave lateral sides and convex apex. Cell wall smooth. Semicells in apical view 3-an-gular with almost straight lateral sides and broadly rounded, often somewhat produced angles. Zygospore globose, covered with large, obtuse, conical projections. Cell length 24-30 μm, cell breadth 21-30 μm.

Ecology: in benthos and tychoplankton of slightly acid, mesotrophic water bodies.

Distribution: Europe, North America. In Europe, *S. groenbladii* is a rare species, only known from Latvia, Poland, Germany and Austria.

Staurastrum haaboeliense Wille

Plate 72: 6-11

Wille 1881, p. 42, pl. 2: 27

Cells broader than long, deeply constricted. Sinus open and acute-angled. Semicells narrowly elliptic-fusiform. Lateral angles produced into short, stout, about parallel processes, tipped with tiny spines. Cell wall provided with acute granules/denticulations, arranged in concentric series around the angles. Semicells in apical view 3(-4)-angular with deeply concave sides and truncate angles. Zygospore unknown. Cell length 15-20(-28) μm, cell breadth (22-)25-35 μm.

Ecology: in benthos and tychoplankton of acidic, oligotrophic water bodies.

Distribution: Eurasia, North America. In Europe, *S. haaboeliense* is particularly known from arc-tic-alpine regions; elsewhere it is very rare.

The original illustration of *S. haaboeliense* in Wille (1880) is rather schematic and but little informative. Therefore, the above concept is mainly based on that in West, West & Carter (1923).

Taxa excluded:
Staurastrum haaboeliense forma *alternatum* Patel et Kumar 1980
S. haaboeliense var. *elevatum* Skuja 1964

Staurastrum habeebense Irénée-Marie   Plate 29: 24-26
Irénée-Marie 1949a, p. 94, text fig. p. 95
Cells longer than broad, slightly constricted. Sinus widely open with obtuse apex. Semicells hexagonal in outline with broadly rounded angles and concave to straight sides. Cell wall smooth. Semicells in apical view 8-angular, the angles but little produced and broadly rounded. Zygospore unknown. Cell length 35-47 µm, cell breadth 26-31 µm.
Ecology: in ephemeral puddles, often on artificial substrate.
Distribution: Europe, North America. In Europe, *S. habeebense* is known from only a few sites in England, Czech Republic, Slovakia and the Netherlands but presumably it is rather widely distributed.
Relevant literature: Coesel & Hindak (2003).

Staurastrum hantzschii Reinsch   Plate 38: 1-6
Reinsch 1867, p. 129, pl. 22D: II
Cells slightly longer than broad, deeply constricted. Sinus open and acute-angled. Semicells broadly elliptic-oval in outline, the poles produced to form short processes which are 2-4-furcate at their apex. Both apically and laterally inserted two other, similar processes are present between each pair of consecutive angles. Cell wall smooth. Semicells in apical view 3(-4)-radiate with straight to concave sides, the angles produced to form short processes with on either side a pair of intramarginal and a pair of marginal, similar processes. Zygospore unknown. Cell length 40-60 µm, cell breadth 30-45 µm.
Ecology: in benthos and tychoplankton of acidic, oligotrophic water bodies.
Distribution: Eurasia, American continents, Africa. In Europe, *S. hantzschii* is of occasional occurrence in central European countries, elsewhere it is rare.

The original figures of *S. hantzschii* by Reinsch (1867) are confusing in that they show three whorls of processes per semicell, versus two whorls in all later publications. Actually, most illustrations of *S. hantzschii* in literature well agree with some of the original figures of *S. intricatum* Delponte. However, those figures (Delponte 1878, pl. 11: 10, 14, 15, 20) in Nordstedt's (1896) authoritative index of desmid taxa are referred to *S. hantzschii*. Possibly, Reinsch's original illustations have to do with an aberrant form. Anyhow, for making up the concept of *S. hantzschii* in the present flora the above-mentioned illustrations in Delponte (1.c.) are taken.

Accounted *Staurastrum hantzschii* (var. *hantzschii*):
Staurastrum hantzschii var. *congruum* (Raciborski) W. et G.S. West 1896
S. hantzschii var. *depauperatum* Gutwiński 1892
S. hantzschii var. *distentum* Grönblad 1962

Taxa excluded:
Staurastrum hantzschii var. *cornutum* W.B. Turner 1892
S. hantzschii var. *japonicum* J. Roy et Bisset 1886 (cf *S. tohopekaligense*)

Staurastrum heimerlianum Lütkemüller

Cells broader than long, deeply constricted. Sinus open and acute-angled. Semicells fusiform. Lateral angles produced into moderately long, parallel or slightly convergent processes, tipped with rather stout spines. Processes towards the base furnished with 2 or 3 (indistinct) concentric series of spines. Also the semicell body provided with a number of distant spines, those at the apex not seldom being emarginate. Semicells in apical view (3-)4(-5)-angular with straight or concave sides, the angles produced to form processes. Zygospore subglobose (somewhat angular) and furnished with long spines that are 1-3 times dichotomous at the apex. Cell length 17-26 µm, cell breadth 27-40 µm.

Ecology: in benthos and tychoplankton of acidic, oligotrophic water bodies.

Distribution: Europe, North America? In Europe, *S. heimerlianum* is locally common in alpine areas, elsewhere it is very rare.

var. *heimerlianum*
Plate 74: 3-8

Lütkemüller 1893, p. 568

Original description: Heimerl 1891, p. 608, pl. 5: 24 (as *Staurastrum cruciatum* Heimerl)

Spines on the processes unequal in size: those inserted about half-way being much stouter than the other ones.

Staurastrum cruciatum Heimerl 1891 is an invalid homonym of *Staurastrum cruciatum* Wolle 1876.

var. *spinulosum* Lütkemüller
Plate 74: 9-11

Lütkemüller 1893, p. 568, pl. 9: 17

All spines (both on the processes and the semicell body) about equal in length.

Taxa excluded:
Staurastrum heimerlianum var. *abundans* Willi Krieger 1932
S. heimerlianum var. *coronatum* Willi Krieger 1932
S. heimerlianum var. *sumatranum* A.M. Scott et Prescott 1961

Staurastrum hexacerum (Ehrenberg) ex Wittrock
Plate 71: 10-16

Wittrock 1872, p. 51

Basionym: *Desmidium hexaceros* Ehrenberg 1834, p. 293

First illustration: Ehrenberg 1838, p. 141, pl. 10: 10

Cells slightly broader than long to about as broad as long, deeply constricted. Sinus open and acute-angled. Semicells elliptic-fusiform (dorsal and ventral margin about equally convex). Lateral angles truncate, usually produced to form short processes. Cells not seldom twisted at the isthmus. Cell wall granulate, granules arranged in concentric series around the angles. Semicells in apical view 3(-4)-angular with concave sides and truncate angles. Zygospore globose, furnished with long spines once or twice bifid at the apex. Cell length 23-30(-41) µm, cell breadth 26-35(-42) µm.

Ecology: in benthos and tychoplankton of slightly acidic to slightly alkaline, mesotrophic water bodies.

Distribution: possibly cosmopolitan. In Europe, *S. hexacerum* is widely distributed.

The original description of *S. hexacerum* in Ehrenberg (1834) is very concise and also the illustration in his 1838 publication is most sketchy. The name of *S. tricorne* Meneghini ex Ralfs, in Ralfs (1848) considered a synonym of *S. hexacerum*, obviously refers to two different species so cannot be properly used, in spite of its nomenclatural priority. In the present flora, for *S. hexacerum* the taxonomic concept in West, West & Carter (1923) is adopted, implying shorter cell processes than in the original concepts of *S. hexacerum* and *S. tricorne*.

Accounted *Staurastrum hexacerum* (var. *hexacerum* forma *hexacerum*):
Staurastrum hexacerum var. *robustum* Kurt Förster 1965
S. hexacerum var. *subdilatatum* Schmidle 1894
S. hexacerum forma *tetragona* Boldt 1885

Taxa excluded:
Staurastrum hexacerum var. *aversum* W. et G.S. West 1897
S. hexacerum forma *alternans* Wille 1879 (cf *S. dispar*)
S. hexacerum var. *ornatum* Borge 1894
S. hexacerum var. *reductum* Hodgetts 1926
S. hexacerum var. *semicirculare* Wittrock 1872 (cf *S. dispar*)

Staurastrum hibernicum W. West   Plate 33: 1-3
W. West 1892, p. 177, pl. 23: 6
Synonym: *Staurastrum orbiculare* var. *hibernicum* (W. West) W. et G.S. West 1912, p. 156
Cells somewhat longer than broad, deeply constricted. Sinus linear with a dilate extremity. Semicells trapeziform in outline, with broadly rounded angles. Cell wall smooth. Semicells in apical view 3-angular with about straight sides and broadly rounded angles. Zygospore unknown. Cell length 42-65 µm, cell breadth 36-56 µm.
Ecology: in oligo-mesotrophic water bodies.
Distribution: cosmopolitan? In Europe, *S. hibernicum* is a widely distributed but rather rare species.

Taxon excluded:
Staurastrum orbiculare var. *hibernicum* forma *quadratum* Therezien 1985

Staurastrum hirsutum Ralfs
Cells (slightly) longer than broad, deeply constricted. Sinus (slightly) open from an acute-angled apex. Semicells trapezoid with broadly rounded angles to almost semicircular/semi-elliptic. Cell wall covered with short spines or acute granules, occasionally grouped in pairs, which are arranged in (obscure) circles around the angles. Semicells in apical view 3-angular with almost straight sides and broadly rounded angles. Zygospore (only known of the nominate variety) globose, provided with stout spines which are repeatedly bi- or threefold furcate. Cell length 35-65 µm, cell breadth 30-55 µm.
Ecology: in benthos of acidic, oligotrophic water bodies.
Distribution: Eurasia, American continents. In Europe, all three varieties are widely distributed and locally common.

S. hirsutum is most polymorphic in respect of cell dimensions as well as density and markedness of cell wall ornamentation. Maybe, several separate species are at issue.

var. *hirsutum*   Plate: 49: 7-10; Plate 50: 1-4
Ralfs 1848, p. 127, pl. 22: 3
Cells usually between 40 and 50 µm in length and between 35 and 45 µm in breadth. Cell wall covered with short, fine spines.

var. *muricatum* (Ralfs) Kurt Förster   Plate 50: 5-8
Förster 1981, p. 247
Basionym: *Staurastrum muricatum* Brébisson ex Ralfs 1848, p. 126, pl. 22: 2
Cells usually between 50 and 65 µm in length and between 40 and 55 µm in breadth. Cell wall covered with conical granules.

105

Accounted *Staurastrum hirsutum* var. *muricatum*:
Staurastrum muricatum var. *demidiatum* Comère 1901

Questionable variety:
S. muricatum var. *trapezicum* Gutwiński 1892 (cf *S. arnellii*)

Taxa excluded:
Staurastrum muricatum var. *australis* Raciborski 1892
S. muricatum var. *subturgescens* Schmidle 1893 (cf *S. turgescens*)
S. muricatum forma *tatrica* Gutwiński 1909 (cf *S. hirsutum* var. *hirsutum*)

var. *pseudarnellii* Coesel et Meesters var. nov.   Plate 50: 9-12
Diagnosis: differs from the nominate variety by cells usually between 35 and 55 μm in length and between 30 and 50 μm in breadth.
Cell wall covered with fine, acute granules, those on the basal angles usually enlarged to minute spines.
Type: Coesel 1997, plate 6: 2 (as *Staurastrum arnellii*)

In Coesel & Meesters (2007) this taxon unjustly is considered to be identical to *S. arnellii* Boldt.

Staurastrum horametrum J. Roy   Plate 51: 1
Roy 1893, p. 238
Roy & Bisset 1894, pl. 3: 2
Cells longer than broad, deeply constricted, hour-glass shaped. Sinus widely open, more or less rectangled. Semicells cuneate with broadly rounded apical angles. Angles with 3 or 4 concentric series of crowded, short spinelets, median part of the semicell flanks without ornamentation. Semicells in apical view 3-4-angular with slightly concave sides and broadly rounded angles, with a circle of spinelets around the centre and 3-4 concentric series of spinelets around each of the angles. Zygospore unknown. Cell length 57-65 μm, cell breadth 48-59 μm.
Ecology: in acidic, oligotrophic water bodies.
Distribution: Scotland.

S. horametrum is a questionable species. Apart from Roy's original description, no other reliable records are known. Maybe, it concerns a deviant form of *S. asperum*.

Taxa excluded:
Staurastrum horametrum var. *faroensis* Børgesen 1901 (cf *S. asperum*)
S. horametrum var. *minus* Gutwiński 1896
S. horametrum var. *orientale* A.M. Scott et Prescott 1961

Staurastrum hystrix Ralfs   Plate 43: 13-19
Ralfs 1848, p. 128, pl. 22: 5
Cells slightly longer than broad to about as long as broad, deeply constricted. Sinus rather widely open from an acute apex. Semicells oblong-oval with a concave to almost straight apex. Each of the lobes provided with some stout, often curved terminal spines and a circle of similar spines just proximal of it. Cell wall otherwise smooth. Semicells in apical view 3-4-angular with concave to almost straight sides and (very) broadly rounded angles; spines confined to the region of the angles. Zygospore more or less globular with a number (ca 10) of truncate, short-pillar-shaped protrusions which are crowned by a whorl of stout spines. Cell length 27-40 μm, cell breadth 28-38 μm.
Ecology: in benthos and tychoplankton of acidic, oligotrophic water bodies.

Distribution: Eurasia, American continents. Within Europe, *S. hystrix* is widely distributed.

Accounted *Staurastrum hystrix* (var. *hystrix* forma *hystrix*):
Staurastrum. hystrix var. *lithuanica* Raciborski 1889a
S. hystrix var. *pannonicum* Lütkemüller 1900
S. hystrix var. *papillifera* Lemmermann 1896
S. hystrix forma *recurvatum* Prescott et A.M. Scott 1942

Taxa excluded:
Staurastrum hystrix var. *brasiliense* Grönblad 1945
S. hystrix var. *floridense* A.M. Scott et Grönblad 1957
S. hystrix var. *granulatum* S. Ricci 1990
S. hystrix var. *paucispinosum* Schmidle 1893 (cf *S. quadrispinatum*)
S. hystrix var. *polyspina* Børgesen 1890
S. hystrix var. *robustum* Willi Krieger 1950
S. hystrix var. *tessulare* Nordstedt 1883

Staurastrum inconspicuum Nordstedt   Plate 114: 9-17
Nordstedt 1873, p. 26, pl. 1: 11
Cells about as long as broad, deeply constricted. Sinus widely open, almost semicircular, not seldom with a minute, median incision. Semicells quadrangular to cup-shaped in outline, the apical
angles extended into short, arm-like processes which are first directed obliquely outward and
then, about halfway their length, abruptly narrowed and directed obliquely upwards. At the joints
of the processes encircling series of minute, acute granules may be developed, just like at the
truncate tip. Semicells in apical view (3-)4(-6)-angular with concave sides, angles extending into
short, articulate processes. Zygospore oval-elliptical, smooth-walled. Cell length 12-20(-30) µm,
cell breadth 12-20(-30) µm.
Ecology: in benthos and plankton of acidic, oligotrophic water bodies.
Distribution: presumably cosmopolitan. In Europe, *S. inconspicuum* is widespread.

S. inconspicuum, mainly characterized by its relatively short, abruptly bended and narrowed processes for the remaining part is most polymorhic. The varieties known from literature appear to be interconnected by such a gradual
range of intermediate forms that their separate taxonomic status does not seem to be justified. Only the taxonomic
position of *S. inconspicuum* forma *gracilior*, described by Maskell (1889) from New Zealand, is open to some doubt
in view of its consistently 3-radiate cells which, in apical view, are marked by distinctly convex sides of the semicell
body. Cells of *S. inconspicuum* may be united to form short filaments (see Pl. 114: 20).

Accounted *Staurastrum inconspicuum* (var. *inconspicuum*):
Staurastrum inconspicuum var. *abbreviatum* Raciborski 1885
S. inconspicuum var. *crassum* Gay 1884
S. inconspicuum var. *planctonicum* G.M. Smith 1924

Questionable non-European taxon:
Staurastrum inconspicuum forma *gracilior* Maskell 1889

Taxon excluded:
Staurastrum inconspicuum var. *minor* Schmidle 1898

Staurastrum inelegans W. et G.S. West
Plate 25: 3-5

W. & G.S. West 1905, p. 501, pl. 7: 11, 12

Synonym: *Staurodesmus inelegans* (W. et G.S. West) Teiling 1948

Cells about as long as broad, deeply constricted. Sinus widely open with an obtuse angle giving rise to a somewhat elongated isthmus. Semicells obversely triangular with (slightly) convex sides in the middle, the angles somewhat produced and more or less acuminated. Cell wall smooth. Semicells in apical view 3-angular with sides that are convex in the middle and with produced angles that are acuminate to narrowly rounded. Zygospore unknown. Cell length 50-60 µm, cell breadth 47-62 µm.

Ecology: in plankton of acidic, oligotrophic water bodies.

Distribution: very rare, only reported from Scotland and Sweden.

Accounted *Staurastrum inelegans* (var. *inelegans*):
Staurastrum inelegans var. *obtusum* Lundberg 1931

Staurastrum informe Grönblad
Plate 88: 3-4

Grönblad 1920, p. 67, pl. 3: 75-76

Cells about as long as broad, deeply constricted. Sinus widely open from an acute-angled to rectangular apex. Semicells lunate in outline, the semicell body bowl-shaped, the apical angles produced into upwardly curved processes. Processes tipped with some three rather stout spines and towards the semicell body furnished with a spiralling series of denticulations. Semicell body itself smooth-walled. Semicells in apical view 3-angular. Zygospore unknown. Cell length 46-53 µm, cell breadth 46-53 µm.

Ecology: in plankton of oligotrophic water.

Distribution: only known from a site in Finland.

According to Grönblad (1938), the illustration in his 1938 paper would be more reliable than the original one in his 1920 paper.

Staurastrum insigne P. Lundell
Plate 29: 20-23

P. Lundell 1871, p. 58, pl. 3: 25

Cells longer than broad, moderately constricted. Sinus widely open, subrectangled. Semicells rhomboid in outline with rounded angles and a produced apex. Cell wall smooth. Semicells in apical view (4-)5-angular. Zygospore unknown. Cell length (17-)26-32 µm, cell breadth (14-)20-25 µm.

Ecology: in benthos of acidic, oligotrophic water bodies.

Distribution: Europe, North America, Asia. In Europe, *S. insigne* is a rare species, confined to arctic-alpine regions.

Accounted *Staurastrum insigne* (var. *insigne* forma *insigne*):
Staurastrum insigne forma *groenlandica* Borge 1894

Staurastrum iversenii Nygaard
Plate 111: 12-13

Nygaard 1949, p. 96, text fig. 49

Cells broader than long, deeply constricted. Sinus V-shaped. Semicell body cup-shaped, the apical angles produced to form long, slender, divergent processes tipped with four stout spines and towards the semicell body furnished with a spiralling series of denticulations. Semicell body in frontal view ornamented with two or three transversal series of acute granules/denticulations: an apical series of bidentate verrucae, a subapical series of short spines/denticulations and a possible median series of acute granules. Semicells in apical view 2-radiate, the elliptic semicell body on

either side furnished with a marginal series of some four dentations (the two in the middle being spine-like) and an intramarginal series of some four bidentate verrucae. Zygospore unknown. Cell length 38-47 µm, cell breadth 70-76 µm.
Ecology: in plankton of oligo-mesotrophic water.
Distribution: only known from a lake in Denmark.

Staurastrum johnsonii W. et G.S. West   Plate 103: 1-4
W. & G.S. West 1896, p. 266, pl. 17: 16
Cells (slightly) broader than long, deeply constricted. Sinus V-shaped. Semicell body vase-shaped with inflated base, the apical angles produced to form long, slender processes. Processes tipped with some three, rather short spines and towards the semicell body furnished with spiralling series of denticulations, the dorsal and ventral ones of which being (much) better developed than the lateral ones. Semicell body at the apex with two transversal series of emarginate verrucae, just above the isthmus generally two concentric whorls of acute granules/denticulations. Semicells in apical view 2(-3)-radiate with either an ellipsoid to fusiform or a triangular semicell body, rather abruptly passing into the processes. Semicell body with two parallel series of verrucae along each of the sides (one marginal and one intramarginal series). Zygospore unknown. Cell length (40-)55-90(-100) µm, cell breadth (60-)80-110(-125) µm.
Ecology: in (tycho)plankton of acidic, oligo-mesotrophic water bodies.
Distribution: Europe, North America. In Europe, *S. johnsonii* is only known from Scandinavian countries where it is widely distributed and may be locally common.

Records of *S. johnsonii* from Europe are mostly labeled var. *perpendiculatum* Grönblad, a variety characterized by upwardly curved processes. However, as shown by Smith (1924) and Scott & Grönblad (1957) projection of the processes may vary, even within a single population.

Accounted *Staurastrum johnsonii* (var. *johnsonii*):
Staurastrum johnsonii var. *coloradense* Cushman 1904
S. johnsonii var. *depauperatum* G.M. Smith 1924
S. johnsonii var. *evolutum* A.M. Scott et Grönblad 1957
S. johnsonii forma *parvum* G.M. Smith 1924
S. johnsonii var. *perpendiculatum* Grönblad 1920
S. johnsonii var. *uralense* Woronichin 1930

Taxa excluded:
Staurastrum johnsonii var. *altior* F.E. Fritsch et F. Rich 1937
S. johnsonii var. *bifurcatum* A.M. Scott et Grönblad 1957
S. johnsonii var. *granulatum* Irénée-Marie 1957 (cf *S. bicorne*)
S. johnsonii var. *sparsidentatum* Nygaard 1977 (= *S. sparsidentatum* Nygaard, in Ostenfeld & Nygaard 1925)
S. johnsonii forma *spinosa* Couté et Rousselin 1975
S. johnsonii var. *triradiatum* G.M. Smith 1924a

Staurastrum julicum Pevalek   Plate 24: 3
Pevalek 1925, p. 81, text fig. 20
Cells somewhat longer than broad, rather deeply constricted. Sinus widely open from a subrectangular apex. Semicells broadly elliptic, the lateral sides furnished with a mamilla. Semicells in apical view 4-angular with slightly convex sides and mamillate angles. Zygospore unknown. Cell length 17.5 µm, cell breadth 13.5-15.5 µm.
Ecology: in benthos/tychoplankton of oligo-mesotrophic water body.
Distribution: only known from a site in Slovenia.

S. julicum is a somewhat questionable species, being recorded but a single time. Teiling (1967: 580) accounts it *Staurodesmus brevispina* (= *Staurastrum brevispina*) but in our opinion it is too different from that species to justify synonymizing.

Staurastrum kanitzii Schaarschmidt　　　　　　　　　　　　　　　　　　Plate 43: 12
Schaarschmidt 1883, p. 273, fig. 16
Cells slightly broader than long, deeply constricted. Sinus open and acute-angled. Semicells elliptic-rhomboid, the lateral angles provided with two stout, divergent spines. Semicell apex with two likewise stout spines between each pair of consecutive angles. Cell wall otherwise smooth. Semicells in apical view 3-angular with slightly concave sides, the angles rounded and furnished with two divergent spines. Within each marginan an additional pair of stout spines. Zygospore unknown. Cell length excluding spines 16 µm, cell breadth excluding spines 20 µm.
Ecology: in acidic, oligotrophic water.
Distribution: only known from a site in Hungary.

Staurastrum keuruense Coesel et Meesters nom. et stat. nov.　　　　　　Plate 31: 1
Synonym: *Staurastrum grande* var. *angulosum* Grönblad 1920, p. 66, pl. 3: 107, 108
Cells about as long as broad, deeply constricted. Sinus linear at its apex, then opening widely. Semicells about trapeziform with broadly rounded basal angles. Cell wall smooth, thickened at the basal angles. Semicells in apical view 3-angular with slightly concave sides and rounded angles. Zygospore unknown. Cell length 61 µm, cell breadth 57 µm.
Eology: in oligo-mesotrophic water body.
Distribution: only known from a site at Keuru (Finland).

Staurastrum kobelianum Schröder　　　　　　　　　　　　　　　　　　Plate 33: 15
Schröder 1919, p. 267, pl. 2: 23
Cells as long as broad, deeply constricted. Sinus open from an acute-angled apex. Semicells rectangular, the basal and apical angles broadly rounded and furnished with a conical granule. Cell wall otherwise smooth. Semicells in apical view 3-angular with slightly concave sides and broadly truncate angles bordered on either side by a granule. Zygospore unknown. Cell length 14-15 µm, cell breadth 14-15 µm.
Ecology: in acidic, oligotrophic water body.
Distribution: only known from a site in Silesia.

Staurastrum kouwetsii Coesel　　　　　　　　　　　　　　　　　　　　Plate 48: 1-5
Coesel 1996, p. 22, fig. 18, 19
Cells about as long as broad, deeply constricted. Sinus open from an acute-angled apex. Semicells subelliptic-rhomboid or subpyramidal with broadly rounded angles, equably covered with rather stout spines arranged in concentric series around the angles. Semicells in apical view 3-angular with slightly concave sides and broadly rounded angles. Zygospore globose and furnished with stout, double-furcate spines. Cell length 40-48 µm, cell breadth 43-48 µm.
Ecology: in benthos and tychoplankton of oligo-mesotrophic, acidic water bodies.
Distribution: unclear, because of possible confusion with other species. In Europe, *S. kouwetsii* is widely distributed.
Relevant literature: Kouwets (1987).

S. kouwetsii resembles *S. brebissonii* but differs by stouter and less densely inserted cell wall spines. *S. kouwetsii* often has been wrongly labeled *S. subbrebissonii* or *S. pilosum*.

Staurastrum laeve Ralfs
Plate 37: 6-14

Ralfs 1848, p. 131, pl. 23: 10

Cells about as broad as long or somewhat broader than long, deeply constricted. Sinus widely open, apex acute-angled to about rectangled. Semicells obversely triangular with (slightly) convex sides. Apical angles produced to form pairs of stout, arm-like, diverging processes which are bifurcate at the ends and are all arranged in the same horizontal plane (incidentally, however, at the semicell apex one or more additional processes may be present). Cell wall smooth. Semicells in apical view 3-radiate, each radius being bilobed thus giving rise to a 6-angular semicell with strongly concave sides. Zygospore globose, provided with a number of stout spines which are double-bifurcate at the apex. Cell length 20-30(-40) μm, cell breadth 23-30(-50) μm.

Ecology: in benthos and tychoplankton of acidic, oligo-mesotrophic water bodies.

Distribution: presumably cosmopolitan. In Europe, *S. laeve* is locally not rare in its appropriate habitat.

Forma *supernumeraria* Nordstedt, characterized by a variable number (1-3) additional, apically orientated processes per semicell (see our Pl. 37: 15) does not deserve a separate taxonomic status considering that most of such cells appear to be dichotypic (Grönblad 1947, Kouwets 1987).

Accounted *Staurastrum laeve* (var. *laeve* forma *laeve*):
Staurastrum laeve var. *latidivergens* Scott et Grönblad 1957
S. laeve var. *major* Bourrelly et Couté 1991
S. laeve forma *supernumeraria* Nordstedt 1873

Taxon excluded:
Staurastrum laeve var. *clevei* Wittrock 1869 (= *S. clevei* (Wittrock 1869) Roy 1893

Staurastrum laevispinum Bissett

Cells about as broad as long or somewhat broader than long, deeply constricted. Sinus widely open, more or less rectangled. Semicells obversely triangular to bowl-shaped. Semicell body at the apical angles gradually passing into diverging to parallel processes which attenuate towards their apices. Apices of the processes usually obtusely rounded, but sometimes emarginate or even bifid. Cell wall smooth. Semicells in apical view 3(-4)-angular with strongly concave sides. Zygospore unknown. Cell length 15-35 μm, cell breadth 17-45 μm.

Ecology: in benthos and plankton of acidic, oligotrophic water bodies.

Distribution: recorded from most of the continents, but only incidentally. In Europe, *S. laevispinum* is of rare occurrence, the nominate variety being known from Great Britain, Ireland and Austria, var. *compactum* only from a single site in Finnish Lappland.

S. laevispinum resembles *S. brachiatum* in most of its characteristics. Actually, the main difference is in the shape of the processes when seen in frontal view: in *S. laevispinum* distinctly tapering towards the end, in *S. brachiatum* broadening at the end because of diverging terminal teeth. In view, however, of the variability of this character in the affiliated species *S. subnudibranchiatum* (see Pl. 35: 14-17) and the high morphological variability in *S. brachiatum* it is thinkable that *S. laevispinum* has to be considered a form of *S. brachiatum*.

var. *laevispinum*
Plate 35: 7-9

Bissett 1884, p. 195, pl. 5: 5

Semicell processes entire at the end.

Accounted *Staurastrum laevispinum* var. *laevispinum* (forma *laevispinum*):
Staurastrum laevispinum forma *major* W. et G.S. West 1902

S. laevispinum var. *subbrachiatum* G.S. West 1905
S. laevispinum var. *tropicum* W. et G.S. West 1907

var. ***compactum*** (Grönblad) Coesel et Meesters comb. nov.   Plate 35: 10-13
Basionym: *Staurastrum brachiatum* var. *compactum* Grönblad 1942, p. 40, pl. 4: 5-9
Semicell processes bifid at the end, the two teeth lying in a horizontal plane so that they can be
best observed in top view.

Taxa excluded:
Staurastrum laevispinum var. *abbreviata* F.E. Fritsch et F. Rich 1937
S. laevispinum forma *sydneyensis* Raciborski 1892

Staurastrum lanceolatum W. Archer
Cells about as broad as long or a little broader than long, deeply constricted. Sinus open and acute-
angled. Semicells elliptic-lanceolate in outline, the lateral angles usually attenuate/mamillate. Cell
wall smooth. Semicells in apical view 3(-4)-angular with more or less concave sides and attenuate/
mamillate angles. Zygospores, known from the nominate variety, globose and furnished with numerous acute spines. Cell length 15-30 µm, cell breadth 15-30 µm.
Ecology: in benthos and tychoplankton of acidic, oligotrophic water bodies.
Distribution: only known with certainty from Europe and North America. In Europe, both var.
lanceolatum and var. *compressum* are of rare occurrence, var. *compressum* being reported in particular from atlantic regions, var. *lanceolatum* also from arctic-alpine regions.

Staurastrum lanceolatum may be confused with *Staurodesmus mucronatus* or *Staurodesmus dickiei*. Last-mentioned species are characterized by semicell angles that are provided with a short spine or acute mucro instead of a
blunt mamilla.

var. ***lanceolatum***   Plate 24: 4-7
Archer 1862, p. 79, pl. 2: 16-22
Synonym: *Staurodesmus lanceolatus* (Archer) Croasdale 1957
Sinus widely opening from its acute apex. Apical cell margin convex.

Accounted *Staurastrum lanceolatum* var. *lanceolatum* (forma *lanceolatum*):
Staurastrum lanceolatum var. *rotundatum* Messikommer 1951
S. lanceolatum forma *tetragona* Wille 1879

var. ***compressum*** W. et G.S. West   Plate 24: 8-11
W. & G.S. West 1894, p. 11, pl. 1: 22
Synonym: *Staurodesmus lanceolatus* var. *compressus* (W. et G.S. West) Teiling 1967
Sinus near its apex narrow and often sublinear, then opening widely outward. Apical cell margin
more or less truncate.

Taxon excluded:
Staurastrum lanceolatum subspec. *perparvulum* Nordstedt 1885 (cf *S. sibiricum*)

Staurastrum lapponicum (Schmidle) Grönblad   Plate 53: 5-8
Grönblad 1926, p. 29, pl. 2: 106, 107
Basionym: *Staurastrum punctulatum* var. *muricatiforme* forma *lapponica* Schmidle 1898, p. 57, pl. 3: 5
Cells about as long as broad, deeply constricted. Sinus widely open from an acute-angled apex.
Semicells ellipsoid to subrhomboid in outline. Cell wall beset with fine granules, near the an-

gles arranged in strikingly regular, concentric series. Semicells in apical view 3(-4)-angular with slightly concave sides and broadly rounded angles. Zygospore unknown. Cell length 32-44 μm, cell breadth 31-42 μm.
Ecology: in benthos and tychoplankton of slightly acidic, mesotrophic water bodies.
Distribution: Eurasia, American continents, Africa. In Europe, *S. lapponicum* is widely distributed and locally rather common.

Accounted *Staurastrum lapponicum* (var. *lapponicum*):
Staurastrum lapponicum var. *ellipticum* (Wille) Grönblad 1926

Taxa excluded:
Staurastrum lapponicum forma *depressum* Jackson 1971
S. lapponicum var. *flaccum* Coesel 1987

Staurastrum lenzenwegeri Coesel et Meesters stat. et nom. nov.   Plate 112: 4-6
Synonym: *Staurastrum octodontum* Skuja var. *tetrodontum* A.M. Scott et Grönblad forma *torta* Schwarz et Lenzenweger 1999, p. 75, fig. 3
Cells about as long as broad, deeply constricted and usually twisted at the isthmus. Sinus broadly U-shaped. Semicell body rectangular with concave apex, the basal angles furnished with a spine, the apical angles produced to form long, divergent processes. Processes tipped with some four stout spines and towards the semicell body furnished with a spiralling series of pronounced denticulations. Semicells in apical view 2-radiate, the elliptic semicell body gradually tapering into the processes. Zygospore unknown. Cell length 45-50 μm, cell breadth 48-54 μm.
Ecology: in plankton of eutrophic water.
Distribution: only known from a site in northern Germany.

S. lenzenwegeri might be related to *S. bibrachiatum* forma *excavatum* Brook 1982.

Staurastrum leptocladum Nordstedt
Cells deeply constricted. Sinus V- or U-shaped. Semicell body vase-shaped, often with inflated base, the apical angles produced to form long, slender, curved processes. Processes bifurcate at the end and towards the semicell body furnished with series of denticulations/short spines. Semicell body at the apex ornamented with some verrucae or spines, just above the isthmus usually with a whorl of granules, otherwise smooth-walled. Semicells in apical view 2-radiate with a rhomboid semicell body, rather abruptly passing into the processes. Zygospore unknown. Cell length 30-105 μm, cell breadth 70-150 μm.
Ecology: in (tycho)plankton of acidic, oligotrophic water bodies.
Distribution: known of all continents, but the main point of distribution is in the tropics. In Europe, *S. leptocladum* is a rare species. The nominate variety is only known from Sweden, var. *cornutum* also from Poland and France.
Relevant literature: Grönblad 1926.

var. *leptocladum*   Plate 106: 7-9
Nordstedt 1870, p. 228, pl. 4: 57
Semicell body in apical view with a series of intramarginal verrucae along each of the sides.

Accounted *Staurastrum leptocladum* var. *leptocladum*:
Staurastrum leptocladum var. *insigne* W. et G.S. West 1896

var. ***cornutum*** Wille   Plate 107: 1-3

Wille 1884, p. 19, pl. 1: 39

Semicell body in apical view with a pair of spines diagonally disposed in the centre.

Accounted *Staurastrum leptocladum* var. *cornutum*:
Staurastrum leptocladum var. *smithii* Grönblad 1945
S. leptocladum var. *rollii* Palamar-Mordvintseva 1964

Non-European variety:
var. ***denticulatum*** G.M. Smith
Smith 1922, p. 351, pl. 11: 14
Semicell body in apical view with a truncate elevation on either side.

Questionable non-European taxa:
Staurastrum leptocladum forma *africanum* G.S. West 1907
S. leptocladum var. *inerme* Kurt Förster 1969
S. leptocladum var. *nudum* Prescott 1936
S. leptocladum var. *parispinuliferum* Kurt Förster 1969
S. leptocladum var. *subinsigne* A.M. Scott et Prescott 1957

Taxa excluded:
Staurastrum leptocladum var. *canadense* Lowe 1924
S. leptocladum var. *coronatum* A.M. Scott et Grönblad 1957
S. leptocladum var. *curvatum* Lowe 1924
S. leptocladum var. *simplex* F.E. Fritsch et F. Rich 1937
S. leptocladum var. *sinuatum* Wolle 1883

Staurastrum levanderi Grönblad

Cells about as long as broad or a little broader than long, deeply constricted. Sinus V-shaped. Semicell body cup-shaped, the apical angles produced to form rather long, divergent processes tipped with three short spines or teeth and towards the semicell body furnished with a spiralling series of denticulations. Semicell body in frontal view ornamented with three transversal series of acute granules: an apical, a subapical and an isthmial series. Granules for the most part packed up into small groups (up to 3 granules). Semicells in apical view 2-radiate, the semicell body furnished with both a marginal and an intramarginal series of dentate verrucae. Zygospore unknown. Cell length 24-40 µm, cell breadth 30-55 µm.

Ecology: in euplankton of eutrophic water bodies.

Distribution: Eurasia, presumably also South America. In Europe, both the nominate variety and var. *hollandicum* are rare, var. *levanderi* being only known from one site in Finland and one site in France, var. *hollandicum* from two sites in the Netherlands.

var. ***levanderi***   Plate 111: 6-8

Grönblad 1938, p. 54, fig. 1: 9

Semicell processes relatively thick, tipped with short spines.

Accounted *Staurastrum levanderi* var. *levanderi*:
Staurastrum levanderi var. *capdeviellei* Tell et Couté 2004

var. ***hollandicum*** (Coesel et Joosten) Coesel et Meesters stat. et comb. nov.   Plate 111 : 9-11

Basionym: *Staurastrum hollandicum* Coesel et Joosten 1996, p. 12, fig. 7-11

Semicell processes relatively thin, tipped with minute dentations.

Staurastrum longipes (Nordstedt) Teiling

Cells usually slightly broader than long, deeply constricted. Sinus V-shaped. Semicells about cup-shaped, the apical angles produced to form long, slender, divergent processes, tipped with some four stout spines. Processes furnished with a spiralling series of dentations. Semicell body smooth-walled. Semicells in apical view 3(-5)-radiate. Zygospore unknown. Cell length (45-)60-75(-85) μm, cell breadth (60-)70-100(-140) μm.

Ecology: in euplankton of acidic, oligotrophic water bodies.

Distribution: Eurasia, American continents, Australia. In Europe, *S. longipes* is common in the north-western atlantic zone, elsewhere rare.

var. *longipes* Plate 107: 4-6; Plate 108: 1-4

Teiling 1946, p. 80

Basionym: *Staurastrum paradoxum* var. *longipes* Nordstedt 1873, p. 35, fig. 17

Semicell body cuneate-campanulate.

Accounted *Staurastrum longipes* var. *longipes*:

Staurastrum longipes var. *longibrachiatum* (Teiling) Skuja 1964

var. *contractum* Teiling Plate 108: 5-7

Teiling 1946, p. 81, fig, 24, 37

Semicell body bowl-shaped.

Taxa excluded:

Staurastrum longipes var. *evolutum* (W. et G.S. West) Thomasson 1955

S. longipes var. *parallelum* Grönblad 1948 (cf *S. paradoxum*)

Staurastrum longispinum (Bailey) W. Archer

Cells exclusive of spines about as long as broad or somewhat longer than broad, deeply constricted. Sinus widely open from its acute-angled apex. Semicells bowl-shaped with (slightly) convex apex. Apical angles provided with two (sometimes three) stout spines of varying length, projected obliquely outward, lying in the same vertical plane and running about parallel. Cell wall smooth. Semicells in apical view 3-angular with slightly concave sides and broadly rounded angles. Zygospore unknown. Cell length excluding spines (50-)80-120 μm, cell breadth excluding spines (60-)80-100 μm. Length of spines 7-30(-45) μm.

Ecology: in plankton of acidic, oligotrophic water bodies.

Distribution: North America, Eurasia, Australia, Africa. In Europe, *S. longispinum* is of very rare occurrence, except for the the British Isles and Sweden.

var. *longispinum* Plate 39: 5; Plate 40: 1-2

Archer in Pritchard 1861, p. 743

Basionym: *Didymocladon longispinum* Bailey 1851, p. 36, pl. 1: 17

The two spines at each of the apical angles are either parallel or converging.

Accounted *Staurastrum longispinum* var. *longispinum* (forma *longispinum*):

Staurastrum longispinum var. *bidentatum* (Wittrock) W. et G.S. West 1903

S. longispinum var. *depressum* Playfair 1908

S. longispinum forma *lundellii* Cedergren 1932

S. longispinum var. *minor* Evens 1949

Non-European variety:
var. ***praelongum*** A.M. Scott et Grönblad
A.M. Scott et Grönblad 1957, p. 41, pl. 29: 2-4
The two spines at each of the apical angles distinctly diverging.

Staurastrum maamense W. Archer   Plate 59: 9-10
Archer 1869, p. 200
First illustration: Cooke 1887, pl. 53: 3
Cells a little longer than broad, deeply constricted. Sinus linear with a dilated extremity. Semicells semicircular to rounded trapezoid. Cell wall provided with three series of crenel-like verrucae, running from apex to semicell base along each of the semicell lobes; cell wall otherwise smooth. Semicells in apical view 3-angular with concave sides and broadly truncate, tricrenate angles. Zygospore unknown. Cell length 32-43 µm, cell breadth 30-35 µm.
Ecology: in benthos and tychoplankton of acidic to neutral, oligo-mesotrophic water bodies.
Distribution: rare; only known from Europe and North America. Within Europe, distribution is strictly confined to atlantic regions.

Taxa excluded:
Staurastrum maamense forma *atypicum* Magnotta 1935 (= *S. magnottae* Scott et Grönblad 1957, cf *S. cornutum*)
S. maamense var. *atypicum* (Magnotta) Skuja 1948 (cf *S. cornutum*)
S. maamense var. *pulchellum* Couté et Rousselin 1975

Staurastrum magdalenae Børgesen et Ostenfeld   Plate 88: 1-2
Børgesen & Ostenfeld 1903, p. 618, text fig. 148
Cells somewhat broader than long, deeply constricted. Sinus widely open with acute-angled to subrectangular apex. Semicell body cup-shaped with slightly concave apex, the apical angles produced to form rather long, divergent processes often incurved at their end. Processes tipped with 4 spines and towards the semicell body furnished with concentric series of about evenly large spines, the proper semicell body smooth-walled. Semicells in apical view 3-radiate with smooth-walled semicell body and spiny processes. Zygospore unknown. Cell length 40-55 µm, cell breadth 65-70 µm.
Ecology: in plankton of mesotrophic water.
Distribution: only known from a site in the Faeröes.

Staurastrum manfeldtii Delponte
Cells (slightly) broader than long, deeply constricted. Sinus V-shaped. Semicell body cylindric to cup-shaped, the apical angles produced to form (rather) long, more or less parallel processes. Processes tipped with 3-4 short spines and towards the semicell body furnished with concentric series of granules or denticulations, the ventral and particularly the dorsal ones being (much) better developed than the lateral ones. Dorsal series of process denticulations continuing along the apex of the semicell body as emarginate verrucae. Semicell body just above the isthmus often with a separate series of granules. Semicells in apical view (2-?)3(-4)-radiate with slightly convex to slightly concave sides, the angles produced to form processes. Along each side an intramarginal series of emarginate verrucae. Zygospore globose, furnished with long spines which are furcate at the apex. Cell length (35-)40-65(-85) µm, cell breadth (45-)60-100(-130) µm.
Ecology: in benthos and tychoplankton of oligo-mesotrophic, slightly acidic to slightly alkaline water bodies.
Distribution: cosmopolitan. In Europe, var. *manfeldtii* is widely distributed and of common occurrence, var. *annulatum* and var. *splendidum* are less common and mainly recorded from mountainous regions whereas var. *productum* is only recorded a few times (from Scotland, Ireland, Finland and the Netherlands).

S. manfeldtii is a highly polymorphic desmid. In Delponte's (1878) original description both cylindrical and cup-shaped semicell bodies are depicted. Contrary to Teiling (1947) we are of opinion that *S. manfeldtii* cannot be reliably distinguished from *S. sebaldi* var. *ornatum* (Coesel 1992). Actually, *S. manfeldtii* takes a position in between *S. sebaldi* and *S. pingue*. Description of the zygospore (of *S. manfeldtii* forma Rich 1935) is rather schematic and might refer to another species.
Relevant literature: Coesel 1996a.

var. *manfeldtii*   Plate 95: 4-7; Plate 96: 1-3
Delponte 1878, p. 64, pl. 13: 6-19
Semicell body cylindric to cup-shaped, just above the isthmus usually provided with series of granules, either only under the base of each process or in the form of a (double) ring round the base of the semicell. Semicell body in apical view with smooth margins.

Accounted *Staurastrum manfeldtii* var. *manfeldtii*:
Staurastrum manfeldtii var. *ornatum* Palamar-Mordvintseva 1981
S. manfeldtii var. *parvum* Messikommer 1942

var. *annulatum* W. et G.S. West 1902   Plate 96: 4-5
W. & G.S. West 1902, p. 56, pl. 1: 30-31
Semicell body cylindric, inflated at the base and provided with a supra-isthmial, double ring of granules. Semicell body in apical view with smooth margins.

var. *productum* (W. et G.S. West 1905) Coesel et Meesters comb. nov.   Plate 97: 2-3
Basionym: *Staurastrum sebaldi* var. *productum* W. et G.S. West 1905, p. 504, pl. 7: 24
Synonym: *Staurastrum productum* (W. et G.S. West) Coesel 1996a, p. 103, fig. 27-29
Semicell body cup-shaped, processes slightly converging. Semicell body in apical view with smooth margins whereas the intramarginal verrucae are flattened, arranged in a remarkably even row and towards the end of the processes gradually decrease in size.

var. *pseudosebaldi* (Wille) Coesel et Meesters stat. et comb. nov.   Plate 98 : 1-4
Basionym: *Staurastrum pseudosebaldi* Wille 1881, p. 45, pl. 2: 30
Semicell body cylindric to cup-shaped, just above the isthmus usually provided with series of granules, either only under the base of each process or in the form of a (double) ring round the base of the semicell. Margins of the semicell body in apical view dentate or spinous.

Accounted *Staurastrum manfeldtii* var. *pseudosebaldi*:
Staurastrum pseudosebaldi var. *elongatum* Messikommer 1957
S. pseudosebaldi var. *polygranulata* Capdevielle et Couté 1980
S. pseudosebaldi var. *simplicius* W. West 1892

var. *splendidum* (Messikommer) Coesel   Plate 97: 1
Coesel 1996a, p. 102, fig. 5-6
Basionym: *Staurastrum gracile* var. *splendidum* Messikommer 1928, p. 209, pl. 9: 14
Semicell body cylindric, inflated at the base and provided with a compound supra-isthmial wart under the base of each process. Semicell body in apical view with two marginal dentations on each side, whereas the intramarginal emarginate verrucae tend to be arranged in a circle.

Taxa excluded:
Staurastrum manfeldtii var. *africanum* Hodgetts 1926
S. manfeldtii var. *bispinatum* W.B. Turner 1892

S. manfeldtii var. *fluminense* Schumacher 1961
S. manfeldtii var. *pinnatum* W.B. Turner 1892
S. manfeldtii var. *planctonicum* Lütkemüller in Grönblad 1942 (cf *S. pingue* var. *planctonicum*)
S. manfeldtii forma *spinulosa* Lütkemüller 1900

Staurastrum margaritaceum Ralfs   Plate 77: 1-10
Meneghini ex Ralfs 1848, p. 134, pl. 21: 9
Basionym: *Pentasterias margaritacea* Ehrenberg 1838, p. 144, pl. 10: 15
Cells a little longer than broad, rather deeply constricted. Sinus widely open, V-shaped at its apex. Semicells campanulate to cup-shaped, the (sub)apical angles produced to form short, truncate, slightly inflexed processes (sometimes almost completely reduced). Processes towards their apex stepwise attenuated and furnished with concentric series of granules. Similar granules, sometimes enlarged to emarginate verrucae, extending across the semicell apex. Semicells at the base not seldom with a supraisthmial whorl of granules. Semicells in apical view 4-6(-9?)-radiate, the angles produced into short, thick-set processes. Zygospore globose, furnished with long spines that are once or twice dichotomous at the apex. Cell length (18-)25-35(-38)µm, cell breadth (17-)20-25 (-32) µm.
Ecology: in benthos and tychoplankton of acidic, oligotrophic water bodies.
Distribution: Eurasia, American continents, New Zealand. In Europe, *S. margaritaceum* is a widely distributed and common species, particularly in *Sphagnum* bogs.

Accounted Staurastrum *margaritaceum* (var. *margaritaceum* forma *margaritaceum*):
Staurastrum margaritaceum var. *alpinum* Schmidle 1894a
S. margaritaceum var. *coronulatum* W. West 1890
S. margaritaceum var. *formosum* Lütkemüller 1900
S. margaritaceum forma *nana* Messikommer 1962
S. margaritaceum var. *subcontortum* W. et G.S. West 1897

Taxa excluded:
Staurastrum margaritaceum forma *amoyensis* Skvortsov 1928
S. margaritaceum var. *cruciatum* Playfair 1913
S. margaritaceum var. *elegans* C.C. Jao 1948
S. margaritaceum var. *gracilius* A.M. Scott et Grönblad 1957
S. margaritaceum var. *hirtum* Nordstedt 1880
S. margaritaceum var. *inornatum* W.B. Turner 1892
S. margaritaceum var. *robustum* W. et G.S. West 1897 (cf *S. crenulatum*)
S. margaritaceum var. *rugosus* J.H. Evans 1962
S. margaritaceum var. *subdivergens* Hinode 1959
S. margaritaceum var. *subtile* Boldt 1885 (cf *S. boreale*)
S. margaritaceum var. *subtile* forma *ornata* Boldt 1885 (cf *S. pertyanum*)
S. margaritaceum var. *subwillsii* Cedercreutz et Grönblad 1936 (cf *S. crenulatum*)
S. margaritaceum var. *tenuibrachum* Messikommer 1956
S. margaritaceum var. *trigonum* Manguin 1937 (cf *S. punctulatum* var. *pygmaeum*)
S. margaritaceum var. *truncatum* Boldt 1888 (cf *S. striolatum*)

Staurastrum megalonotum Nordstedt   Plate 64: 1
Nordstedt 1875, p. 35, pl. 8: 38
Cells nearly as long as broad, deeply constricted. Sinus widely open and acute-angled. Semicells subhexagonal fusiform with concave apical and upper lateral sides, the angles ending in a serrate spine. Apical margin with a pair of serrate protrusions, similar to those at the apical angles. Cell

wall furthermore provided with sharp denticulations, arranged in concentric series around the angles. Semicells in apical view 4-angular with concave sides, the angles ending in acute, serrate protrusions; each margin with 2 marginal and 2 intramarginal protrusions, similar to those at the angles. Cell length 51-55 µm, cell breadth 53-57 µm. Zygospore unknown.
Ecology: in benthos of slightly acidic, oligo-mesotrophic water bodies.
Distribution: Spitzbergen.

As far as known, there are, apart from the original description by Nordstedt (1875), no other reliable records of this species. All other records under the name of *S. megalonotum* refer to other species, mostly *S. monticulosum*. The relationship between *S. megalonotum* and *S. monticulosum* is undeniable and, maybe, *S. megalonotum* better is to be considered a variety of *S. monticulosum*. However, to decide in that question, more reliable records of *S. megalonotum* are needed.

Taxa excluded:
Staurastrum megalonotum forma *hastatum* Lütkemüller 1893 (cf *S. monticulosum*)
S. megalonotum var. *nordstedtii* Kurt Förster 1963 (cf *S. monticulosum*)
S. megalonotum var. *obtusum* Hastings 1892
S. megalonotum var. *waltheri* Nygaard 1977

Staurastrum meriani Reinsch
Plate 57: 2-8

Reinsch 1867, p. 125, pl. 23: D: I: 1-11
Cells about twice as long as broad, slightly constricted. Sinus a V- or U-shaped incision. Semicells obversely campanulate in outline, with concave to straight lateral margins and a (slightly) convex apex. Angles broadly rounded, the apical ones often a little produced. Cell wall, except for the isthmial region, evenly granulated; granules near the apical angles in concentric series, otherwise without a definite arrangement. Semicells in apical view (3-)4-6-angular with broadly rounded angles and concave to straight sides. Zygospore pumpkin-shaped, with 10-12 marginal crenations. Cell length 35-50 µm, cell breadth 20-28 µm.
Ecology: in benthos of acidic to neutral, oligo-mesotrophic water bodies, also sub-atmophytic on wet substrates.
Distribution: only known from Europe and the American continents. Typical upland species; locally not rare in European arctic and (sub)alpine regions.

The characteristic shape of the zygospore suggests a relationship with *S. alternans* and allied species.

Accounted *Staurastrum meriani* (var. *meriani* forma *meriani*):
Staurastrum meriani forma *campanulata* Ducellier 1918
S. meriani forma *constricta* Teodoresco 1907
S. meriani forma *rotundata* Borge 1892

Staurastrum micron W. et G.S. West
Cells about as broad as long or a little broader than long, deeply constricted. Sinus V-shaped. Semicell body bowl-shaped, the apical angles produced to form rather stout, diverging processes. Processes slightly dilate at the apex, tipped with 3-4 minute spines, towards the semicell body furnished with some two concentric series of short spines/dentations. Semicells in apical view (2-)3(-4)-radiate with concave to straight sides, the angles produced to form short processes. Zygospore unknown. Cell length (12-)15-20(-25) µm, cell breadth (12-)15-25(-30) µm.
Ecology: in benthos and tychoplankton of acidic, oligo-mesotrophic water bodies.
Distribution: Eurasia, North America, Africa. In Europe, var. *micron* is widely distributed but rather rare, var. *spinulosum* is only known from the Netherlands and Finland.

S. *micron* var. *perpendiculatum* as originally described by Grönblad (1920) as S. *iotanum* var. *perpendiculatum* most likely is a triradiate form of S. *tetracerum*. The algae pictured by Brook (!959d) under the name of S. *micron* var. *perpendiculatum*, however, do refer to S. *micron* (var. *micron*) indeed.

var. *micron*   Plate 92: 9-14

W. & G.S. West 1896a, p. 159, pl. 4: 50-51
Semicell body (almost) smooth-walled.

Accounted *Staurastrum micron* var. *micron*:
Staurastrum micron var. *angolense* W. et G.S. West 1897b
S. *micron* forma *biradiata* Irénée-Marie 1938
S. *micron* var. *granulatum* Schmidle 1898

var. *spinulosum* Coesel   Plate 92: 15-17

Coesel 1996, p. 22, fig. 24-25
Semicell body spinous, in apical view with two prominent spines projecting from each margin.

Taxa excluded:
Staurastrum micron forma *major* Brook 1959d (cf S. *paradoxum*)
S. *micron* var. *perpendiculatum* (Grönblad) Brook 1959d (cf S. *tetracerum*)
S. *micron* var. *unidentatum* Hirano 1959

Staurastrum micronoides Coesel et Joosten   Plate 92: 5-8

Coesel & Joosten 1996, p. 15, fig. 12-14
Cells about as long as broad, deeply constricted. Sinus widely open, acute-angled. Semicell body cup-shaped (triangular), the apical angles produced to form rather long, diverging processes. Processes usually curved upwards, slightly dilated at the apex, tipped with some four spines, towards the semicell body furnished with two or three (indistinctly) concentric series of short spines/dentations. Semicells in apical view 3-radiate with straight or slightly concave sides, the angles produced to form processes, the semicell body with two intramarginal (bi)dentate verrucae on each side. Zygospore unknown. Cell length 21-30 μm, cell breadth 24-30 μm.
Ecology: in plankton of eutrophic, pH-circumneutral water bodies.
Distribution poorly known because of possible confusion with morphologically resembling species. S. *micronoides* is of occasional occurrence in the Netherlands (from where it was originally described) but presumably it is rather widely distributed within Europe.

S. *micronoides* might be identical to given desmid forms labeled S. *floriferum* (e.g. in Grönblad 1938, fig. 2: 3)

Staurastrum miedzyrzecense (B. Eichler) Coesel et Meesters stat. nov.   Plate 101: 4-8

Basionym: *Staurastrum grallatorium* var. *miedzyrczecense* B. Eichler 1896, p. 132, pl. 4: 51
Synonym: *Staurastrum saltans* var. *miedzyrczecense* (B. Eichler) Cedercreutz et Grönblad 1936, p. 7, pl. 2: 26-30
Cells broader than long, deeply constricted. Sinus widely open from an acute-angled (to almost closed) apex. Semicells cup-shaped with an elevated, truncate apex the angles of which are furnished with a marked spine. Upper lateral sides produced to form rather long, parallel to convergent processes tipped with 2-3 short spines and towards the semicell body furnished with a spiralling series of denticulations. Semicells in apical view 2-radiate with an elliptic semicell body passing into the processes. Semicell body with four divergent, intramarginal spines placed in the corners. Zygospore unknown. Cell length 28-47 μm, cell breadth 32-57 μm.

Ecology: in benthos and tychoplankton of acidic, oligo-mesotrophic water bodies.
Distribution: Europe. *S. miedzyrzecense* is a rare species only known from sites in Poland, Finland, France and Belgium.

Cedercreutz et Grönblad (1936) rightly state that the taxon under discussion, originally described by Eichler (1896) as a variety of *S. grallatorium* Nordstedt, shows much more resemblance to *S. saltans* Joshua (described from Burma) than to *S. grallatorium* (described from Brazil). However, also *S. saltans* distinctly differs from *S. miedzyrzecense* in that it has larger cell dimensions, a supristhmial whorl of granules and remarkably large spines at the end of the processes.

Staurastrum minimum Coesel   Plate 91: 4-7
Coesel 1996, p. 23, fig. 20-23

Cells about as long as broad, deeply constricted. Sinus V-shaped. Semicell body cup-shaped (triangular), the apical angles produced to form rather long, divergent processes. Processes with slightly undulate/dentate margin and tipped with some minute dentations. Semicells in apical view 3-radiate with about straight sides, the angles produced to form processes. Cell wall very delicate. Zygospore unknown. Cell length 16-20 μm, cell breadth 18-23 μm.
Ecology: in oligotrophic, acidic water bodies.
Distribution: only recorded from the Netherlands and France, but most likely (much) more widely distributed.

S. minimum is an inconspicuous *Staurastrum* species. Except for its small cell dimensions and delicate cell wall, hardly any specific morphological features can be distinguished. It might be compared with *S. tenuissimum* W. et G.S. West (1895), described from Madagascar, which is characterized by somewhat longer, more slender processes.

Staurastrum minutissimum Reinsch
Cells about as long as broad, slightly constricted. Sinus widely open and obtuse-angled. Semicells quadrate-cuneate with broadly rounded, truncate, or retuse apical angles. Cell wall smooth. Semicells in apical view (?2-)3-5-angular, with broadly rounded, truncate or retuse angles and slightly concave to straight sides. Zygospore unknown. Cell length 8-17 μm, cell breadth 7-17 μm.
Ecology: in benthos of acidic, oligotrophic water bodies, also subatmophytic in wet moss cushions.
Distribution: Europa, N. America. In Europe, *S. minutissimum* is only incidentally recorded, mainly from arctic regions.
Relevant literature: West & West (1912).

S. minutissimum, particularly its var. *convexum*, sometimes is hard to be distinguished from *S. sibiricum*.

var. *minutissimum*   Plate 29: 7-8
Reinsch 1867, p. 140, pl. 23A: II: 3-8
Synonym: *Staurodesmus minutissimus* (Reinsch) Teiling 1967
Cells with concave apices, usually somewhat longer than broad.

var. *convexum* W. et G.S. West   Plate 29: 9-15
W. & G.S. West 1912, p. 131, pl. 119: 2
Cells with straight or slightly convex apices, usually somewhat broader than long.

Accounted *Staurastrum minutissimum* var. *convexum* (forma *convexum*):
Staurastrum. minutissimum forma *tetragona* Nordstedt 1875

Taxa excluded:
Staurastrum. minutissimum forma *trigona major* Wille 1879 (cf *S. sibiricum*)
S. minutissimum forma *trigona minor* Wille 1879 (cf. *S. sibiricum*)
S. minutissimum var. *constrictum* W. West 1892a (cf. *S. sibiricum*)
S. minutissimum forma *trigona* Nordstedt 1875 (cf *S. sibiricum*)

Staurastrum monticulosum Ralfs   Plate 63: 9-15
Brébisson ex Ralfs 1848, p. 130, pl. 34: 9
Cells about as long as broad or slightly longer than broad, deeply constricted. Sinus widely open and acute-angled. Semicells subtrapeziform with concave to straight lateral sides, the basal angles truncate with two spines lying in the same vertical plane, the apical angles acute and provided with a simple or a bifurcate spine. Apical margin with a pair of intramarginal spines, similar to those at the apical angles. Cell wall furthermore provided with a variable number of small granules, arranged in concentric series around the angles. Some additional spines between each pair of consecutive basal angles may occur. Semicells in apical view 3(-4)-angular with slightly convex to almost straight sides and subacute angles provided with two spines (one above the other). Inside each margin two spinate projections; also at the margin itself some short spines/dentations may occur. Cell length (27-)40-57 µm, cell breadth (26-)35-50 µm. Zygospore unknown.
Ecology: in benthos and tychoplankton of acidic, oligo-mesotrophic water bodies.
Distribution: Europe, North America. In Europe, *S. monticulosum* is widely distributed but mainly confined to boreal and (sub)alpine regions.
Relevant literature: Peterfi (1972).

Accounted *Staurastrum monticulosum* (var. *monticulosum*):
Staurastrum monticulosum var. *diplacanthum* (De Notaris) Nordstedt in De Toni 1889
S. monticulosum var. *groenlandicum* Grönblad 1920
S. monticulosum var. *inermius* Irénée-Marie 1952
S. monticulosum var. *minor* Lhotsky 1949

Taxa excluded:
Staurastrum monticulosum var. *alaskanum* Hirano 1968 (cf *S. pungens*)
S. monticulosum var. *bidens* G.S. West 1907
S. monticulosum var. *bifarium* Nordstedt 1873
S. monticulosum var. *pulchrum* W. et G.S. West 1902 (cf *S. furcatum* var. *aciculiferum*)
S. monticulosum var. *rhomboideum* Boldt 1885
S. monticulosum var. *sanctae-annae* Schaarschmidt 1883 (cf *S. furcatum*)
S. monticulosum var. *vallesiacum* Viret 1909 (cf. *S. pungens*)

Staurastrum multinodulosum Grönblad   Plate 109: 1-3
Grönblad 1926, p. 30, pl. 3: 113, 114
Cells 2-radiate, slightly broader than long to about as broad as long, deeply constricted. Sinus V-shaped. Semicell body in frontal view campanulate in outline, often with a short-cylindrical base Apical angles produced to form long, slender, divergent processes which usually are distinctly curved upwards. Processes tipped with some four minute spines and towards the semicell body furnished with a spiralling series of denticulations. Apex of the semicell body in frontal view marked by a pair of emarginate verrucae. Semicell body in lateral view with a number of short, parallel, transversal series of small denticulations ranging from the base of the process down to the semicell base. Wall of the semicell body otherwise quite smooth. Semicell body in apical view elliptic-rectangular with a pair of intramarginal, emarginate verrucae on either side. Zygospore unknown. Cell length 45-90 µm, cell breadth 50-95 µm.

Ecology: in plankton of meso-eutrophic, alkaline water bodies.
Distribution: Eurasia, Australia. In Europe, only known from some sites in Poland, Italy and Finland but presumably more widely distributed.

Staurastrum muticum Ralfs   Plate 31: 2-11
Brébisson ex Ralfs 1848, p. 125, pl. 21: 4
Cells about as long as broad, deeply constricted. Sinus open, obtusely rounded at the apex. Semicells subelliptic, subpyramidal to semicircular in outline. Cell wall smooth. Semicells in apical view (?2-)3(-4)-angular with concave sides and broadly rounded angles. Zygospores globose, covered with large, conical projections. Cell length 15-45 μm, cell breadth 15-40 μm.
Ecology: in benthos and plankton of acidic or circumneutral, oligo-mesotrophic water bodies.
Distribution: presumably cosmopolitan, in Europe widespread.

The zygospore represented in our Pl. 31: 8 is reproduced from Homfeld (1929) and in its shape essentially deviates from *S. muticum* zygospores as illustrated in Ralfs 1848 (furnished with bifurcate spines) and Salisbury 1936 (covered with simple teeth). On account of the adhering semicells, Homfeld's illustration seems to be the most reliable. For that matter, Dick (1923) depicts a similar zygospore type. However, it is well thinkable that *S. muticum* in its actual concept has to be considered a catch-all taxon involving different species, maybe even triradiate facies of *Cosmarium* species (Kouwets 1991).

Accounted *Staurastrum muticum* (var. *muticum* forma *muticum*):
Staurastrum muticum var. *depressum* (Nägeli) Nordstedt 1880
S. muticum var. *ellipticum* Wolle 1884
S. muticum forma *minor* Rabenhorst 1868
S. muticum var. *minor* Wolle 1884
S. muticum var. *parvum* Prescott, in Prescott, Bicudo & Vinyard 1982

Taxa excluded:
Staurastrum muticum forma *bieneanum* (Rabenhorst) Reinsch 1867a (= *S. bieneanum* Rabenhorst 1862)
S. muticum var. *convexum* (Scott et Prescott) Croasdale, in Croasdale, Flint & Racine 1994
S. muticum var. *polygonum* Grönblad 1945
S. muticum forma *punctatum* Compère 1967
S. muticum var. *subcurtum* (Nordstedt) Croasdale, in Croasdale, Flint & Racine 1994 (= *S. coarctatum* var. *subcurtum*)
S. muticum var. *subsphaericum* Børgesen 1894 (= *S. ellipticum*)
S. muticum var. *substriolatum* Raciborski 1890
S. muticum var. *victoriense* G.S. West 1905

Staurastrum mutilatum P. Lundell   Plate 57: 12
Lundell 1871, p. 74, pl. 5: 3
Cells longer than broad, slightly constricted. Sinus V-shaped. Semicells hexagonal in outline with broadly rounded angles, the lateral angles somewhat produced. Semicell wall provided with some five transversal series of compound verrucae: one supraisthmial series at the base, one series in the midregion passing the lateral angles, and three series in the upper part of the semicell. Semicells in apical view 4-angular with broadly rounded angles. Zygospore unknown. Cell length 57-63 μm, cell breadth 42-44 μm.
Ecology: in benthos of oligotrophic water bodies.
Distribution: Extremely rare, only known from a site near Uppsala (Sweden).

The record of *S. mutilatum* in Hirano 1953 likely refers to another (newly to be described) species.

Staurastrum myrdalense (K. Strøm) Coesel et Meesters stat. nov.   Plate 31: 12-15
Basionym: *Staurastrum bieneanum* var. *myrdalense* K. Strøm 1926, p. 229, pl. 5: 20
Cells somewhat longer than broad, deeply constricted. Sinus widely open from its acute-angled apex. Semicells elliptic in outline. Cell wall smooth. Semicells in apical view 3-angular with almost straight sides and broadly rounded angles. Zygospore unknown. Cell length 30-40 μm, cell breadth 25-35 μm.
Ecology: in benthos and tychoplankton of acidic, oligotrophic water bodies.
Distribution: Eurasia, North America. In Europe, *S. myrdalense* is a rare species, known from Norway, Switzerland, Germany and the Netherlands.

Staurastrum natator W. West   Plate 106: 4-6
W. West 1892a, p. 183, pl. 23: 14
Cells slightly broader than long to about as broad as long, deeply constricted. Sinus V- or U-shaped. Semicell body cup-shaped, the apical angles produced to form rather long, divergent processes. Processes tipped with some three short spines and towards the semicell body furnished with concentric series of acute granules or minute denticulations. Semicell body at the apex with three large mucronate or emarginate verrucae, just above the isthmus generally with one or two concentric whorls of acute granules. Face of the semicell body with a prominent central protuberance ornamented with a circle of granules. Semicells in apical view 2(-3)-radiate, with either a rhomboidal or a triangular semicell body passing into the processes. Semicell body with a prominent protuberance in the middle of each margin and a series of three verrucae within each margin. Zygospore unknown. Cell length (40-)50-65(-75) μm, cell breadth (55-)65-75(-85) μm.
Ecology: in tychoplankton of acidic, oligo-mesotrophic water bodies.
Distribution: Eurasia, North America. In Europe, *S. natator* is confined to the UK, Scandinavia and Austria. Especially in arctic regions it may be locally rather common.

The picture of *S. natator* by Smith (1924) who examined type material of this species collected by W. West from Ireland differs from the original drawing by West (1892a) in the presence of an isthmial whorl of granules and a less pronounced central inflation of the semicell body. Obviously, those characteristices are subject to variation.

Accounted *Staurastrum natator* (var. *natator*):
Staurastrum natator var. *arctoum* Schmidle 1898
S. natator var. *boldtii* Grönblad 1920
S. natator var. *crassum* W. et G.S. West 1896
S. natator var. *triquetrum* Grönblad 1920

Taxa excluded:
Staurastrum natator ssp. *dimazum* Lütkemüller 1910 (= *S. dimazum*)
S. natator var. *rhomboideum* N. Carter 1935

Staurastrum neglectum G.S. West   Plate 95: 1-3
G.S. West 1909, p. 70, pl. 3: 12
Cells slightly broader than long, deeply constricted. Sinus V-shaped, hardly visible. Semicell body vase-shaped, the lower part almost cylindrical, the apical angles produced to form gradually attenuating processes. Processes slightly divergent to slightly convergent, tipped with minute spines and furthermore provided with minute granules (arranged in concentric series). Cells usually twisted at the isthmus. Semicells in apical view 3-angular, the margins concave between the processes which are encircled with about 7 series of acute, minute granules. Zygospore said to be glo-

bose with slender processes that are 2 or 3 times forked at the apex. Cell length (?12-)18-26(-30?) μm, cell breadth (?20-)24-35(-49?) μm.
Ecology: in benthos and tychoplankton of circumneutral oligo-mesotrophic water bodies.
Distribution: Europe, Australia, North America. In Europe, *S. neglectum* is very rare, only recorded from the British Isles and Finland.

S. neglectum was originally described from southern Australia (West 1909). According to the latter author it would be identical to *S. tricorne* var. β in Ralfs (1848). To our mind it is questionable however, in view of the rather stout spines tipping the semicell processes, whether the zygospore pictured in Ralfs (1848, pl. 34: 8d) really belongs to *S. neglectum*.

Staurastrum novae-semliae Wille　　　　　　　　　　　　　Plate 52: 1-4
Wille 1879, p. 51, pl. 13: 58
Cells slightly longer than broad to as long as broad, rather deeply constricted (breadth of isthmus $^1/_2$ cell breadth). Sinus open, acute-angled. Semicells semicircular-ellipsoid, the basal angles curved downwards. Cell wall at the basal angles thickened, elsewhere provided with broad, flattened verrucae. Semicells in apical view 3-angular with straight to slightly convex sides and broadly rounded angles. Zygospore unknown. Cell length 28-34 μm, cell breadth 24-34 μm.
Ecology: in benthos of shallow, oligotrophic water bodies.
Distribution: Asia, Europe. *S. novae-semliae* is a rare, arctic species, in Europe only known from Lappland.

Accounted *Staurastrum novae-semliae* (var. *novae-semliae*):
Staurastrum novae-semliae var. *crassangulatum* Grönblad 1942

Staurastrum nygaardii Coesel et Meesters stat. et nom. nov.　　　　Plate 112: 7-9
Synonym: *Staurastrum smithii* Teiling var. *verrucosum* Nygaard 1949, p. 85, text fig. 40 a-d
Cells about as long as broad, deeply constricted, not seldom twisted at the isthmus. Sinus V-shaped. Semicell body bowl-shaped, the apical angles produced to form long, slender, divergent processes. Processes tipped with some four rather stout spines and towards the semicell body furnished with a spiralling series of dentations. Centre of semicell body with two verrucae (protuberances) positioned in the same horizontal plane; otherwise smooth-walled. Semicell in apical view 2-radiate. Semicell body with two dentate protuberances on either side. Zygospore unknown. Cell length 62-68 μm, cell breadth 54-70 μm.
Ecology: in plankton of eutrophic water.
Distribution: only known from a lake in Denmark.

S. nygaardii in its morphology links up with given forms of *S. tetracerum*. However, to avoid a further increase of polymorphism in the last-mentioned species, we prefer to render the taxon under discussion the status of a separate species.

Staurastrum obscurum Coesel　　　　　　　　　　　　　　Plate 29: 16-19
Coesel 1996, p. 23, fig. 34-42
Cells about as long as broad or slightly longer than broad, deeply constricted. Sinus open, acute-angled. Semicells in outline subhexagonal-subrhomboid with broadly rounded angles, the lateral angles often slightly produced. Cell wall faintly undulate, otherwise smooth. Semicells in apical view 4-angular, the angles of one semicell more or less alternating with those of the other. Zygospore unknown. Cell length 35 –39 μm, cell breadth 27-40 μm.
Ecology: in benthos of acidic, oligotrophic water bodies.
Distribution: only known from a site in the Netherlands.

Staurastrum ophiura P. Lundell

Cells distinctly broader than long, deeply constricted. Sinus widely open with subacute to U-shaped apex. Semicells bowl-, cup- or vase-shaped with convex apex, the apical angles produced to form long, slender, parallel or slightly converging, arm-like processes. Cell wall of the processes ornamented with many (at least 10) concentric series of denticulations and some 3 teeth at the tip. Cell wall of the semicell body with a ring of papillae or acute granules at the base, and a crown of conical granules or emarginate verrucae at the top, cell wall for the remaining part smooth. Semicells in apical view circular with a whorl of (4-5?)6-8(-9) radiating processes and a central ring of granula/verrucae equal in number to and alternating with the processes. Zygospore unknown. Cell length (50-)65-90 µm, cell breadth 100-170 µm.

Ecology: in plankton of acidic, oligotrophic water bodies.

Distribution: Europe, North America and Japan. In Europe, most of the records of var. *ophiura* originate from Great Britain, Ireland and Scandinavia; var. *subcylindricum* is only known from Sweden. Outside the atlantic region *S. ophiura* is very rare.

var. *ophiura*   Plate 116: 3; Plate 117: 1

Lundell 1871, p. 69, pl. 4: 7

Semicell body bowl- or cup-shaped, about as long as broad.

Accounted *Staurastrum ophiura* var. *ophiura*:

Staurastrum ophiura var. *cambricum* W. et G.S. West 1894

var. *subcylindricum* Borge 1939   Plate 117: 2-3

Borge 1939, p. 23, fig. 14

Semicell body vase-shaped, distinctly longer than broad.

Questionable non-European taxa:

Staurastrum ophiura var. *pentacerum* Wolle 1884

S. ophiura var. *tetracerum* Wolle 1884

Taxa excluded:

Staurastrum ophiura var. *coronatum* Scott 1950

S. ophiura var. *horizontale* Scott et Prescott 1958

S. ophiura var. *horridum* Scott 1950

S. ophiura var. *longiradiatum* Scott 1950

S. ophiura var. *minor* Couté et Rousselin 1975

S. ophiura var. *minus* Prescott et Scott 1942

S. ophiura var. *nanum* Woodhead et Tweed 1960

S. ophiura var. *perornatum* Grönblad 1945

Staurastrum orbiculare Ralfs   Plate 32: 1-5

Ehrenberg ex Ralfs 1848, p. 125, pl. 21: 5h

Cells somewhat longer than broad, deeply constricted. Sinus linear with a dilate extremity. Semicells subsemicircular in outline. Cell wall smooth. Semicells in apical view 3-angular with slightly concave sides and broadly rounded angles. Zygospore unknown. Cell length (31-)50-62 µm, cell breadth (24-)43-52 µm.

Ecology: in benthos and tychoplankton of acidic to neutral, oligo-mesotrophic water bodies.

Distribution: not well known because most records refer to varieties that should be accounted other species. In Europe, *S. orbiculare* is a widely distributed but rather rare species.

Relevant literature: West & West (1912).

Cell wall punctation in *S. orbiculare* as reproduced in literature varies in density and markedness. Dense patterns of coarse dots, indicative of a scrobiculate cell wall, presumably refer to other species.

Questionable non-European taxa:
Staurastrum orbiculare var. *borgei* Prescott 1966
S. orbiculare var. *minor* Prescott 1936
S. orbiculare var. *waterbergense* Claassen 1961

Taxa excluded:
Staurastrum orbiculare var. *aculeatum* Playfair 1910
S. orbiculare var. *angulatum* Kaiser 1919 (= *S. crassangulatum*)
S. orbiculare var. *bieneanum* (Rabenhorst) Rabenhorst 1868 (= *S. bieneanum*)
S. orbiculare var. *crassiusculum* Förster 1964
S. orbiculare var. *denticulatum* Nordstedt 1870 (cf *S. bidentulum* Grönblad 1945)
S. orbiculare var. *depressum* J. Roy et Bisset 1886 (= *S. ralfsii* var. *depressum*)
S. orbiculare var. *extensum* Nordstedt 1873 (= *S. extensum*)
S. orbiculare var. *germinosum* Playfair 1908
S. orbiculare var. *granulosum* Playfair 1910
S. orbiculare var. *hibernicum* (W. West) W. et G.S. West 1912 (= *S. hibernicum*)
S. orbiculare var. *maxima* Van Oye 1953
S. orbiculare var. *messikommeri* Hirano 1950
S. orbiculare var. *planctonicum* Playfair 1912
S. orbiculare var. *protractum* Playfair 1912 (cf *S. bieneanum*)
S. orbiculare forma *punctata* Gutwiński 1892
S. orbiculare var. *quadratum* Schmidle 1897
S. orbiculare var. *ralfsii* W. et G.S. West 1912 (= *S. ralfsii*)
S. orbiculatum forma *subangulata* Messikommer 1927
S. orbiculare var. *subdenticulatum* Bourrelly et Couté 1991
S. orbiculare var. *subralfsii* Grönblad 1960 (anomalous form of *S. ralfsii*?)
S. orbiculare var. *truncatum* Prescott 1940
S. orbiculare var. *verrucosum* Wille 1881

Staurastrum oxyacanthum W. Archer

Cells slightly broader than long to about as broad as long, deeply constricted. Sinus widely open and acute-angled. Semicells cup-shaped, the apical angles produced to form fairly long, parallel or slightly converging processes. Processes tipped with short spines and towards the semicell body furnished with acute granules or denticulations being arranged in concentric series but on the dorsal and ventral side often better developed than on the lateral sides. Semicell body also furnished with spines or denticulations but predominantly or even exclusively at the apex. Semicells in apical view 3(-4)-radiate with almost straight sides, the angles produced to form processes. Zygospore of the nominate variety globose, furnished with slender spines which are double-furcate at the end. Cell length (26-)30-40(-53) μm, cell breadth (36-)45-55(-68) μm.
Ecology: in benthos and tychoplankton of slightly acidic, meso-oligotrophic water bodies.
Distribution: Eurasia, American continents. In Europe, *S. oxyacanthum* is widely distributed, var. *oxyacanthum* being rather common, var. *polyacanthum* rather rare.

var. *oxyacanthum*   Plate 80: 1-5
Archer 1860, p. 78, pl. 7: 1-2
Processes slightly convergent. Semicell body in apical view with a pair of stout, intramarginal spines projecting on each side.

Accounted *Staurastrum oxyacanthum* var. *oxyacanthum*:
S. oxyacanthum var. *croasdaleae* Hirano 1968
S. oxyacanthum var. *patagonicum* Borge 1901

var. *polyacanthum* Nordstedt   Plate 80: 10-13
Nordstedt 1885, p. 11, pl. 7: 9
Processes slightly convergent. Semicell body in apical view with quite a number of both intramarginal and marginal spines on each side.

Accounted *Staurastrum oxyacanthum* var. *polyacanthum* (forma *polyacanthum*):
Staurastrum oxyacanthum var. *polyacanthum* forma *fallax* Růžička 1972

var. *sibiricum* (Boldt) De Toni   Plate 80: 6-9
De Toni 1889, p. 1220
Basionym: *Staurastrum oxyacanthum* subsp. *sibiricum* Boldt 1885, p. 119, pl. 6: 40-41
Processes parallel to slightly divergent. Semicell body in apical view with a pair of stout, intramarginal spines projecting on each side.

Staurastrum oxyrhynchum J. Roy et Bisset 1886
Cells about as long as broad, deeply constricted. Sinus open, acute-angled. Semicells elliptic with crenulate margins, at the lateral-basal angles with a short, stout spine pointing downwards, at the top with an apical and a parallel subapical series of (emarginate) verrucae/granules. Semicells in apical view 3-angular with straight to slightly concave sides, the rounded angles furnished with a short spine. Parallel to each side two curved series of verrucae/granules reaching from angle to angle. Zygospore unknown. Cell length 26-30 μm, cell breadth 25-30 μm.
Ecology: in acidic, oligo-mesotrophic water bodies.
Distribution: very rare. The nominate variety of *S. oxyrhynchum* is not known from Europe but only from Japan. Var. *truncatum* is only known from a site in Böhmerwald (present Czech Republic).

var. *oxyrhynchum*   Plate 61: 9
Roy et Bisset 1886, p. 238, pl. 268: 6.
Series of granules that reach from angle to angle are only present at the apex of the semicell. Semicells in apical view with about smooth margins.

var. *truncatum* (Lütkemüller) Coesel et Meesters stat. nov.   Plate 61: 10
Basionym: *Staurastrum oxyrhynchum* subsp. *truncatum* Lütkemüller 1910, p. 499, pl. 3: 19-21
Next to the apical series of granules there is also a median series of granules extending from angle to angle. Semicells in apical view with dentate margins.

Staurastrum pachyrhynchum Nordstedt
Cells about as long as broad, deeply constricted. Sinus open, acute-angled to subrectangled. Semicells elliptic, rhomboid, obversely triangular, or subsemicircular in outline. Lateral angles broadly rounded, usually slightly produced and marked by a strongly thickened cell wall. Cell wall smooth. Semicells in apical view 3-5-angular with concave sides and broadly rounded angles. Zygospore unknown. Cell length 22-45 μm, cell breadth 22-45 μm.
Ecology: in benthos of acidic, oligotrophic water bodies.
Distribution: Europe, North America; records from other continents for the most part are questionable. In Europe, var *pachyrhynchum* is locally not rare in arctic-alpine regions, var. *convergens* is only known from a few sites in Poland and Austria, var. *polonicum* only from a site in Poland.

S. pachyrhynchum is a fairly polymorphic species. Main differentiating characteristic on species level is the incrassate cell wall at the lateral angles. However, expression of that feature is variable. Specimens in which the cell wall is hardly thickened may be confused with *S. subpygmaeum* and some other species.

var. *pachyrhynchum* Plate 28: 1-7
Nordstedt 1875, p. 32, pl. 8: 34
Synonym: *Staurodesmus pachyrhynchus* (Norstedt) Teiling 1967
Semicells in frontal view elliptic, rhomboid, or obversely triangular in outline. Lateral angles directed horizontally or with a faint upward tilt.

Accounted *Staurastrum pachyrhynchum* var. *pachyrhynchum* (forma *pachyrhynchum*):
Staurastrum pachyrhynchum var. *ellipticum* Skuja 1964
S. pachyrhynchum forma *kossinskajae* Croasdale 1962

var. *convergens* Raciborski Plate 28: 8-9
Raciborski 1889, p. 98, pl. 7: 14
Synonym: *Staurodesmus pachyrhynchus* var. *convergens* (Raciborski) Teiling 1967
Semicells in frontal view subsemicircular. Lateral angles downturned.

Accounted *Staurastrum pachyrhynchum* var. *convergens*:
Staurastrum pachyrhynchum var. *uhtuense* Grönblad 1921

var. *polonicum* (B. Eichler et Gutwiński) Coesel et Meesters comb. nov. Plate 28: 10
Basionym: *Staurastrum pseudopachyrhynchum* var. *polonicum* B. Eichler et Gutwiński 1895, p. 174, pl. 5: 46
Semicells in frontal view anvil-shaped. Lateral angles directed horizontally.

Taxa excluded:
Staurastrum pachyrhynchum var. *minor* Skuja 1964 (cf *S. subpygmaeum*)
S. pachyrhynchum var. *nothogeae* Skuja 1976
S. pachyrhynchum var. *pseudopachyrhynchum* (Wolle) Croasdale 1965
S. pachyrhynchum var. *pusillum* Cedergren 1938 (cf *S. sibiricum*)
S. pachyrhynchum var. *tenerum* Grönblad 1942 (cf *S. subpygmaeum*)
Staurodesmus pachyrhynchus var. *pseudobacillaris* (Grönblad) Förster 1972

Staurastrum paradoxoides Coesel et Meesters stat. et nom. nov. Plate 90: 11-13
Synonym: *Staurastrum paradoxum* var. *reductum* Coesel 1996a, p. 25, fig. 48-51
Cells about as broad as long, deeply constricted. Sinus widely open and acute-angled. Semicells cup-shaped, the apical angles produced to form very short, about parallel processes. Processes tipped with some 4 spines and towards the semicell body furnished with acute granules arranged in some concentric series. Semicell body also furnished with some granules but almost exclusively on the apex. Semicells in apical view 3-radiate with about straight sides. Within each margin a series of distant granules. Zygospore unknown. Cell length 25-30 µm, cell breadth 28-32 µm.
Ecology: in acidic, oligotrophic water bodies.
Distribution: unclear because of possible confusion with other species (e.g., *S. polymorphum*). In Europe, *S. paradoxoides* is only known from a few sites in the Netherlands but presumably it is not that rare.

Staurastrum paradoxum Ralfs
Plate 90: 1-10

Ralfs 1848, p. 138, pl. 23: 8

Cells somewhat broader than long to about as broad as long, deeply constricted. Sinus widely open and acute-angled. Semicells campanulate or cup-shaped, the apical angles produced to form rather long, stout, (slightly) diverging processes. Processes tipped with 3-4 spines and towards the semicell body furnished with acute granules being arranged in concentric series. Semicell body also furnished with (sparse) granules but almost exclusively on the apex. Semicells in apical view 3-4-radiate with about straight sides, the angles produced to form processes. Within each margin of the semicell body a single series of scattered granules. Zygospore more or less globose, furnished with long spines that are variably furcate. Cell length 30-50 µm, cell breadth 30-60 µm.
Ecology: in benthos and tychoplankton of acidic, oligotrophic water bodies.
Distribution: unclear because of diverse taxonomic conceptions. In Europe, *S. paradoxum* is widespread.

Brook (1959a, 1959b) suggested to abandon the species *S. paradoxum* because of lack of type material, inadequacy of the original description and the confusing number of forms referred to this species. However, the illustrations in Ralfs (1848) making part of the first valid publication, where clarity is concerned leave but little to be desired. Characteristic features are, among others, the rather stocky, diverging processes tipped with relatively stout spines. We do not agree with Brook (l.c.) that those figures are likely to refer to a reduced form of *S. anatinum*. Actually, the taxonomic confusion with respect to the species under discussion is largely to be traced back to the flora of West, West & Carter (1923) illustrating quite a series of obviously different species under the name of *S. paradoxum*.

Taxa excluded:
Staurastrum paradoxum var. *aequabile* Skuja 1956 (cf *S. saltator* var. *pendulum*)
S. paradoxum var. *biradiatum* B.M. Griffiths 1925 (cf *S. chaetoceras*)
S. paradoxum var. *chaetoceras* Schröder in Zacharias 1898 (= *S. chaetoceras*)
S. paradoxum var. *cingulum* W. et G.S. West 1903 (= *S. cingulum*)
S. paradoxum var. *depressum* W.B. Turner 1892
S. paradoxum var. *diacanthum* (A. Lemaire) Homfeld 1929 (= *S. diacanthum*)
S. paradoxum var. *evolutum* (W. et G.S. West) N. Carter, in West, West & Carter 1923 (cf *S. tetracerum*)
S. paradoxum var. *fusiforme* Boldt 1885 (= *S. natator* var. *boldtii* Grönblad 1920)
S. paradoxum var. *longibrachiatum* Teiling 1946 (cf *S. longipes*)
S. paradoxum var. *longipes* Nordstedt 1873 (= *S. longipes*)
S. paradoxum var. *nodulosum* W. West 1892a (cf *S. tetracerum*)
S. paradoxum var. *osceolense* Wolle 1885
S. paradoxum var. *parvum* (W. West) N. Carter, in West, West & Carter 1923 (cf *S. minimum*)
S. paradoxum var. *perornatum* Playfair 1913
S. paradoxum var. *reductum* Coesel 1996 (= *S. paradoxoides*)
S. paradoxum var. *tosnense* Skorikow 1904
S. paradoxum var. *zonatum* Woodhead et Tweed 1960

Staurastrum pelagicum W. et G.S. West
Plate 69: 8

W. & G.S. West 1902, p. 46, pl. 2: 26-27

Cells a little broader than long, deeply constricted. Sinus open and acute-angled. Semicells about bowl-shaped, the apical angles produced to form short solid processes that are deeply bifurcate at the end. Cell wall distantly granulate, the granules around the angles arranged in concentric series. Semicells in apical view 3-angular with slightly concave sides, angles terminating in short, smooth-walled, bifurcate processes. Zygospore unknown. Cell length (?35-)45-55 µm, cell breadth (?50)65-75 µm.
Ecology: in plankton of slightly acidic oligotrophic lakes.

Distribution: badly known because of possible confusion with *Staurastrum avicula*. In Europe it has been recorded from Scotland, Ireland, Iceland, Sweden and Switzerland.
Relevant literature: West, West & Carter (1923).

S. pelagicum is a somewhat questionable species. Actually, it seems to be in a direct evolutionary line with *S. avicula* var. *avicula* and *S. avicula* var. *planctonicum*.

Staurastrum pentasterias Grönblad   Plate 75: 1-3
Grönblad 1963, p. 30, pl. 5: 35
Cells slightly broader than long to about as broad as long, deeply constricted. Sinus widely open and acute-angled (the apex often V-shaped). Semicells cup-shaped to ellipsoid, the apex often a little produced, the apical/lateral angles produced to form rather short to moderately long, about parallel processes tipped with short spines. Cell wall of the processes with concentric series of acute granules/denticulations. Semicell body at the base with a supraisthmial whorl of granules, otherwise (almost) smooth-walled. Semicells in apical view (4-)5-radiate, the angles produced to form processes; semicell body with a central circle of emarginate verrucae. Zygospore unknown. Cell length 29-34(-38) μm, cell breadth 32-38 μm.
Ecology: in benthos and tychoplankton of oligo-mesotrophic water bodies.
Distribution: poorly known because of possible confusion with resembling species. Records only known from Europe (Sweden, Germany, Austria, France).

S. pentasterias is a rather ill-defined species. Records could also refer to 5-radiate forms of other species (e.g., *S. gracile*, *S. boreale*, *S. crenulatum*).

Staurastrum pertyanum Coesel et Meesters nom. nov.   Plate 75: 4-7
Synonym: *Phycastrum asperum* Perty 1849, p.174
Staurastrum ornatum W.B. Turner var. *asperum* (Perty) Schmidle 1896, p. 36, pl. 16: 22
Cells about as broad as long or slightly broader than long, deeply constricted. Sinus V-shaped, then widely open. Semicells cup-shaped, the apical angles produced to form rather short to moderately long, slightly convergent processes. Processes tipped with minute spines/denticulations and to the semicell body furnished with a few concentric series of acute granules. At the dorsal base of each process a pair of short, denticulate, accessory processes. Semicell body at the base with a supraisthmial whorl of granules, otherwise (almost) smooth-walled. Semicells in apical view 5-radiate. At the base of each arm-like process two short, dentate, diverging accessory processes. Zygospore unknown. Cell length 24-29 μm, cell breadth 28-35 μm.
Ecology: in benthos of oligotrophic water bodies.
Distribution: Eurasia, North America. In Europe it is a rare species, particularly known from arctic-alpine regions.

The original diagnosis of this taxon by Perty (1849) is only based on an apical cell view without any illustration, so little informative. Consequently, our above-given diagnosis is primarily based on data provided by Schmidle (1896). *S. pertyanum* may be confused with *S. subpinnatum* Schmidle [= *S. pinnatum* Turner var. *subpinnatum* (Schmidle) W. et G.S. West], a (sub)tropical taxon that is characterized by larger, predominantly 6-radiate cells with a campanulate semicell body.

Staurastrum petsamoense Järnefelt   Plate 89: 1-3
Järnefelt 1934, p. 208, fig. 18
Cells slightly broader than long, deeply constricted. Sinus widely open and acute-angled. Semicells bowl-shaped with a slightly convex apex, the apical angles produced to form short, stout, parallel or slightly divergent processes. Processes tipped with 3-5 (rather short) spines. Cell wall

furnished with coarse, conical granules/denticulations arranged in concentric series around the processes, the larger granules on the semicell body often being disposed in pairs (emarginate verrucae). Flanks of the semicell just below the apical series of granules usually with an unsculptured zone. Semicells in apical view 3-angular with slightly convex to slightly concave sides, the angles produced to form rather short processes. Intramarginal granules on the semicell body usually partly in the form of (more or less circular) emarginate verrucae. Zygospore unknown. Cell length (?40-)70-85 µm, cell breadth (?50-)90-125 µm.
Ecology: in plankton of oligotrophic lakes.
Distribution: Europe, North America. In Europe, *S. petsamoense* is confined to arctic-alpine regions. In central Europe it is very rare, but in northern Europe it may be locally rather common (Skuja 1964).

Most records of small-sized forms of *S. petsamoense* probably belong to other species, e.g., *S. gracile, S. boreale* or *S. proboscideum*.

Taxa excluded:
Staurastrum petsamoense var. *biverticillatum* Nygaard 1954 (cf *S. boreale?*)
S. petsamoense var. *minus* (Messikommer) Thomasson 1957 (cf *S. gracile?*)

Staurastrum pileolatum Ralfs

Plate 58: 1-7

Brébisson ex Ralfs 1848, p. 215, pl. 35: 22
Cells distinctly longer than broad, slightly constricted. Sinus a V- or U-shaped incision. Semicells rectangular-subcampanulate in outline, with concave lateral margins and a strongly retuse apex. Angles broadly rounded, the apical ones more or less conical. Cell wall in the apical region provided with granules or verrucae, arranged in concentric series around the angles. At the base of the semicell a ring of emarginate verrucae, separated from the apical ornamentation by an unsculptured zone. Semicells in apical view 3(-4?)-angular with rounded angles and almost straight sides. Zygospore described as globular, furnished with a few stout spines, 2-3-fid at the apex. Cell length 35-45 µm, cell breadth 19-25 µm.
Ecology: in benthos of acidic, oligo-mesotrophic water bodies.
Distribution: upland species, only known with certainty from Europe and Japan. In Europe, *S. pileolatum* is pretty rare. Records mainly refer to montane regions in western and central Europe.
Relevant literature: West & West (1912).

S. pileolatum differs from *S. capitulum* by its upwardly directed apical angles; in addition, granules are never spine-like. The below-described varieties of *S. pileolatum* are interconnected by transitional forms. *S. pileolatum* var. *brasiliense*, as originally described under the name of *S. amoenum* var. *brasiliense* by Børgesen (1890) from Brasil, presumably refers to another species. The only description of *S. pileolatum* zygospores is by Roy (1893) who did not add any illustration and stated that the spores in question got lost.

Accounted *Staurastrum pileolatum* (var. *pileolatum*):
Staurastrum pileolatum var. *cristatum* Lütkemüller 1893

Questionable non-European taxon:
Staurastrum pileolatum var. *brasiliense* (Børgesen) Lütkemüller 1893

Staurastrum pingue Teiling

Cells somewhat broader than long to about as broad as long, deeply constricted. Sinus V-shaped. Semicell body cylindric to cup-shaped or vase–shaped (with inflated base), the apical angles produced to form long, slender, divergent to more or less parallel processes. Processes tipped with 3-4 (small) spines and towards the semicell body furnished with a series of granules or denticulations, the ventral and particularly the dorsal ones being (much) better developed than the lateral ones. Semicell body just above the isthmus often with a variable number of granules. Semicells in apical view 3(-5)-radiate with about straight sides, the angles produced to form slender processes. Along each of the sides of the semicell body two emarginate intramarginal verrucae (often reduced). Zygospore unknown. Cell length 30-110 μm, cell breadth 40-110 μm.

Ecology: in plankton of meso-eutrophic, slightly acidic to alkaline water bodies, may cause blooms in eutrophic waters.

Distribution: presumably cosmopolitan. In Europe, both var. *pingue* and var. *planctonicum* are widely distributed and locally common.

var. *pingue*
Plate 103: 5-7; Plate 104: 1-3

Teiling 1942, p. 66, fig. 3-5

Cells relatively small (breadth of isthmus < 10 μm), processes about straight and generally tipped with 4 spines.

Accounted *Staurastrum pingue* var. *pingue*:
Staurastrum pingue var. *tridentata* Nygaard 1949

var. *planctonicum* (Teiling) Coesel et Meesters stat. et comb. nov.
Plate 104: 4-9; Plate 105: 1-4

Basionym: *Staurastrum planctonicum* Teiling 1946, p. 77, fig. 30-32

Cells relatively large (breadth of isthmus ≥ 10 μm), processes not seldom curved (semicells with 'shrugged shoulders') and often tipped with three spines.

Accounted *Staurastrum pingue* var. *planctonicum*:
Staurastrum planctonicum var. *bullosum* Teiling 1946
S. planctonicum var. *laeviceps* Grönblad 1960
S. planctonicum var. *ornatum* (Grönblad) Teiling 1947

Taxa excluded:
Staurastrum pingue var. *evolutum* (W. et G.S. West) Prescott, in Prescott, Bicudo & Vinyard 1982
S. planctonicum var. *bulbosum* (W. West) Thomasson 1960 (= *S. bulbosum*)

Staurastrum platycerum Joshua
Plate 92: 1-4

Joshua 1886, p. 643, pl. 24: 1-2

Cells broader than long, deeply constricted. Sinus widely open, obtuse-angled. Semicells cup-shaped with an elevated apex, the subapical angles produced to form long, more or less parallel, arm-like processes. Processes tipped with some 3 stout spines and towards the semicell body furnished with a spiralling series of dentations/spines. Semicell body smooth-walled (except for some possible apical granules). Semicells in apical view 4-5(-6)-radiate. Zygospore unknown. Cell length 34-36 μm, cell breadth 83-95(-160) μm.

Ecology: in plankton of oligotrophic water bodies.

Distribution: Asia, Europe. In Europe, *S. platycerum* (sensu Grönblad 1938) is only known from some sites in Finland.

When describing *S. platycerum* from Burma, Joshua (1886) provided only some apical cell views as an illustration. Therefore there is doubt about the true identity of *S. platycerum*. The above-given diagnosis is mainly based on Grönblad's (1938) concept. The large, dorsal spines about half-way the processes, shown in his fig. 3: 6, likely are part of the natural variability in this species. Anyhow, there is no ground to consider this form identic to *S. sexangulare* var. *dentatum* Playfair as Grönblad (1938, p. 62) does. Possibly, *S. platycerum* is closely affiliated to *S. rotula* Nordstedt (in particular its var. *smithii* Thomasson), a common species in tropical regions.

Taxon excluded:
Staurastrum platycerum var. *dentatum* (Playfair) Grönblad 1938 (cf *S. sexangulare*)

Staurastrum podlachicum B. Eichler et Gutwiński   Plate 65: 9-12
Eichler & Gutwiński 1894, p. 175, pl. 5: 49

Cells about as as long as broad, deeply constricted. Sinus widely open from an acute-angled apex, or narrowly open and more or less linear in its apical part and then widely opening outward. Semicells hexagonal to ellipsoid with (slightly) produced lateral angles. Lateral angles furnished with two minute spines lying in the same vertical plane. Cell wall of the semicell lobes provided with acute granules or minute spines, in principle arranged in more or less regular concentric series around the angles, but usually much reduced. Semicells in apical view 3-angular with slightly concave to almost straight sides and acute angles. Zygospore unknown. Cell length 27-44 μm, cell breadth 26-46 μm.
Ecology: in benthos and tychoplankton of acidic, oligotrophic water bodies.
Distribution: Europe, New Zealand. In Europe, *S. podlachicum* has only incidentally been recorded (Poland, Finland, France, the Netherlands).

S. podlachicum is a poorly-known species with a variably elaborated cell wall sculpturing. Possibly, it is a mere reduction form of, e.g., *S. forficulatum*.

Staurastrum pokljukense Pevalek   Plate 33: 16
Pevalek 1925, p. 81, text fig. 19

Cells slightly broader than long, deeply constricted. Sinus linear with a dilate extremity. Semicells rectangular with broadly rounded angles and straight to slightly convex sides. Cell wall smooth. Semicells in apical view 3-angular with broadly rounded or somewhat domed angles and concave sides. Zygospore unknown. Cell length 10 μm, cell breadth 11-12 μm.
Ecology: in oligo-mesotrophic water bodies.
Distribution: only known from a site in Slovenia.

S. pokljukense is a somewhat questionable species. It looks like a small form of *S. quadratulum* but would differ in the outline of the semicell angles in apical view, being described as 'broadly rounded or on two sides somewhat flattened so that each angle has three small corners.'

Staurastrum polonicum Raciborski

Cells longer than broad, (rather) deeply constricted. Sinus widely open, apex acute-angled to about rectangled (rarely a V-shaped invagination). Semicells hexagonal in outline with broadly rounded, more or less produced angles. Cell wall, except for the isthmial region, evenly granulated; granules near the apical angles in concentric series, otherwise without a definite arrangement. Semicells in apical view stellate, the margins concave between 7-9 broadly rounded angles; within the series of 7-9 marginal lobules an additional circle of 3-7 intramarginal lobules. Zygospore unknown. Cell length 45-65 μm, cell breadth 35-50 μm.
Ecology: in benthos of acidic to neutral, oligo-mesotrophic water bodies.
Distribution: rare species, only known from (sub)alpine regions in Europe (var. *polonicum*) and North America (var. *coronulatum*). In Europe, var. *polonicum* is known from montane regions in southern Germany, Poland and Austria.

var. *polonicum*   Plate 57: 9-11

Raciborski 1884, sep. p. 17, pl. 1: 10

Isthmus relatively broad (ca $^1/_2$ cell breadth). Margins between the lateral and apical lobules only slightly concave. Semicell apex (in frontal view) broadly truncate and almost straight.

Non-European variety:

var. *coronulatum* Prescott

Prescott 1936, p. 140, pl. 15: 11-15

Isthmus relatively narrow (ca $^1/_3$ cell breadth). Margins between the lateral and apical lobes deeply concave. Semicell apex distinctly concave.

Staurastrum polymorphum Ralfs   Plate 71: 17-23 ; Plate 72: 1-5

Brébisson ex Ralfs 1848, p. 135, pl. 34: 6

Cells slightly broader than long, deeply constricted. Sinus widely open and acute-angled. Semicells broadly fusiform to cup-shaped, the lateral/apical angles usually produced to form short, stout processes, the processes tipped with some four short, stout spines. Cell wall furnished with acute granules (minute spines) arranged in concentric series around the angles. Semicells in apical view 3-4(-6?)-angular with (slightly) concave sides and truncate angles; along the sides with intramarginal series of (paired) granules. Zygospore globose, furnished with stout, furcate spines. Cell length (19-)25-30 μm, cell breadth (28-) 35-40(-45) μm.

Ecology: in benthos and tychoplankton of slightly acidic to slightly alkaline, mesotrophic water bodies.

Distribution: Eurasia, American continents? In Europe, *S. polymorphum* is rather widely distributed but much less common than often suggested.

S. polymorphum, presumably also because of its invitatory epitheton, has become a notorious catch-all species. According to Ralfs (1848) a relevant differentiating feature with respect to the morphologically resembling *S. hexacerum* is in the markedly spinous nature of the processes. Grönblad (1921) who examined specimens in Wittrock & Nordstedt's collection of exsiccatae figured cells conformable to that characteristic. Next to that, his illustrations of cells in apical view show distinct intramarginal series of (paired) granules along the semicell sides, which could be an important additional differentiating feature.

Relevant literature: Grönblad 1921, 1960.

Accounted *Staurastrum polymorphum* (var. *polymorphum*, forma *polymorphum*):

Staurastrum polymorphum forma *obesa* Heimerl 1891

S. polymorphum var. *pygmaeum* Grönblad 1921

S. polymorphum var. *spinosa* E. Larsen 1907

Taxa excluded:

Staurastrum polymorphum var. *bornensis* Capdevielle1978

S. polymorphum var. *cinctum* Messikommer 1963

S. polymorphum var. *coronatum* Steinecke 1916

S. polymorphum var. *divergens* Nygaard 1949 (cf *S. paradoxum*)

S. polymorphum forma *eductum* Pevalek 1925 (cf *S. paradoxum*)

S. polymorphum var. *groenbladii* Hirano 1948

S. polymorphum var. *hinganicum* Noda et Skvortsov 1969

S. polymorphum forma *intermedia* Wille 1879 (cf *S. hexacerum*)

S. polymorphum var. *itirapinense* C.E.M. Bicudo 1967

S. polymorphum forma *monstrosa* Wille 1879 (cf *S. hexacerum*)

S. polymorphum var. *munitum* W. West 1892

S. *polymorphum* var. *pentagonum* F.L. Harvey 1892
S. *polymorphum* var. *pusillum* W. West 1912 (cf S. *inflexum*)
S. *polymorphum* var. *simplex* W. et G.S. West 1905 (cf S. *paradoxum*)
S. *polymorphum* var. *subgracile* Wittrock 1872
S. *polymorphum* var. *verruciferum* Grönblad 1960
S. *polymorphum* var. *waldense* Dick 1926

Staurastrum polytrichum (Perty) Rabenhorst — Plate 46: 1-3
Rabenhorst 1868, p. 214
Basionym: *Phycastrum polytrichum* Perty 1852, p. 210, pl. 16: 24
Cells about as long as broad or a little longer than broad, deeply constricted. Sinus widely open from an acute-angled apex. Semicells elliptic-oval, equally covered with stout spines which are arranged in obscure circles round the angles. Semicells in apical view 3-angular with almost straight sides and broadly rounded angles. Zygospore unknown. Cell length (55-)65-80(-90) μm, cell breadth (50-)60-80(-85) μm.
Ecology: in benthos of slightly acidic, mesotrophic water bodies.
Distribution: Europe, American continents. In Europe, S. *polytrichum* is a rather rare species, mainly occurring in montane regions.

The zygospore depicted by Cushman (1905) under the name of S. *polytrichum* var. *readingense* (see also West, West & Carter 1923, pl. 136: 11) obviously belongs to another, as yet unidentified species.

Accounted *Staurastrum polytrichum* (var. *polytrichum* forma *polytrichum*):
Staurastrum polytrichum forma *biseriatum* Kaiser 1926
S. *polytrichum* var. *ornatum* Taft 1945

Questionable non-European variety:
Staurastrum polytrichum var. *brasiliense* (Grönblad) Coesel 1987

Taxa excluded:
Staurastrum polytrichum var. *alpinum* Schmidle 1895 (cf S. *teliferum*)
S. *polytrichum* var. *readingense* Cushman 1905
S. *polytrichum* var. *reniforme* J. Sampaio 1923

Staurastrum proboscideum (Ralfs) W. Archer — Plate 76: 1-6
Archer in Pritchard 1861, p. 742
Basionym: *Staurastrum asperum* var. *proboscideum* Brébisson ex Ralfs 1848, p. 139, pl. 23: 12a,b
Cells about as long as broad, deeply constricted. Sinus widely open and acute-angled. Semicells broadly fusiform, the lateral angles usually produced to form (very) short, stout, parallel processes, the processes tipped with short spines or acute granules. Cell wall furnished with coarse, conical granules, arranged in concentric series around the angles, those at the apex usually emarginate. Flanks of the semicells just below the apical series of granules sometimes with an unsculptured zone. Semicells at the base with a supraisthmial whorl of (simple) granules. Semicells in apical view 3(-4)-angular with straight to concave sides and truncate angles; along the sides with intra-marginal series of emarginate granules. Zygospore not known with certainty. Cell length 35-50 μm, cell breadth 40-55 μm.
Ecology: in benthos and tychoplankton of slightly acidic, mesotrophic water bodies.
Distribution: Eurasia, North America. In Europe, S. *proboscideum* is widespread but only of occasional occurrence.

The picture of an (obviously immature) zygospore in Williamson (1997) might refer to another species.

Taxa excluded:
Staurastrum proboscideum var. *altum* Boldt 1885 (cf *S. sebaldi*)
S. proboscideum var. *brasiliense* Børgesen 1890
S. proboscideum forma *javanica* Nordstedt 1880
S. proboscideum var. *productum* Messikommer 1942 (cf *S. boreale*)
S. proboscideum var. *subglabrum* W. West 1890
S. proboscideum var. *subjavanicum* Nygaard 1926

Staurastrum pseudopelagicum W. et G.S. West   Plate 70: 1-6
W. & G.S. West 1903, p. 547, pl. 18: 1-3
Cells slightly broader than long to about as broad as long, deeply constricted. Sinus widely open from an acute-angled apex. Semicells triangular to semicircular, the apical angles produced to form (fairly) short diverging processes ending in (1-)2(-4) stout, diverging spines. Cell wall covered with granules (varying from very distinct to almost invisible) arranged in concentric series around the processes and continuing onto (part of) the semicell body. Semicells in apical view (2-)3-radiate with slightly concave to slightly convex sides, the angles produced to form short processes; the inner series of granules on the semicell body often with a pair of more or less prominent spines on each of the sides. Zygospore unknown. Cell length (30-)40-55(-70) µm, cell breadth (40-)50-80(-90) µm.
Ecology: in plankton of acidic, oligo-mesotrophic water bodies.
Distribution: Europe, North America. In Europe, *S. pseudopelagicum* is a rare species, mainly confined to atlantic regions.
Relevant literature: Brook (1957).

Accounted *Staurastrum pseudopelagicum* (var. *pseudopelagicum* forma *pseudopelagicum*):
Staurastrum pseudopelagicum var. *bifurcatum* Huitfeld-Kaas 1906
S. pseudopelagicum forma *biradiata* Grönblad 1938
S. pseudopelagicum var. *minor* G.M. Smith 1924
S. pseudopelagicum var. *ornatum* Korshikov 1941
S. pseudopelagicum var. *tumidum* G.M. Smith 1924

Taxon excluded:
Staurastrum pseudopelagicum var. *spinosa* Thomasson 1952 (cf *S. furcatum*)

Staurastrum pseudopisciforme B. Eichler et Gutwiński   Plate 120: 7-8
Eichler & Gutwiński 1895, p. 175, pl. 5: 50
Cells somewhat broader than long, deeply constricted. Sinus open, acute-angled. Semicells subelliptic-hexagonal, the lateral angles produced to form relatively long, divergent processes that are strikingly bifurcate at the end (like a fishtail). A couple of similar or somewhat shorter processes is also present on the apex in between each pair of angles. Cell wall at the base of the semicell lobes with some concentric series of granules. Semicell body in apical view 3-radiate with about straight sides, the angles ending in a stout spine. On either side of each angle a bi- or trifurcate intramarginal spine projecting beyond the margin. Zygospore unknown. Cell length 32-42 µm, cell breadth 38-61µm.
Ecology: in acidic, oligotrophic water bodies.
Distribution: Europe. Distribution unclear because of possible confusion with other species. Original records are from Poland and Finland.

S. pseudopisciforme is a questionable species. West & West (1895a) consider it a variety of *S. arcuatum*. On the contrary, the figure of *S. pseudopisciforme* in Coesel & Meesters (2007) might rather refer to *S. furcatum*.

Accounted *Staurastrum pseudopisciforme* (var. *pseudopisciforme*):
Staurastrum pseudopisciforme var. *dimidio-minor* Grönblad 1920

Staurastrum pseudotetracerum (Nordstedt) W. et G.S. West   Plate 91: 1-3
W. & G.S. West 1895, p. 79, pl. 8: 39
Basionym: *Staurastrum contortum* var. *pseudotetracerum* Nordstedt 1887, p. 157
First illustration: Nordstedt 1888a, p. 37, pl. 4: 9
Cells about as long as broad, deeply constricted. Sinus widely open, about rectangled. Semicell body about triangular in outline, the apical angles produced to form rather short, divergent processes that tend to curve upwards. Processes tipped with three or four minute spines and towards the semicell body furnished with concentric series of small granules. Semicell body itself for the most part smooth-walled. Semicells in apical view 3-(4-) radiate with slightly concave sides, the poles attenuated into processes. Zygospore unknown. Cell length 18-27 µm, cell breadth 19-30 µm.
Ecology: in tychoplankton of meso-eutrophic water bodies.
Distribution: uncertain because of ready confusion with other species, possibly cosmopolitan. Records from Europe are widely spread but relatively low in number.

S. pseudotetracerum could be confused with reduction forms of triradiate *S. chaetoceras*.

Staurastrum punctulatoides Coesel et Meesters stat. et nom. nov.   Plate 57: 1
Synonym: *Staurastrum kjellmanii* var. *rotundatum* W. et G.S. West 1896a, p. 158, pl. 4: 46
Staurastrum inflatum W. et G.S. West 1912, p. 191
Cells distinctly longer than broad, moderately constricted. Sinus very widely open, obtuse-angled. Semicells obovate-subcircular. Cell wall beset with coarse granules without a definite disposition. Semicells in apical view quadrate-circular. Zygospore unknown. Cell length 43 µm, cell breadth 25-27 µm.
Ecology: in oligotrophic water.
Distribution: only known from a site in northern England.

S. punctulatoides is a poorly known species. Originally described as a variety of *S. kjellmanii*, West & West (1912) rendered it the status of a separate species under the name of *S. inflatum*. As, however, that latter name appears to be used already by Bernard (1908) for a Javanese desmid species, *S. inflatum* W. et G.S. West had to be renamed again.

Staurastrum punctulatum Ralfs   Plate 56: 5-19
Brébisson ex Ralfs 1848, p. 133, pl. 22: 1
Cells longer than broad (sometimes about as long as broad), deeply constricted. Sinus widely open. Semicells subrhomboid (dorsal and ventral margins about equally convex) with rounded lateral angles. Cell wall granulate, the granules around the angles arranged in more or less distinct concentric series. Semicells in apical view 3(-5)-angular with slightly concave to slightly convex sides and narrowly to broadly rounded angles. Zygospore globose, furnished with stout spines that are double-furcate at the apex and arise from a mamillate base. Cell length (20-)25-40(-50) µm, cell breadth (17-)20-35(-45) µm.
Ecology: in benthos and tychoplankton of acidic, oligotrophic water bodies.
Distribution: possibly cosmopolitan. In Europe, *S. punctulatum* is widely distributed and common.

S. punctulatum is a frequently recorded desmid taxon. However, many records are doubtful, presumably referring to other, superficially resembling species. Even with respect to those forms in the present flora that are considered to belong to *S. punctulatum* several ones possibly refer to separate species. Particularly some coarsely granulate forms from arctic regions such as depicted by Wille (1879) and Förster (1965), see our Pl. 56: 8 and 19, might be related to *S. punctulatoides*.

Despite the high degree of polymorphy shown in our figures of *S. punctulatum* we did not succeed in linking a particular morphotype to any infraspecific taxon as diagnosed in its original description. Consequently, no separate varieties are distinguished.

Accounted *Staurastrum punctulatum* (var. *punctulatum* forma *punctulatum*):
Staurastrum punctulatum forma *contorta* Schmidle 1896
S. punctulatum var. *kjellmanii* Wille, in Bergh & Lotken 1887
S. punctulatum forma *minor* W. et G.S. West 1912
S. punctulatum var. *pygmaeum* (Ralfs) W. et G.S. West 1912
S. punctulatum var. *subproductum* W. et G.S. West 1912

Taxa excluded:
Staurastrum punctulatum var. *coronatum* (Schmidle) W. et G.S. West 1912
S. punctulatum forma *crassa* Gay 1891
S. punctulatum forma *elliptica* M. Lewin 1888 (cf *S. lapponicum*)
S. punctulatum var. *muricatiforme* Schmidle 1898
S. punctulatum var. *pygmaeum* forma *obtusum* (Wille) Prescott, in Prescott, Bicudo & Vinyard 1982
S. punctulatum var. *pygmaeum* forma *trilineata* (W. West) W. et G.S. West 1912
S. punctulatum var. *striatum* W. et G.S. West 1912 (= *S. striatum*)
S. punctulatum var. *subdilatatum* Kurt Förster 1963
S. punctulatum var. *submuricatiforme* C.C. Jao 1949
S. punctulatum var. *subrugulosum* Raciborski 1885 (cf *S. turgescens*)
S. punctulatum var. *triangulare* C.C. Jao 1949
S. punctulatum var. *turgescens* (De Notaris) Rabenhorst 1868 (= *S. turgescens*)

Staurastrum pungens Ralfs   Plate 63: 1-8
Brébisson ex Ralfs 1848, p. 130, pl. 34: 10

Cells about as long as broad or slightly longer than broad, deeply constricted. Sinus widely open and acute-angled. Semicells subelliptic-cuneate in outline, the lateral-apical angles ending in a large, obliquely upwardly projected spine. Apical margin provided typically with two accessory spines between each pair of consecutive angles. Cell wall further provided with a variable number of small granules, arranged in concentric series around the angles. Semicells in apical view 3-angular with almost straight sides, angles ending in a stout spine. Within each margin an additional pair of stout spines. Zygospore globose, furnished with long spines which are forked at the apex. Cell length 36-46 μm, cell breadth 33-43 μm.

Ecology: in benthos and tychoplankton of acidic, oligo-mesotrophic water bodies.

Distribution: Europe, North America. In Europe, *S. pungens* is a widely distributed, but rare species.

In the original description of *S. pungens*, in Ralfs (1848), the cell is characterized by a 'smooth front, and in the 1923 flora of West, West & Carter there is talk of a 'minutely punctulate' cell wall. Yet, there is hardly any doubt that the cell wall of this species normally is furnished with a number of concentric series of granules, or minute dentations around the angles. Irregularities in number, position and shape of the accessory spines have been described by several authors, see our Pl. 63: 5-7.

Accounted *Staurastrum pungens* (var. *pungens*):
Staurastrum pungens var. *sublunatum* Homfeld 1929

Taxa excluded:
Staurastrum pungens var. *madagascariense* Bourrelly et Couté 1991

Staurastrum pyramidatum W. West   Plate 47: 1-5
W. West 1892a, p. 179
Original description and figure: W. West 1890, p. 294, pl. 5: 14 (as *S. muricatum* var. *acutum*)
Cells slightly longer than broad, deeply constricted. Sinus linear, closed or narrowly open for the greater part. Semicells trapezoid with broadly rounded angles to almost semicircular/semi-elliptic. Cell wall densely beset with short, solid spines (sometimes locally doubled) arranged in circles around the angles. Semicells in apical view 3-angular with almost straight sides and rounded angles. Zygospore globose, provided with stout spines which are repeatedly bi- or threefold furcate. Cell length (50-)60-80 µm, cell breadth 50-70 µm.
Ecology: in benthos of slightly acidic, oligo-mesotrophic water bodies.
Distribution: Europe, North America. In Europe, *S. pyramidatum* is widely distributed, particularly in mountainous areas, but rather rare.

The original figure of *S. pyramidatum* in West (1890) differs from later illustrations of that species by a less dense configuration of the cell wall spines. The present concept of this species is based on that provided by West, West & Carter (1923).

Accounted *Staurastrum pyramidatum* (var. *pyramidatum* forma *pyramidatum*):
Staurastrum pyramidatum var. *coilon* W. et G.S. West 1894
S. pyramidatum forma *longispina* Manguin 1939

Taxon excluded:
Staurastrum pyramidatum var. *bispinosum* Schmidle 1898 (cf *S. arnellii*)

Staurastrum quadrangulare Ralfs   Plate 43: 5-9
Brébisson ex Ralfs 1848, p. 128, pl. 22: 7, pl. 34: 11
Cells about as long as broad, deeply constricted. Sinus widely open from its acute-angled apex. Semicells in general outline trapezoid to almost rectangular. Apical angles provided wirh two spines, lying in the same horizontal plane. Basal angles provided with a variable number of spines, arranged in various patterns. Cell wall smooth. Semicells in apical view (3-)4(-5)-angular with concave to straight sides. Zygospore unknown. Cell length (20-)25-35 µm, cell breadth (20-)25-35 µm.
Ecology: in benthos and tychoplankton of acidic, oligotrophic water bodies.
Distribution: Eurasia, American continents, Africa. In Europe, *S. quadrangulare* is widely distributed but not common.

S. quadrangulare is rather variable in the ornamentation of the basal angles. Based on that, quite a number of different varieties have been described. The significance of most of these varieties may be seriously doubted. In Ralfs (1848) two rather different illustrations are provided. Unfortunately, none of them clearly shows the ornamentation pattern of the basal semicell angles. A number of exclusively tropical varieties probably be better considered separate species.

Accounted *Staurastrum quadrangulare* var. *quadrangulare*:
Staurastrum quadrangulare var. *alatum* Wille 1884
S. quadrangulare var. *americanum* Raciborski 1892

S. quadrangulare var. *armatum* W. et G.S. West 1896
S. quadrangulare var. *attenuatum* Nordstedt 1870
S. quadrangulare var. *contectum* (W.B. Turner) Grönblad 1945
S. quadrangulare var. *sexcuspidatum* B. Eichler 1867

Questionable non-European varieties:
Staurastrum quadrangulare var. *longispina* Børgesen 1890
S. quadrangulare var. *prolificum* Croasdale, in Scott, Grönblad & Croasdale 1965
S. quadrangulare var. *sanctipaulense* C.E.M. Bicudo 1967
S. quadrangulare var. *setigerum* Grönblad 1945

Taxa excluded:
Staurastrum quadrangulare var. *ornatum* Bourrelly et Couté 1991
S. quadrangulare var. *subarmatum* Claassen 1961

Staurastrum quadratulum Coesel et Meesters stat. et nom. nov.   Plate 33: 14
Synonym: *Staurastrum orbiculare* var. *quadratum* Schmidle 1897, p. 24, text fig. 4: 1
Cells as long as broad, roughly quadrate in outline, deeply constricted. Sinus linear with a dilate extremity. Semicells rectangular with broadly rounded angles and straight to slightly convex sides. Cell wall smooth. Semicells in apical view 3-angular with broadly rounded angles and concave sides. Zygospore unknown. Cell length 24 μm, cell breadth 24 μm.
Ecology: in oligo-mesotrophic water bodies.
Distribution: only known from some sites in Germany.

Staurastrum quadrispinatum W.B. Turner   Plate 43: 1-4
W.B. Turner 1886, p. 35, pl. 1: 4
Cells somewhat longer than broad, deeply constricted. Sinus widely open from its acute-angled apex. Semicells hexagonal to trapezoid. Apical angles provided with two stout spines, lying in the same horizontal plane. Basal angles provided with 1-3 stout spines, also lying in one and the same horizontal plane. Cell wall smooth. Semicells in apical view 3(-4)-radiate with broadly truncate angles and concave to almost straight sides. Zygospore unknown. Cell length (30-)35-45(-55) μm, cell breadth (25-)30-40(-50) μm.
Ecology: in benthos and tychoplankton of acidic, oligotrophic water bodies.
Distribution: Europe, American continents. In Europe, *S. quadrispinatum* is of rare occurrence, known from Wales, Austria and Romania.
Relevant literature: Péterfi 1973.

Accounted *Staurastrum quadrispinatum* (var. *quadrispinatum* forma *quadrispinatum*):
Staurastrum quadrispinatum forma *minus* Kossinskaja 1949
S. quadrispinatum var. *spicatum* (W. et G.S. West) A.M. Scott et Grönblad 1957
S. quadrispinatum var. *transsylvanicum* S. Péterfi 1943

Staurastrum ralfsii (W. et G.S. West) Coesel et Meesters stat. nov.
Cells longer than broad to about as long as broad, deeply constricted. Sinus linear with a dilate extremity. Semicells pyramidal in outline with broadly rounded angles. Cell wall smooth. Semi-cells in apical view 3-angular with distinctly concave sides and broadly rounded angles. Zygospore globose and furnished with numerous simple, acute spines.
Ecology: in benthos and tychoplankton of acidic to neutral, oligo-mesotrophic water bodies.
Distribution: presumably cosmopolitan. In Europe, both var. *ralfsii* and var. *depressum* are wide-spread and locally common.

var. *ralfsii*   Plate 32: 6-10

Basionym: *Staurastrum orbiculare* var. *ralfsii* W. et G.S. West 1912, p. 156, pl. 124: 12-16

Cells distinctly longer than broad. Cell length 29-41(-61) μm, cell breadth 22-36(-50) μm.

Accounted *Staurastrum orbiculare* var. *ralfsii* (forma *ralfsii*):
Staurastrum orbiculare forma *major* W.West 1892

Taxa excluded:
Staurastrum orbiculare var. *ralfsii* forma *maius* Förster 1964
S. orbiculare var. *ralfsii* forma *punctata* Manguin 1939

var. *depressum* (J. Roy et Bisset) Coesel et Meesters comb. nov.   Plate 32: 11-18

Basionym: *Staurastrum orbiculare* var. *depressum* J. Roy et Bisset 1886, p. 237, pl. 268: 14

Cells about as long as broad. Semicells depressed, Cell length 22-27 μm, cell breadth 19-28 μm.

Zygospores as illustrated by Hegde & Bharati (1983) in number and stoutness of spines distinctly differ from those pictured by Coesel (1997) which might indicate that different taxa are at issue.

Staurastrum reductum (Messikommer) Coesel et Meesters stat. nov.   Plate 105: 5-7

Basionym: *Staurastrum dimazum* var. *reductum* Messikommer 1927, p. 345, pl. 2: 15

Cells (slightly) broader than long, deeply constricted. Sinus V-shaped. Semicell body cylindric, the apical angles produced to form rather long, divergent processes which not rarely are more or less curved upwards. Processes tipped with some three rather stout spines and towards the semicell body furnished with spiralling series of marked dentations. Dorsal series of process denticulations continuing along the apex of the semicell body as pronounced, emarginate verrucae. Semicell body just above the isthmus often with a separate series of acute granules. Semicells in apical view 2-radiate. Along the sides of the semicell body an intramarginal series of emarginate verrucae. Zygospore unknown. Cell length 40-65 μm, cell breadth 55-80 μm.
Ecology: in plankton of mesotrophic, circumneutral water bodies.
Distribution: Eurasia. In Europe, *S. reductum* is known from Switzerland, Austria and the Netherlands, but presumably wider distributed.

S. reductum is a biradiate desmid obviously belonging to the affinity group of *S. manfeldtii*. Actually, in its frontal outline it is more or less intermediate between *S. manfeldtii* and *S. pingue* (var. *planctonicum*).

Staurastrum retusum W.B. Turner

Cells about as long as broad, deeply constricted. Sinus usually closed, more or less linear with a (slightly) dilated extremity; less often, however, widely opening from an acute-angled apex. Semicells more or less trapeziform, with straight to slightly concave apex and (broadly) rounded angles. Cell wall scrobiculate, particularly at the angles. Semicells in apical view 3-angular with concave sides and broadly rounded angles. Zygospore presumably globose with a distinctly scrobiculate wall and furnished with simple, acute spines. Cell length 15-30 μm, cell breadth 15-30 μm.
Ecology: in benthos and tychoplankton of acidic, oligo-mesotrophic water bodies.
Distribution: Eurasia, Australia, possibly also N. America. Main point of distribution is in the Indo-Malaysian/Northern Australian region. In Europe it is a rare species; var. *retusum* is known from Scotland, Germany, Poland and Hungary, var. *hians* only from Germany and Poland. Zygospores are only described of var. *retusum*, from N. Australia.

The original description of *S. retusum* in Turner (1892) is rather poor. The present concept of *S. retusum* is based on that in West & West (1912) who stress the marked angular cell wall punctulation to be a relevant characteristic of this species. Particularly in tropical specimens the scrobiculation may be such pronounced that the outline of the cell angles is finely crenate rather than smooth. In view of this, it is not certain whether var. *boreale* W. et G.S. West is justly accounted the species under discussion, the cell wall of this variety being described as 'smooth' (West & West 1912). From experience, however, we know that the cell wall scrobiculation in question, even in tropical specimens, may be reduced to such a degree that it is only observed at close examination. It is not quite sure that the zygospores described by Coesel & Dingley (2005) refer to *S. retusum*. The adhering cells figured by those authors, on closer consideration, could be identified as *S. pseudoretusum* Hegde (1986) rather than *S. retusum*.

var. *retusum*   Plate 33: 4-11
Turner 1892, p. 104, pl. 13: 13
Sinus closed for the larger part, linear or slightly undulate.

Accounted *Staurastrum retusum* var. *retusum*:
Staurastrum retusum var. *boreale* W. et G.S. West 1905a
S. retusum var. *concavum* Noda et Skvortzov 1969
S. retusum var. *nitidulum* Hinode 1965
S. retusum var. *punctulatum* B. Eichler et Gutwiński 1895

var. *hians* (B. Eichler et Gutwiński) Coesel et Meesters stat. nov.   Plate 33: 12-13
Basionym: *Staurastrum retusum* var. *punctulatum* forma *hians* B. Eichler et Gutwiński 1895, p. 174, pl. 5: 45
Sinus widely opening from an acute apex.

Although Růžička (1972: 480) is of opinion that forma *hians* only slightly differs from the nominate forma of *S. retusum* we think its identity more questionable. Unfortunately, only a few records are known so that it is hard to assess its taxonomic status.

Staurastrum rhabdophorum Nordstedt   Plate 59: 3-4
Nordstedt 1875, p. 36, pl. 8: 40
Cells longer than broad, slightly constricted. Sinus a U-shaped incision. Semicells subquadrate in outline, with slightly concave lateral margins and truncate to slightly convex apices. Apical angles broadly rounded, basal angles about rectangled. Cell wall at the base of the semicell with a broad ring of longitudinally directed, granulated ridges. Just below the apex a similar series of granulate-denticulate verrucae. Also the apical margin itself provided with variably pronounced, emarginate verrucae. Semicells in apical view subcircular, the entire margin furnished with emarginate verrucae and with a subcircular series of verrucae within the margin. Zygospore unknown. Cell length 45-65 μm, cell breadth 30-40 μm.
Ecology: subatmophytic on moist rocks, also in benthos of small, pH-circumneutral alpine pools.
Distribution: rare arctic species, only known from Europe and North America.

Staurastrum ricklii Huber-Pestalozzi   Plate 52: 5
Huber-Pestalozzi 1928, p. 690, pl. 13: 10
Cells slightly longer than broad, deeply constricted. Sinus linear with a dilate extremity. Semicells trapeziform with broadly rounded angles. Cell wall coarsely granulate, granules rather far apart and arranged in concentric series around the angles. Semicells in apical view 3-angular with about straight sides and broadly rounded angles. Zygospore unknown. Cell length 41-45 μm, cell breadth 33 μm.
Ecology: in tychoplankton of oligo-mesotrophic water.
Distribution: only known from a lake in Corsica.

S. ricklii is a somewhat questionable species as it might refer to a triradiate *Cosmarium* species (*C. margaritiferum*?)

Staurastrum saltator Grönblad

Cells slightly broader than long, deeply constricted. Sinus V-shaped. Semicell body cup-shaped, the apical angles produced to form rather long and slender, divergent processes which are curved downwards towards their end. Processes tipped with 3-4 spines and towards the semicell body furnished with a series of denticulations, the dorsal ones being (much) better developed than the lateral ones. Semicell body provided with some (sub)apical denticulations, otherwise smooth. Semicells in apical view 3-radiate with about straight sides, the angles produced to form processes. Zygospore unknown. Cell length 30-32 µm, cell breadth 42-54 µm.
Ecology: in plankton of meso-eutrophic water bodies.
Distribution: the nominate variety is only known from a site in Finland, var. *pendulum* only from a site in Denmark.

S. saltator, in particular its var. *pendulum*, might be a mere, somewhat deviating form of *S. pingue*.

var. ***saltator*** Plate 88: 7
Grönblad 1938, p. 56, fig. 2: 4
Semicell body in apical view with 2 intramarginal denticulations on each side. Processes tipped with 4 spines.

var. ***pendulum*** (Nygaard) Coesel et Meesters stat. et comb. nov. Plate 88: 8-9
Basionym: *Staurastrum pendulum* Nygaard 1949, p. 99, text fig. 52
Semicell body in apical view with 3 intramarginal and 3 marginal denticulations on each side. Processes tipped with 3 spines.

Accounted *Staurastrum saltator* var. *pendulum*:
Staurastrum pendulum var. *pinguiforme* Croasdale 1958

Staurastrum scabrum Ralfs Plate 51: 6-10
Ralfs 1848, p. 214, pl. 35: 20
Cells about as long as broad, deeply constricted. Sinus open from an acute-angled apex or linear in its apical part and then open. Semicells trapezoid with broadly rounded angles. Cell wall furnished with acute or conical granules arranged in concentric series around the basal angles. Granules near the angles simple, those along the flanks are usually emarginate or merged into compound verrucae. Semicells in apical view 3-angular with almost straight sides and rounded angles, along the sides a marginal and an intramarginal series of emarginate verrucae. Zygospore described to be angular-globose, furnished with short, stout spines which are 2- to 4-fid at the apex. Cell length 27-40 µm, cell breadth 26-38 µm.
Ecology: in benthos of oligotrophic, acidic water bodies.
Distribution: Eurasia, North America. In Europe, *S. scabrum* is widely distributed and particular in arctic-alpine rgions rather common.

The original figure of *S. scabrum* in Ralfs (1848) is fairly poor, a much more informative illustration is provided in Wittrock & Nordstedt (1893). The zygospore of *S. scabrum* is described by Roy (1893) without illustration. The figure in Grönblad (1921) is without adhering semicells that could prove its correct assignment.

Taxon excluded:
Staurastrum scabrum forma *boldtii* Croasdale 1973

Staurastrum schroederi Grönblad   Plate 29: 5

Grönblad 1926, p. 32, pl. 2: 93, 94

Cells slightly longer than boad, rather deeply constricted. Sinus V-shaped. Semicells campanulate, the apical angles broadly rounded and somewhat produced (capitate). Cell wall smooth. Semicells in apical view 3-angular with concave sides and broadly rounded angles. Zygospore unknown. Cell length 23 μm, cell breadth 16 μm.

Ecology: in benthos of oligotrophic, acidic water.

Distribution: only known from a site in Silesia (Poland).

S. schroederi is a questionable species. It might refer to an anomalous form of some other smooth-walled *Staurastrum* species (e.g., *S. pachyrhynchum*) or to a triraduate form of some *Cosmarium* (e.g., *C. capitulum* var. *groenlandicum* Børgesen).

Staurastrum sebaldi Reinsch   Plate 98: 5-7

Reinsch 1867, p. 133, pl. 24D: I: 1-3

Cells about as long as broad, deeply constricted. Sinus V-shaped. Semicell body cup-shaped with convex dorsal margin, the angles produced to form rather short, converging processes. Processes tipped with (2-)3-4(-6) short spines and towards the semicell body furnished with (indistinct) concentric series of acute granules or denticulations, the dorsal ones being much more pronounced than the ventral and lateral ones. Dorsal series of process denticulations continuing on the apex of the semicell body as (furcate) spines. Semicell body just above the isthmus usually with a separate series of granules. Semicells in apical view 3(-4)-radiate with about straight sides, the angles produced to form processes. Along the sides an intramarginal series of (furcate) spines. Zygospore unknown. Cell length 73-85 μm, cell breadth 69-100 μm.

Ecology: in benthos of oligo-mesotrophic, slightly acidic water bodies.

Distribution: Eurasia, American continents. In Europe, *S. sebaldi* (in the above conception) is a rare species, mainly recorded from mountainous regions.

The original illustration of *S. sebaldi* by Reinsch (1867) shows a rather uniform, dense setting with stout spines in the upper half of the semicell body corresponding with a distinct marginal series of spines when seen in top-view. Remarkably, none of the later illustrations of this species fits Reinsch's original description so that the concept of *S. sebaldi* is quite unclear. The above-provided diagnosis is based on figures in West, West & Carter (1923) and describes a taxon that shows relationship with *S. traunsteineri* on the one hand and with *S. manfeldtii* on the other. Relevant literature: West, West & Carter (1923), Coesel (1992).

Accounted *Staurastrum sebaldi* (var. *sebaldi*):
Staurastrum sebaldi var. *altum* (Boldt) W. et G.S. West 1896
S. sebaldi var. *jarynae* Gutwiński 1892
S. sebaldi var. *spinosum* Wolle 1883

Taxa excluded:
Staurastrum sebaldi var. *brasiliense* Børgesen 1890 (cf. *S. manfeldtii*)
S. sebaldi var. *corpulentum* A.M. Scott et Grönblad 1957 (cf *S. traunsteineri*)
S. sebaldi var. *depauperatum* Boldt 1885 (cf *S. petsamoense*)
S. sebaldi var. *gracile* Messikommer 1927
S. sebaldi forma *groenlandica* Børgesen 1894 (cf. *S. manfeldtii*)
S. sebaldi var. *impar* Croasdale 1958
S. sebaldi var. *multispinosum* A.M. Scott et Grönblad 1957 (cf *S. manfeldtii* var. *pseudosebaldi*)
S. sebaldi forma *orientalis* W. et G.S. West 1907 (cf *S. manfeldtii* var. *pseudosebaldi*)
S. sebaldi var. *ornatum* Nordstedt 1873 (cf *S. manfeldtii*)

S. sebaldi var. *productum* W. et G.S. West 1905 (= *S. manfeldtii* var. *productum*)
S. sebaldi var. *quarternum* F.L. Harvey 1892
S. sebaldi var. *raciborski* (Gutwiński) Gronblad 1945
S. sebaldi var. *spinosum* S. Peterfi 1964 (cf *S. manfeldtii*)
S. sebaldi var. *traunsteineri* (Hustedt) Beck-Mannagetta 1931 (= *S. traunsteineri*)
S. sebaldi var. *triangularis* Bongale 1989 (cf *S. manfeldtii* var. *pseudosebaldi*)
S. sebaldi var. *ventriverrucosum* A.M. Scott et Prescott 1961

Staurastrum setigerum Cleve

Cells about as long as broad, deeply constricted. Sinus widely open from an acute-angled apex. Semicells (sub)elliptical, provided with a number of long spines, arranged chiefly at the angles, with a variable number on the faces; the spines at the angles usually distinctly longer and stouter than those on the faces in between. Semicells in apical view 3-angular with slightly concave to slightly convex sides, the angles provided with some long, stout spines, the sides and the centre of the apex with a variable number of additional spines, usually more delicate in nature, whether or not arranged in distinct patterns. Zygospore unknown. Cell length 60-80(-100) μm, cell breadth 65-85(-125) μm.
Ecology: in (tycho)plankton of acidic, oligotrophic water bodies.
Distribution: Eurasia, American continents, Africa. In Europe, var. *setigerum* is widely distributed but of rare occurrence. Var. *longirostre* is confined to tropical regions in South America.

S. setigerum is most variable in number, size and arrangement of the spines. Small-sized forms with angular spines hardly larger than the facial ones may be confused with *S. teliferum*.

var. **setigerum**   Plate 45: 4-9
Cleve 1864, p. 490, pl. 4: 4
Length of angular spines less than half the breadth of the semicell body.

Accounted *Staurastrum setigerum* var. *setigerum*:
Staurastrum setigerum var. *apertum* Palamar-Mordvintseva 1961
S. setigerum var. *brevispinum* G.M. Smith 1924
S. setigerum var. *cristatum* A.M. Scott et Grönblad 1957
S. setigerum var. *minor* Schmidle 1898a
S. setigerum var. *occidentale* W. et G.S. West 1896
S. setigerum var. *pectinatum* W. et G.S. West 1896
S. setigerum var. *subvillosum* Grönblad 1945
S. setigerum var. *tristichum* Nygaard 1926

Non-European variety:
var. **longirostre** Grönblad
Grönblad 1945, p. 30, pl. 13: 265
Length of angular spines more than half the breadth of the semicell body.

Accounted *Staurastrum setigerum* var. *longirostre*:
S. setigerum var. *reductum* Grönblad 1945
S. setigerum var. *spinellosum* Kurt Förster 1969

Taxa excluded:
Staurastrum setigerum forma *alaskanum* Croasdale 1957 (cf *S. teliferum*)
S. setigerum var. *alaskense* Irénée-Marie et Hilliard 1963 (cf *S. polytrichum*)

S. setigerum var. *argentinensis* Couté et Tell 1981
S. setigerum forma *major* Prescott, C.E.M. Bidudo et Vinyard 1982
S. setigerum var. *nyansae* Schmidle 1898a
S. setigerum var. *sublongirostre* Thomasson 1957

Staurastrum sexangulare (Bulnheim) P. Lundell

Cells somewhat broader than long to about as broad as long, deeply constricted. Sinus widely open. Semicells subfusiform to cuneate in outline, the angles deeply cleft to form an upper and a lower process. Rarely, small, accessory processes are developed or just the reverse, the upper series of processes is largely reduced. Processes tipped with (2-)3-4 spines, towards their base furnished with concentric or spiralling series of acute granules/denticulations. Semicell body often with a subapical, transversal series of granules, otherwise smooth-walled. Semicells in apical view (3-)5-7(-8) radiate, the angles produced to form tapering processes, with an upper process arising from the base of each lower process, the processes of upper and lower whorls rarely being exactly superimposed. Semicell body often with an intramarginal, more or less closed, circular series of granules. Zygospore unknown. Cell length (30-)60-100 µm, cell breadth (50-)80-125 µm.
Ecology: in (tycho)plankton of (slightly) acidic, oligotrophic water bodies.
Distribution: Eurasia, Australia, North America?, South America (var. *brasiliense*). In Europe, *S. sexangulare* is a rare species, mainly confined to boreal-atlantic regions.

Particularly in the Indo-Malaysian/northern Australian region *S. sexangulare* is a rather common species and most variable in number and size of the cell processes (Ling & Tyler 2000).

var. *sexangulare*   Plate 118: 1-4

Lundell 1871, p. 71, pl. 4: 9
Basionym: *Didymocladon sexangularis* Bulnheim 1861, p. 51, pl. 9A: 1
Cells relatively slender (breadth of istmus $^1/_5 - ^1/_{10}$ of total cell breadth). Semicell body subfusiform in outline.

Accounted *Staurastrum sexangulare* var. *sexangulare*:
S. sexangulare var. *asperum* Playfair 1910
S. sexangualare forma *australica* Schmidle 1896a
S. sexangulare var. *coronatum* Hinode 1959
S. sexangulare var. *crassum* W.B Turner 1892
S. sexangulare var. *dentatum* Playfair 1910 (anomaly?)
S. sexangulare var. *gemmescens* Playfair 1910 (anomaly?)
S. sexangulare var. *incurvum* Borge 1896
S. sexangulare var. *intermedium* W.B Turner 1892
S. sexangulare var. *productum* Nordstedt 1888a
S. sexangulare var. *subglabrum* W. et G.S. West 1902a
S. sexangulare var. *supernumerarium* W. et G.S. West 1903

Non-European variety:
var. *brasiliense* Grönblad
Grönblad 1945, p. 30, pl. 13: 270
Cells relatively stout (breadth of isthmus $^1/_3 - ^1/_4$ of total cell breadth). Semicell body cuneate in outline.

Taxa excluded:
Staurastrum sexangulare var. *attenuatum* W.B. Turner 1892
S. sexangulare var. *bidentatum* Gutwiński 1902 (cf *S. elegans* Borge 1896)
S. sexangulare var. *platycerum* (Joshua) Playfair 1910 (= *S. platycerum* Joshua 1886)

Staurastrum sexcostatum Ralfs
Plate 75: 12-16

Brébisson ex Ralfs 1848, p. 129, pl. 23: 5

Cells somewhat longer than broad to about as long as broad, rather deeply constricted. Sinus widely open, V-shaped at its apex. Semicells broadly elliptic-oval, the lateral angles produced to form short, stout, parallel processes (sometimes almost completely reduced). Processes broadly truncate at the end and furnished with some concentric series of stout, conical granules. Similar granules, usually enlarged to emarginate verrucae, extending onto the semicell apex. Semicells at the base with a supraisthmial whorl of granules. Semicells in apical view (5-)6(-8)-radiate, the angles produced into short, thick-set processes. Apex with curved series of intramarginal granules following the outline of the semicell. Zygospore unknown. Cell length (30-)40-50(-65) μm, cell breadth (25-)35-45(-50) μm.

Ecology: in benthos of shallow, often ephemeral, oligo-mesotrophic pools.

Distribution: Eurasia, North America. In Europe, *S. sexcostatum* is widely distributed but only of occasional occurrence.

Accounted *Staurastrum sexcostatum* (var. *sexcostatum*):
Staurastrum sexcostatum var. *depauperatum* Gutwiński 1896
S. sexcostatum var. *ornatum* (Nordstedt) Kurt Förster 1963
S. sexcostatum subsp. *productum* W. West 1892
S. sexcostatum var. *truncatum* Raciborski 1885

Staurastrum sibiricum Borge
Plate 26: 5-11

Borge 1891, p. 9, fig. 4

Synonym: *Staurodesmus sibiricus* (Borge) Croasdale 1962

Cells about as long as broad, rather deeply constricted. Sinus open, subrectangled. Semicells obversely subtriangular. Apical angles narrowly rounded or acuminated, often with an incrassate cell wall. Cell wall smooth. Semicells in apical view 2-angular (forma *ovalis*) or 3-(-5?) angular, with narrowly rounded angles and convex (forma *ovalis*) or almost straight sides. Zygospore unknown. Cell length 13-20(-25?) μm, cell breadth 15-21(-24?) μm.

Ecology: in plankton and benthos of acidic, oligotrophic water bodies, also subatmophytic on dripping rocks.

Distribution: Europe, Asia and North America. In Europe, *S. sibiricum* is a rare species, only recorded from arctic regions.

Accounted *Staurastrum sibiricum* (var. *sibiricum* forma *sibiricum*):
Staurastrum sibiricum var. *crassangulata* Børgesen 1894
S. sibiricum forma *ovalis* Borge 1891
S. sibiricum forma *trigona* W. et G.S. West 1896a

Taxa excluded:
Staurastrum sibiricum var. *baffinensis* (Whelden) Prescott, in Prescott, Bicudo & Vinyard 1982
S. sibiricum var. *occidentale* W. et G.S. West 1896

Staurastrum simonyi Heimerl

Cells about as long as broad, deeply constricted. Sinus open and acute-angled. Semicells elliptic to semicircular, the lateral/basal angles provided with 2-4 spines. Apical margin with two spines between each pair of consecutive angles. Cell wall further provided with a variable number of acute granules or minute spines, arranged in more or less regular concentric series around the angles. Semicells in apical view 3(-4)-angular with concave to straight sides. The angles may be acute, rounded or truncate and are furnished with one or more, usually rather stout spines. Within each margin an additional pair of marked spines. Zygospore described as globose and furnished with conical spines which are singularly forked at the apex. Cell length 18-28 µm, cell breadth 20-30 µm.

Ecology: in benthos and tychoplankton of acidic, ologotrophic water bodies.

Distribution: Eurasia, North America. In Europe, all three below-described varieties of *S. simonyi* are widely distributed and locally fairly common.

S. simonyi is a most variable species, its extreme forms often are found under different species names. As for the zygospore there is only one description, by Homfeld (1929), without any illustration.

Relevant literature: West, West & Carter (1923), Kouwets (1988).

var. *simonyi*   Plate 62: 7-12
Heimerl 1891, p. 607, pl. 5: 23

Semicells elliptic in outline. Lateral angles provided with two spines, lying in the same vertical plane. The four spines at the apical margin well developed; remaining ornamentation in the form of acute granules. Semicells in apical view with acute angles.

Accounted *Staurastrum simonyi* var. *simonyi*:
Staurastrum simonyi var. *elegantius* Grönblad 1920
S. simonyi var. *gracile* Lütkemüller 1893
S. simonyi var. *spinosius* Grönblad 1948a

var. *semicirculare* Coesel   Plate 62: 17-20
Coesel 1996, p. 25, fig. 27-28

Semicells semicircular in outline. Basal angles provided with a variable number of spines, the most basal one of which is often (much) longer than the other ones. The four spines at the apical margin not seldom reduced (hardly larger than the granules elsewhere on the cell wall). Semicells in apical view with rounded angles.

var. *sparsiaculeatum* (Schmidle) Hirano   Plate 62: 13-16
Hirano 1953, p. 208, pl. 2: 7
Basionym: *Staurastrum sparsiaculeatum* Schmidle 1896, p. 60, pl. 16: 20 ('*Staurastrum sparese-aculeatum*')

Semicells elliptic in outline. Lateral angles provided with 3-4 spines, lying in different planes. Next to the four apical spines there are shorter or longer spines elsewhere on the cell wall. Semicells in apical view with broadly rounded to truncate angles, furnished with stout spines.

Accounted *Staurastrum simonyi* var. *sparsiaculeatum*:
Staurastrum simonyi var. *convergens* A.M. Scott et Grönblad 1957

Staurastrum sinense Lütkemüller
Cells about as long as broad or somewhat longer than broad, (rather) deeply constricted. Sinus widely open, obtuse-angled with the apex minutely acuminate. Semicells in frontal view some-what table-shaped, with a rather abrupt transition between a columnar basal part and a flattened upper part with broadly rounded corners. Cell wall at the lateral lobes provided with a number of concentric series of fine granules, otherwise quite smooth. Semicells in apical view (3-)4-angular with concave sides and produced, broadly rounded to truncate angles. Zygospore described as smooth-walled, spherical, with crenate margins. Cell length (15-)20-25(-35) µm, cell breadth (14-)20-25(-33) µm.

Ecology: in benthos and tychoplankton of acidic, oligo-mesotrophic water bodies.

Distribution: possibly cosmopolitan. In Europe, the only reliably documented record (referring to the nominate variety) is by Borge (1906) from Sweden.

The main difference with the closely allied species *S. striolatum* is in the limitation of cell wall granulation to the radiating semicell lobes. If this difference would appear to be only gradual in nature, *S. sinense* better would be synonymized with *S. striolatum*. The only zygospore illustration of *S. sinense* is by Patel et Kumar (1980). In spite of the poor quality of this illustration it is quite clear that the morphology of this zygospore agrees with that of the related species *S. striolatum*, *S. dilatatum* and *S. alternans*.

var. *sinense*   Plate 54: 11-12
Lütkemüller 1900a, p. 124, pl. 6: 39, 40
Synonym: *Staurastrum disputatum* var. *sinense* (Lütkemüller) W. et G.S. West 1912
Cells deeply constricted, breadth of isthmus ca $^1/_3$ of cell breadth. Semicell lobes distinctly produced.

Accounted *Staurastrum sinense* var. *sinense*:
Staurastrum sinense var. *extensum* (Borge) Compère 1983

Non-European variety?:
var. *insigne* (Raciborski) Compère
Compère 1983, p. 157
Basionym: *Staurastrum dilatatum* var. *insigne* Raciborski 1892, p. 388, pl. 7: 13
Synonym: *Staurastrum disputatum* W. et G.S. West 1912, p. 176
Cells rather deeply constricted, breadth of isthmus ca $^1/_2$ of cell breadth. Semicell lobes only slightly produced.

Taxon excluded:
Staurastrum disputatum var. *annulatum* Rich 1939

Staurastrum smithii Teiling   Plate 112: 10-14
Teiling 1946, p. 82
Synonym: *S. contortum* G.M. Smith 1924, p. 98, pl. 76: 17-20 (invalid homonym of *S. contortum* Delponte 1878)
Cells slightly broader than long, very deeply constricted and strongly twisted at the isthmus. Sinus widely open with obtuse apex. Semicell body in outline cup-shaped with concave apex, the apical angles gradually produced to form long, divergent processes. Processes tipped with some three minute spines and towards the semicell body furnished with a spiralling series of undulations/denticulations. Semicell body smooth-walled. Semicells in apical view 2(–3?)-radiate with straight or slightly convex sides, the poles attenuated into processes. Zygospore unknown. Cell length 36-64 µm, cell breadth 44-66 µm.
Ecology: in plankton of meso-eutrophic water bodies.
Distribution: Eurasia, American continents, Australia. In Europe, *S. smithii* is rare, only known with certainty from Sweden, the Netherlands, Austria and Hungary.

S. smithii hardly can be distinguished from certain reduction forms of *S. bibrachiatum* (Grönblad & Scott 1955, Brook 1982) so possibly is identic to that latter species (in which case the name of *bibrachiatum* would have priority).

Taxon excluded:
Staurastrum smithii var. *verrucosum* Nygaard 1949 (= *S. nygaardii*)

Staurastrum spetsbergense (Nordstedt) Coesel et Meesters stat. nov.   Plate 58: 11-14
Basionym: *S. capitulum* var. *amoenum* forma *spetsbergensis* Nordstedt 1872, p. 39 and 41, pl. 7: 25
Cells distinctly longer than broad, slightly constricted. Sinus a V- or U-shaped incision. Semicells campanulate in outline, with concave lateral margins and a straight apex. Angles rounded or truncate, the apical ones distinctly produced. Cell wall provided with compound verrucae, arranged in

obscure concentric series around the angles and in a supraisthmial whorl at the base of the semi-cell. Semicells in apical view 3-angular with subacute angles and almost straight sides. Zygospore unknown. Cell length 30-45 μm, cell breadth 25-35 μm.
Ecology: in benthos of acidic to neutral, oligo-mesotrophic water bodies, also sub-atmophytic on wet substrates.
Distribution: Eurasia, North America. In Europe, *S. spetsbergense* is occasionally encountered in arctic-alpine areas.

Nordtstedt's original figure as represented in our Plate 58: 11, according to the author himself, would not be quite correct (West & West 1912, p. 126).

Staurastrum spongiosum Ralfs   Plate 60: 1-6
Brébisson ex Ralfs 1848, p. 141, pl. 23: 4
Cells about as long as broad, deeply constricted. Sinus narrowly open from an acute-angled apex (or linear near the apex and then open). Semicells in rough outline trapezoid, provided with short emarginate or furcate processes arranged in obscure, semicircular series around the lobes. Semicells in apical view 3(-4)-angular with slightly convex to slightly concave sides, the angles ending in an emarginate or furcate process, the sides with similar, marginal and intramarginal processes. Zygospore described as to be globose, furnished with numerous spines that are once or twice dichotomous at their apices. Cell length 45-65 μm, cell breadth 40-60 μm.
Ecology: in benthos of slightly acidic, oligo-mesotrophic water bodies.
Distribution: Eurasia, American continents. In Europe, *S. spongiosum* is widely distributed.

The zygospore of *S. spongiosum* was described by Lundell (1871) but without illustration. No other records are known.

Accounted *Staurastrum spongiosum* (var. *spongiosum* forma *spongiosum*):
Staurastrum spongiosum forma *dentatum* S. Peterfi 1964
S. spongiosum forma *depauperata* Raciborski 1890
S. spongiosum var. *griffithsianum* (Nägeli) Lagerheim, in Wittrock & Nordstedt 1886
S. spongiosum var. *perbifidum* W. West 1892a
S. spongiosum forma *spinosa* Irénée-Marie 1938
S. spongiosum var. *splendens* Tarnogradsky 1960

Taxa excluded:
Staurastrum spongiosum var. *americanum* W.B. Turner 1885
S. spongiosum var. *cumbricum* A.W. Bennet 1888

Staurastrum striatum (W. et G.S. West) Růžička   Plate 55: 1-6
Růžička 1957, p. 148, Fig. 3: 53-55
Basionym: *Staurastrum punctulatum* Ralfs var. *striatum* W. et G.S. West 1912, p. 186, pl. 128: 5, 6
Cells about as long as broad, deeply constricted. Sinus widely open, acute-angled. Semicells (sub) rhomboid with rounded, or rounded-truncate lateral angles. Cell wall with rather distant series of delicate granules, arranged in distinct concentric patterns around the lateral angles. Usually also a supraisthmial ring of granules may be distinguished. Semicells in apical view 3-angular with slightly concave sides and rounded, or rounded-truncate angles. Zygospore globose, furnished with long, stout spines that are double-furcate at the apex and arise from a mamillate base. Cell length 25-35 μm, cell breadth 24-36 μm.
Ecology: in benthos and tychoplankton of slightly acidic to slightly alkaline, meso-eutrophic water bodies.

Distribution: Eurasia. In Europe, *S. striatum* is locally rather common.

S. striatum may be readily confused with some other species, e.g., *S. punctulatum* and *S. dispar.*

Staurastrum striolatum (Nägeli) W. Archer   Plate 54: 8-10
Archer in Pritchard 1861, p. 740
Basionym: *Phycastrum striolatum* Nägeli 1849, p. 126, pl. 8A: 3
Cells about as long as broad, deeply constricted. Sinus widely open, acute-angled or obtuse-angled with the apex minutely acuminate. Semicells subtriangular to suboval in outline, the apex concave to almost straight, the basis often somewhat elongate, lateral angles broadly rounded or subtruncate. Cell wall finely granulate, granules arranged in concentric rings around the angles. Semicells in apical view 3(-4)-angular with concave sides and broadly rounded to truncate angles. Zygospore smooth-walled, compressed spherical, in top view with undulate margins. Cell length 19-28 µm, cell breadth 18-28 µm.
Ecology: in benthos and tychoplankton of acidic, oligo-mesotrophic water bodies.
Distribution: possibly cosmopolitan. In Europe, *S. striolatum* is fairly widely distributed but presumably rather rare (see note below).

The taxonomic concept of *S. striolatum* is somewhat questionable. In the original description by Nägeli (1849) the cell wall is characterized as 'unweaponed, at the radii ring-shaped striated'. In the concept by West & West (1912) there is mention of a finely granulated cell wall. In addition to that, the lateral angles are described as rounded or subtruncate. Especially the latter characteristic (subtruncate angles) appears to have played an important role in identification by later authors. According to various illustrations in literature the feature of subtruncate angles usually has been attached greater value than the original characteristic of a concave-straight apical outline. Larger-sized cell forms showing distinctly convex apical margins, however, probably be better accounted *S. dilatatum.*

Accounted *Staurastrum striolatum* (var. *striolatum*):
Staurastrum striolatum forma *brasiliensis* W.B. Turner 1892
S. striolatum var. *divergens* W. et G.S. West 1902a
S. striolatum var. *intermedium* Thérézien 1985
S. striolatum var. *oelandicum* Wittrock 1872

Taxa excluded:
Staurastrum striolatum var. *acutius* Maskell 1889
S. striolatum var. *divergens* forma *major* Irénée-Marie 1958

Staurastrum subboergesenii Grönblad   Plate 111: 1-5
Grönblad 1938, p. 60, fig. 3: 1
Cells about as long as broad, deeply constricted. Sinus open and acute-angled. Semicells cup-shaped, the apical angles produced to form slender, divergent processes which are bi- or trifurcate at the apex. Both apically and laterally inserted usually two more, but much shorter furcate processes are present between each pair of consecutive angles. Cell wall otherwise predominantly smooth. Semicells in apical view 3-4-radiate with straight or concave sides, the angles attenuated into spine-like procsses with on either side of their base a pair of intramarginal and a pair of marginal, short, furcate processes (incidentally, however, for a lesser or greater part reduced). Zygospore unknown. Cell length 35-65 µm, cell breadth 45-80 µm.
Ecology: in acidic, oligotrophic water bodies.
Distribution: Europe, Asia. In Europe it is only known from eastern Finland and Swedish Lappland.

Possibly, *S. subboergesenii* is closely affiliated to the North American species *S. novae-terrae* W.R. Taylor.

Staurastrum subbrebissonii Schmidle   Plate 46: 4-5

Schmidle 1894, p. 554, pl. 28: 15

Cells about as long as broad, deeply constricted. Sinus widely open from an acute-angled apex. Semicells subtrapezoid with broadly rounded angles, evenly covered with rather stout spines which are arranged in obscure circles round the angles. Semicells in apical view 3(-4)-angular with almost straight sides and broadly rounded angles. Zygospore unknown. Cell length 50-80μm, cell breadth 40-80 μm.

Ecology: in benthos of slightly acidic, mesotrophic water bodies.

Distribution: central Europe.

Relevant literature: Kouwets (1987).

S. subbrebissonii is a poorly known, questionable species, somewhat intermediate between *S. hirsutum* and *S. polytrichum*. Actually, most records in literature refer to *S. kouwetsii*.

Taxon excluded:

Staurastrum subbrebissonii var. *hexagonum* Gutwiński 1909 (cf *S. trapezioides*)

Staurastrum subcruciatum Cooke et Wills   Plate 69: 9-12

Cooke & Wills, in Cooke 1887, p. 148, pl. 51: 3

Cells somewhat broader than long, deeply constricted. Sinus widely open from an acute-angled apex. Semicells triangular to elliptic, the lateral sides produced obliquely upwards and ending in two diverging spines which lie in the same vertical plane. Cell wall covered with minute granules which are arranged in concentric series round the angles and are reduced or wanting more remote of them. Semicells in apical view 3(-4)-angular with slightly concave sides and acutely extended angles. Zygospore unknown. Cell length 27-32 μm, cell breadth 33-47 μm.

Ecology: in benthos and tychoplankton of oligo-mesotrophic, acidic to pH-circumneutral water bodies.

Distribution: only known for certain from Eurasia and North America, but possibly wider distributed. In Europe, *S. subcruciatum* is a rare species, mainly confined to atlantic regions.

Accounted *Staurastrum subcruciatum* (var. *subcruciatum*):

Staurastrum subcruciatum var. *demissum* Palamar-Mordvintseva 1961

Questionable taxon:

Staurastrum subcruciatum forma *nana* Lütkemüller 1900

Taxa excluded:

Staurastrum subcruciatum var. *reductum* Bourrelly et Couté 1991

S. subcruciatum var. *trispinatum* Irénée-Marie 1951

Staurastrum subexcavatum (Grönblad) Coesel et Meesters stat. nov.   Plate 114: 1-2

Basionym: *Staurastrum tetracerum* var. *subexcavatum* Grönblad 1921, p. 62, pl. 5: 28-29

Cells broader than long, deeply constricted and usually twisted at the isthmus. Sinus V-shaped. Semicell body bowl-shaped with excavated apex, the apical angles produced to form long, parallel to slightly divergent processes. Processes tipped with some three minute spines and towards the semicell body furnished with a spiralling series of undulations/dentations. Semicells in apical view 2-radiate, the elliptic semicell body rather gradually passing into the processes. Zygospore unknown. Cell length 15-40 μm, cell breadth 35-65 μm.

Ecology: in acidic, oligo-mesotrophic water bodies.
Distribution: unclear because of possible confusion with other species. In Europe it is known from Finland and N.W. Spain.

Staurastrum subexcavatum may be considered an intermediate form between *S. tetracerum* and *S. excavatum* W. et G.S. West 1895, the latter species being described from Madagascar.

Staurastrum sublaevispinum W. et G.S. West   Plate 35: 5-6
W. & G.S. West 1898, p. 314, pl. 18: 20-22
Cells somewhat broader than long, deeply constricted. Sinus widely open with an obtuse angle giving rise to a somewhat elongate isthmus. Semicells obversely triangular with almost straight lateral sides and a concave apex. Semicell body relatively small, almost imperceptibly passing into finger-like, diverging processes which are hardly attenuated towards their obtusely rounded or conical apices. Cell wall smooth. Semicells in apical view 3(-4)-angular with strongly concave sides. Zygospore unknown. Cell length 25-35 µm, cell breadth 30-45 µm.
Ecology: in benthos and plankton of acidic, oligotrophic water bodies.
Distribution: reported from North America, Europe, S.E. Asia and Australia, but only incidentally. In Europe, *S. sublaevispinum* is of very rare occurrence, only known from Scotland and S.W. Finland.

S. sublaevispinum sometimes can hardly be distinguished from *S. laevispinum*, see, e.g., Thomasson & Tyler (1971). The latter authors report also *S. sublaevispinum*-like specimens of *S. brachiatum*. So, possibly, just like *S. laevispinum* also *S. sublaevispinum* may be considered a form of *S. brachiatum*.

Staurastrum subnivale Messikommer   Plate 77: 11-16
Messikommer 1942, p. 170, pl. 18: 5
Cells about as long as broad, deeply constricted. Sinus V-shaped. Semicell body cup-shaped to campanulate, the apical angles produced to form somewhat incurved processes tipped with short spines. Cell wall furnished with dentations, arranged in concentric series around the processes, those at the apex usually emarginate. Flanks of the semicells just below the apical series of granules usually with an unsculptured zone. Semicells at the base with a supraisthmial whorl of (simple) granules. Semicells in apical view 3-angular with about straight sides, the angles produced to form short processes. Sides of the semicell body with an intramarginal series of emarginate dentations. Zygospore globose, furnished with slender, furcate spines. Cell length 26-32 µm, cell breadth 25-38 µm.
Distribution: Europe, North America? Distribution is poorly known because of possible confusion with other species. In Europe, *S. subnivale* has only been recorded from Switzerland and Austria but in our opinion also some records of *S. borgeanum* originating from Sweden, Poland and the Netherlands may be accounted *S. subnivale*.

Questionable non-European taxon:
Staurastrum subnivale forma *alaskanum* Croasdale 1957

Staurastrum subnudibranchiatum W. et G.S. West   Plate 35: 14-17
W. & G.S. West 1905, p. 502, pl. 7: 18, 19
Cells slightly broader than long, deeply constricted. Sinus widely open, more or less rectangled. Semicells bowl-shaped with convex apex and convex lateral sides. Apical angles produced to form arm-like, diverging processes which are obtusely bifid (sometimes obtusely entire) at the ends. Cell wall smooth. Semicells in apical view 4-5-angular with strongly concave sides. Zygospore unknown. Cell length 25-45 µm, cell breadth 33-61 µm.

Ecology: in plankton of oligotrophic soft-water bodies.
Distribution: rare species, only known with certainty from Europe and North America. Within Europe, *S. subnudibranchiatum* is known from Scotland, Finnish Lapland and S.W. France.

S. subnudibranchiatum differs from *S. brachiatum* in a more spherical shape of the semicell body, particularly its convex apex. Yet, intermediate forms are known, e.g., the illustrations of *S. subnudibranchiatum* by Tell (1980) and Vijverman (1991) from Argentine and Papua New Guinea, respectively. The cell wall dots illustrated by Nygaard (1991) obviously refer to gelatinous protrusions excreted by cell wall pores and not to real cell wall granules as suggested by that author.

Accounted *Staurastrum subnudibranchiatum* (var. *subnudibranchiatum*):
Staurastrum subnudibranchiatum var. *incisum* G.M. Smith 1924

Taxon excluded:
Staurastrum subnudibranchiatum var. *latispinum* Irénée-Marie 1957

Staurastrum suborbiculare W. et G.S. West   Plate 32: 19-20
W. & G.S. West 1896a, p. 158, pl. 4: 48
Cells about as long as broad, deeply constricted. Sinus linear, with a dilate extremity. Semicells pyramidate-subsemicircular with rounded angles, slightly convex lateral sides and faintly retuse (concave) apex. Cell wall smooth. Semicells in apical view 3-angular with broadly rounded angles and almost straight sides. Zygospores globose, densely covered with small conical protuberances, each beset with a minute spine. Cell length 35-44 μm, cell breadth 35-38 μm.
Ecology: in benthos and tychoplankton of acidic, oligotrophic water bodies.
Distribution: unclear as reliable identification is mainly on the basis of zygospores. Presumably, *S. suborbiculare* is a rare species. Zygospores are only known from a site in Scotland. Some records of vegetative cells from the American continents are credible.

Staurastrum subosceolense Grönblad   Plate 88: 5-6
Grönblad 1920, p. 79, pl. 3: 103-104
Cells (slightly) broader than long, deeply constricted. Sinus acute-angled to almost rectangular. Semicells cup-shaped (triangular), the apical angles produced to form rather long, (slightly) diverging processes. Processes tipped with 3-4 spines, towards the semicell body furnished with a spiralling series of dentations. Semicells in apical view 3-4-radiate with about straight sides, the angles produced to form processes, the semicell body on each side with two intramarginal emarginate verrucae.
Zygospore unknown. Cell length 30-60 μm, cell breadth 60-75 μm.
Ecology: in oligo-mesotrophic water bodies.
Distribution: Europe. Only known from some sites in Finland and Sweden.

Staurastrum subosceolense is a questionable, poorly-known species that may readily be confused with some other species, such as reduction forms of *S. floriferum*.

Staurastrum subpygmaeum W. West
Cells about as long as broad, (rather) deeply constricted. Sinus open, acute-angled to subrectangled. Semicells cuneate-subrhomboid with (slightly) convex sides, the lateral angles broadly rounded and more or less produced. Cell wall smooth. Semicells in apical view 3-4-angular with convex sides and broadly rounded, produced angles. Zygospore unknown. Cell length (23?-)35-45(-53) μm, cell breadth (23?-)35-45(-52) μm.
Ecology: in benthos and tychoplankton of acidic, oligotrophic water bodies.

Distribution: Europe, American continents, Japan. In Europe, *S. subpygmaeum* var. *subpygmaeum* is widely distributed, but rare; var. *subangulatum* is only known with certainty from Scotland and N.W. Russia, var. *undulatum* only from a site in western Ireland.

var. *subpygmaeum* Plate 28: 11-12
W. West 1892a, p. 178, pl. 23: 8

Synonym: *Staurodesmus subpygmaeus* (W. West) Croasdale 1962

Cells rather deeply constricted (breadth of isthmus $^1/_2 - ^1/_3$ of total cell breadth). Semicells cuneate in outline. Semicell flanks firmly lined.

Accounted *Staurastrum subpygmaeum* var. *subpygmaeum* (forma *subpygmaeum*):

Staurastrum subpygmaeum var. *apertum* V. et P. Allorge 1931

S. subpygmaeum forma *glabra* W. et G.S. West 1894

S. subpygmaeum forma *tetragona* Shirshov 1935

var. *subangulatum* W. et G.S. West 1912 Plate 28: 13
W. & G.S. West 1912, p. 163, pl. 124: 1

Cells deeply constricted. Semicells subrhomboid in outline. Semicell flanks firmly lined.

var. *undulatum* (D.B. Williamson et D.M. John) Coesel et Meesters comb. nov. Plate 28: 14-16
Basionym: *Staurastrum clepsydra* var. *undulatum* D.B. Williamson et D.M. John 2009, p. 130, pl. 29Aa-Ad

Cells deeply constricted. Semicells cuneate in outline. Semicell flanks with some two, variously pronounced undulations.

Taxa excluded:

Staurastrum subpygmaeum var. *ecorne* (B. Eichler et Gutwiński) W. et G.S. West 1895 (= *S. crassimamillatum*)

S. subpygmaeum var. *minus* Scott et Grönblad 1957

S. subpygmaeum var. *spiniferum* Scott et Grönblad 1957

Staurastrum subsphaericum Nordstedt Plate 34: 4-6
Nordstedt 1875, p. 31, pl. 8: 33

Cells distinctly longer than broad, moderately constricted. Sinus widely open, about rectangled at the apex. Semicells subcircular in outline. Cell wall looking rough because of mucilage extrusions ('subgranulate'). Semicells in apical view 3-5(-8)-angular, with straight to slightly convex sides. Zygospore unknown. Cell length 48-59 µm, cell breadth 30-39 µm.

Ecology: in oligo-mesotrophic water bodies.

Distribution: only known from Spitsbergen.

Taxon excluded:

S. subsphaericum forma *americana* Raciborski 1892

Staurastrum suchlandtianum Messikommer Plate 75: 9-11
Messikommer 1942, p. 163, pl. 14: 10, pl. 18: 4

Cells slightly broader than long to about as broad as long, deeply constricted. Sinus widely open and acute-angled. Semicells sub-elliptic to cup-shaped, the apical angles produced to form rather short to fairly long, parallel or slightly converging processes. Processes tipped with short spines and towards the semicell body furnished with acute granules or denticulations being arranged in indistinct concentric series, those on the dorsal side much better developed than those on the other sides. Semicell body also furnished with spines or denticulations but predominantly or even almost exclusively at the apex. Semicells in apical view 3-radiate with almost straight

sides, the angles produced to form processes. At the base of each process two short, bifurcate, diverging spines. Semicell body with 1 or 2 intramarginal bifurcate spines projecting on each side which, however, sometimes may be largely reduced. Zygospore unknown. Cell length 27-30 µm, cell breadth 34-42 µm.
Ecology: in benthos of oligotrophic water bodies.
Distribution: only known from the European alps, rare.

Staurastrum teliferum Ralfs

Cells about as long as broad, deeply constricted. Sinus more or less open from an acute-angled apex. Semicells (sub)elliptical, provided with a number of stout spines, arranged chiefly at the angles with a variable number positioned on the faces. Semicells in apical view 3-angular with (slightly) concave sides and broadly rounded angles. Zygospore (only known of the nominate variety) globular, provided with a number of long, stout spines, bifurcate at the apex. Cell length (35-)40-55(-65) µm, cell breadth 35-50 µm.
Ecology: in (tycho)plankton of acidic to slightly alkaline, oligo-mesotrophic water bodies.
Distribution: Eurasia, American continents, Africa. In Europe, var. *teliferum* is widely distributed and of common occurrence in acidic, oligotrophic water bodies; var. *gladiosum* is less common and seems to prefer circumneutral to slightly alkaline, mesotrophic habitats.

In the original descriptions the most important difference between *S. teliferum* Ralfs and *S. gladiosum* W.B. Turner is in the positioning of the spines: confined to the angles in *S. teliferum*, arranged in series all over the cell surface in *S. gladiosum*. However, in practice this differentiating characteristic appears not to hold out. Usually, next to the angular spines a smaller or larger number of scattered spines on the semicell face may be observed, as is even the case in Ralfs (1848, t. 34: 14), see our Pl. 44: 2.

var. *teliferum*   Plate 44: 1-9
Ralfs 1848, p. 128, pl. 22: 4, pl. 34: 14
Cells sligthly longer than broad, spines more or less concentrated at the lobes.

Accounted *Staurastrum teliferum* var. *teliferum* (forma *teliferum*):
Staurastrum teliferum var. *alpinum* (Schmidle) J. Sampaio 1944
S. teliferum var. *compactum* Kurt Förster 1974
S. teliferum var. *horridum* Lütkemüller 1900
S. teliferum var. *ordinatum* Børgesen 1894
S. teliferum forma *obtusa* W. West 1892a
S. teliferum var. *pecten* (Wolle) Grönblad 1945
S. teliferum var. *subteliferum* (J. Roy et Bisset) Kurt Förster 1970
S. teliferum var. *validum* Max Schmidt 1903

var. *gladiosum* (W.B. Turner) Coesel et Meesters stat. et comb. nov.   Plate 44: 10-13
Basionym: *Staurastrum gladiosum* W.B. Turner 1885, p. 6, pl. 16: 21
Cells slightly broader than long, spines rather evenly distributed over the cell surface.

Accounted *Staurastrum teliferum* var. *gladiosum* (forma *gladiosum*):
Staurastrum gladiosum var. *delicatulum* W. et G.S. West 1900
S. gladiosum forma *ornata* Laporte 1931
S. teliferum var. *subacutangulum* Messikommer 1971
S. teliferum var. *tatricum* Gutwiński 1909

Taxa excluded::
Staurastrum teliferum var. *convexum* Bennett 1886
S. teliferum var. *groenbladii* Kurt Förster 1964
S. teliferum forma *lagoensis* Wille 1884
S. teliferum var. *lagoense* (Wille) Grönblad 1945
S. teliferum var. *longispinum* Grönblad 1945
S. teliferum var. *reniforme* J. Sampaio 1944
S. teliferum var. *transvaalense* Claassen 1961
S. gladiosum forma *curvispinum* Grönblad, in Grönblad, Scott & Croasdale 1964
S. gladiosum var. *longispinum* W.B. Turner 1892
S. gladiosum var. *submuricatum* Beck 1926

Staurastrum tetracerum (Kützing) Ralfs

Cells about as long as broad, deeply constricted and usually twisted at the isthmus. Sinus V-or U-shaped. Semicell body bowl-shaped to rectangular, the apical angles produced to form rather long, divergent processes. Processes tipped with some four minute spines and towards the semicell body furnished with a spiralling series of undulations/denticulations (the dorsal denticulations consequently alternating with the ventral ones); also in the basal part of the semicell body often a transversal series of distant denticulations. Semicells in apical view 2(-4)-radiate with about straight sides, the poles attenuated into processes. Zygospore described to be globose, furnished with long processes once or twice dichotomous at the apex. Cell length (13-)20-30(-50) µm, cell breadth (14-)20-35(-55) µm.

Ecology: in benthos and plankton of both oligotrophic and eutrophic, both acidic and alkaline water bodies, may cause blooms in eutrophic waters.

Distribution: cosmopolitan. In Europe, var. *tetracerum* is common, var. *cameloides* and var. *irregulare* are less common and var. *biverruciferum* is only known from a site in Finland.

S. tetracerum is a polymorphic taxon. Also because of its wide ecological range a taxonomic revision could be desirable. The zygospore of *S. tetracerum* was described by Lundell (1871) but no reliable illustrations are known.

var. ***tetracerum***   Plate 113: 7-17
Ralfs ex Ralfs 1848, p. 137, pl. 23: 7
Basionym: *Micrasterias tetracera* Kützing 1833, p. 602, pl. 19: 83-84
Semicell body without any frontal protuberance.

Accounted *Staurastrum tetracerum* var. *tetracerum*:
Staurastrum tetracerum var. *evolutum* W. et G.S. West 1905a
S. tetracerum forma *tetragona* W. et G.S. West 1897
S. tetracerum forma *trigona* P. Lundell 1871
S. tetracerum var. *undulatum* W. et G.S. West 1895

var. ***biverruciferum*** Grönblad   Plate 113: 18
Grönblad 1921, p. 61, pl. 5: 31-32
Semicell body in frontal view furnished with two median protuberances, each of them being positioned near the base of a process.

var. ***cameloides*** M. Florin 1957   Plate 113: 26-30
Florin 1957, p. 133, fig. 29: 5-8
Synonym: *Staurastrum paradoxum* var. *osceolense* forma *biradiata* Georgevitch 1910, p. 245, text fig. 6

Semicell body without any frontal protuberance but with a prominent inflation at the base of the processes (giving the impression of shrugged shoulders).

var. ***irregulare*** (W. et G.S. West) Brook 1982   Plate 113: 19-25
Brook 1982, p. 263, text fig. 2
Basionym: *Staurastrum irregulare* W. et G.S. West 1894, p. 48, fig. 49-50
Semicell body furnished with a more or less prominent, truncate-denticulate central protuberance.

Accounted *Staurastrum tetracerum* var. *irregulare*:
Staurastrum irregulare var. *spinosum* Willi Krieger et Bourrelly 1956
S. irregulare var. *subosceolense* Grönblad 1945
S. irregulare forma *tenue* Förster 1963

Non-European variety:
var. ***trigranulatum*** W. et G.S. West
W. & G.S. West 1907, p. 219, pl. 15: 19
Semicell body in frontal view furnished with three median tubercles, positioned in a transversal row.

Taxa excluded:
Staurastrum tetracerum var. *bohlinii* Kurt Förster 1974 (cf *S. smithii*)
S. tetracerum var. *maximum* Messikommer 1966 (cf *S. chaetoceras*)
S. tetracerum var. *validum* W. et G.S. West 1897 (cf *S. chaetoceras*)

Staurastrum thomassonii Nygaard   Plate 34: 7-11
Nygaard 1991, 213, pl. 4: 79-81, pl. 7: 120a,b
Cells (slightly) longer than broad, moderately constricted. Sinus widely open, acute-angled to almost rectangled. Semicells subrhomboid with broadly rounded angles. Cell wall smooth. Semicells in apical view 3-angular with straight to slightly concave sides and broadly rounded angles. Zygospore unknown. Cell length 20-29 µm, cell breadth 15-23 µm.
Ecology: in plankton of oligo-mesotrophic water.
Distribution: only known from a lake in Denmark.

Staurastrum tohopekaligense Wolle   Plate 38: 7-10
Wolle 1885, p. 128, pl. 51: 4, 5
Cells slightly longer than broad to somewhat broader than long, deeply constricted. Sinus open and acute-angled. Semicells broadly elliptic in outline, the poles produced to form arm-like processes which are bi- or trifurcate at their apex, not seldom with two other similar processes between each pair of consecutive angular processes, furthermore with a dorsal series of similar processes, two of which project between each pair of consecutive angles. Cell wall smooth. Semicells in apical view 3(-4)-radiate with about straight sides, the angles produced into arm-like processes with on either side a similar process inserted within the margin. Not seldom on either side of each angular process also a pair of marginal processes is present. Zygospore unknown. Cell length (30-)45-90 µm, cell breadth (30-)45-95 µm.
Ecology: in (tycho)plankton of acidic, oligotrophic water bodies.
Distrinution: cosmopolitan. In Europe, *S. tohopekaligense* is rather rare, displaying a predominantly atlantic distribution.
Relevant literature: Ling & Tyler (1995).

Particularly in tropical regions, *S. tohopekaligense* is much variable in both cell dimensions and number of processes. In the original diagnosis of *S. tohopekaligense* by Wolle (1885) is mention of triradiate semicells with two whorls of processes : a lower whorl of 3 and an upper whorl of 6 processes. However, later on also cell forms with a lower whorl of 9 instead of 3 processes, as well as janus forms have been recorded. Large forms with 15 processes per semicell may be confused with the non-European, (sub)tropical species *S. leptacanthum* Nordstedt, the semicell body of which, in apical view, is subcircular rather than triangular.

Accounted *Staurastrum tohopekaligense* (var. *tohopekaligense* forma *tohopekaligense*):
Staurastrum tohopekaligense forma *acuminatum* A.M. Scott et Prescott 1961
S. tohopekaligense var. *brevispinum* G.M. Smith 1924
S. tohopekaligense var. *europaeum* Grönblad 1947
S. tohopekaligense forma *minus* (W.B. Turner) A.M. Scott et Prescott 1961
S. tohopekaligense var. *nonanum* (W.B. Turner) Schmidle1898a
S. tohopekaligense var. *quadrangulare* W. et G.S. West 1895
S. tohopekaligense var. *trifurcatum* W. et G.S. West 1895

Questionable non-European taxon:
Staurastrum tohopekaligense var. *robustum* A.M. Scott et Prescott 1961

Taxa excluded:
Staurastrum tohopekaligense var. *brevispinum* forma *hexagonum* F. Rich 1935
S. tohopekaligense forma *hexagonum* F.Rich 1935
S. tohopekaligense var. *insigne* W. et G.S. West 1902 (cf *S. leptacanthum*)
S. tohopekaligense var. *quadridentatum* C.C. Jao 1949 (cf. *S. hantzschii*)

Staurastrum tortum (Lagerheim et Nordstedt) W. et G.S. West   Plate 26: 12-15
W. & G.S. West 1912, p. 161, pl. 125: 9
Basionym: *Cosmarium tortum* Lagerheim et Nordstedt, in Wittrock, Nordstedt & Lagerheim 1903, p. 16, fig. 1-8
Cells a little bit longer than broad, slightly constricted and conspicuously twisted at the isthmus. Sinus widely open and obtuse-angled. Semicells cuneate with rounded apical angles and retuse apex. Cell wall smooth. Semicells in apical view elliptic or 3-angular with rounded angles and straight sides. Zygospore unknown. Cell length 16-20 µm, cell breadth 14-17 µm.
Ecology: in (tycho)plankton of oligotrophic water bodies.
Distribution: Europe, Asia, Australia, North America. In Europe, only the biradiate form of *S. tortum* is known, from a few sites in Sweden, Great Britain, Portugal and the Netherlands.

The taxonomic status of *S. tortum* is uncertain; it very much resembles both *S. minutissimum* and *S. sibiricum*.

Accounted *Staurastrum tortum* (var. *tortum* forma *tortum*):
Staurastrum tortum forma *trigona* G.S. West 1909

Staurastrum trachytithophorum W. et G.S. West   Plate 56: 3-4
W. & G.S. West 1897, p. 493, pl. 6: 22
Cells about as long as broad, deeply constricted. Sinus widely open with acute-angled apex. Semicells obversely triangular with (slightly) convex sides, domed apex and somewhat produced, rounded, apical angles. Cell wall at the apical angles provided with a number of scattered, acute granules, arranged in some ill-defined concentric series; otherwise smooth. Semicells in apical view 3-angular with faintly convex sides and slightly produced, rounded angles. Zygospore globose, furnished with long, slender spines, 2-4-fid at the apex. Cell length 30-34 µm, cell breadth 29-33 µm.

Ecology: in benthos of acidic, oligotrophic water bodies.
Distribution: Europe, possibly also North America. Apart from the original description from the south of England, hardly any reliable records are known.

Although West & West (1897) state *S. trachytithophorum* to be quite distinct from any other species, most later records are dubious. Therefore it cannot be excluded that *S. trachytithophorum* is an anomalous form of some other species (e.g., *S. acutum* var. *varians*).

Taxon excluded:
Staurastrum trachytithophorum var. *pingue* Grönblad 1960

Staurastrum trapezioides Coesel et Meesters stat. et nom. nov.   Plate 49: 1-5
Synonym: *Staurastrum brebissonii* var. *truncatum* Grönblad 1926, p. 27, pl. 2: 85, 86
Cells about as broad as long or a little broader than long, deeply constricted. Sinus open from an acute-angled apex. Semicells trapeziform with broadly rounded angles, evenly and closely covered with short spines arranged in concentric series around the angles. Semicells in apical view 3-angular with concave sides and broadly rounded angles. Zygospore unknown. Cell length 40-55μm, cell breadth 42-58 μm.
Ecology: in benthos and tychoplankton of mesotrophic, slightly acidic water bodies.
Distribution: Europe. Known from Germany, the Netherlands, Austria and Poland but presumably wider distributed.

Earlier (e.g., Růžička 1972, Coesel & Meesters 2007) the species under discussion was identified as *S. trapezicum*. On closer examination, however, *S. trapezicum* described by Boldt (1888) from Greenland relates to another species.

Staurastrum traunsteineri Hustedt   Plate 99: 1-4
Hustedt 1911, p. 340, fig. 35
Cells broader than long, deeply constricted. Sinus widely open from an acute-angled apex. Semicell body elliptic to broadly fusiform, the lateral angles produced to form strongly converging processes. Processes tipped with (2-)3-4 short spines and towards the semicell body furnished with concentric series of granules. Dorsal series of granules continuing over the apex of the semicell body as flattened, emarginate verrucae. Semicells in apical view 3-radiate with straight to slightly convex sides, the angles produced to form processes. Semicell body in apical view with smooth margins whereas the intramarginal verrucae are arranged in remarkably even rows and gradually decrease in size from the midst of the semicell towards the end of the processes. Zygospore unknown. Cell length 55-70 μm, cell breadth 70-90 μm.
Ecology: in benthos of oligo-mesotrophic water bodies.
Distribution: rare, only recorded from alpine regions in central Europe.

Staurastrum tristichum Elfving   Plate 56: 1-2
Elfving 1881, p. 8, pl. 1: 4
Cells longer than broad, rather deeply contricted. Sinus widely open, obtuse-angled. Semicells subrhomboid with (sub)acuminate lateral angles and broadly rounded apex. Semicell wall at each of the lateral angles furnished with some three concentric series of granules, otherwise smooth. Semicells in apical view 4-angular with concave sides and acuminate angles. Zygospore unknown. Cell length 30-38 μm, cell breadth 21-32(-38) μm.
Ecology: in oligo-mesotrophic water bodies.
Distribution: only known from SW Finland.

S. tristichum is a questionable species, maybe it is just a form of *S. acutum* var. *varians*.

Staurastrum tumidum Ralfs   Plate 23: 5-6; Plate 24: 1-2
Brébisson ex Ralfs 1848, p. 126, pl. 21: 6
Synonym: *Staurodesmus tumidus* (Ralfs) Teiling 1967, p. 578
Cells slightly longer than broad, rather deeply constricted. Sinus widely opening from its acute-angled apex. Semicells broadly elliptic-oval in outline, the lateral sides with a median button-like thickening or mamilla. Semicells in apical view 3(–4)-angular with convex sides and mamillate angles. Zygospore described as ovoid-oblong, with a thick, lamellose wall, furnished with scattered conical papillae. Cell length 110-135 µm, cell breadth 90-125 µm.
Ecology: in benthos of acidic, oligotrophic water bodies.
Distribution: Europe, North America and Japan. In Europe, *S. tumidum* is fairly widespread, but generally rare.

Accounted *Staurastrum tumidum* (var. *tumidum*):
Staurastrum tumidum var. *quadrangulatum* M. Mix 1973

Taxa excluded:
Staurastrum tumidum var. *attenuatum* Borge 1918
S. tumidum var. *bipapillatum* Croasdale, in Grönblad & Croasdale 1971

Staurastrum turgescens De Notaris   Plate 53: 1-4
De Notaris 1867, p. 51, pl. 4: 43
Cells a little longer than broad, deeply constricted. Sinus open, acute-angled, with an acute or narrowly rounded apex. Semicells elliptic-oval in outline. Cell wall densely granulate. Semicells in apical view 3-angular with concave sides and broadly rounded angles. Zygospore said to be 'smooth-walled, compressed spherical, in the broad view with 9-12 marginal undulations'. Cell length 28-42 µm, cell breadth 25-37 µm.
Ecology: in benthos and tychoplankton of acidic, oligo-mesotrophic water bodies.
Distribution: possibly cosmopolitan. In Europe, *S. turgescens*, although possibly widely distributed, is a rather uncommon species.

The original description of *S. turgescens* in De Notaris (1867) is very poor. Our present concept is based on that in West & West (1912). Although the latter authors explicitly state that the cell wall granules in this species have no definite arrangement several later authors illustrate granules disposed in concentric series around the semicell angles. Unfortunately, the taxonomic significance of this feature could not be examined. Description of the zygospore (pumpkin-like as in *S. dilatatum*) is from Archer (1878) who, however, did not provide any illustration so that his observation cannot be checked. *S. turgescens* var. *arcticum* reproduced in West & West (1912) to our mind is better to be considered a species of its own. So far, its possible occurrence in Europe has not been confirmed by an adequate illustration.

Accounted *Staurastrum turgescens* (var. *turgescens* forma *turgescens*):
Staurastrum turgescens forma *minor* G.S. West 1907
S. turgescens var. *sparsigranulatum* Scott et Grönblad 1957

Taxa excluded:
Staurastrum turgescens var. *arcticum* Wille 1879
S. turgescens var. *canadensis* Irénée-Marie 1957

S. turgescens forma *majus* Woodhead et Tweed 1960

Staurastrum uhtuense Grönblad   Plate 114: 18-19
Grönblad 1921, p. 60, pl. 5: 30
Cells about as long as broad or slightly longer than broad, deeply constricted. Sinus a V-shaped, narrowly open to almost closed incision. Semicells trapeziform–rectangular with concave sides, the apical angles produced to form (rather) long, diverging processes. Processes tipped with two spines and towards the semicell body furnished with a spiralling series of denticulations. Semicell body with a supraisthmial series of compound, dentate verrucae, otherwise smooth-walled. Semi-cells in apical view biradiate with a rhomboid-subcircular semicell body and crenulate margins. Zygospore unknown. Cell length 48-62 µm, cell breadth 39-72 µm.
Ecology: in slightly acidic, oligo-mesotrophic water bodies.
Distribution: Europe, Asia. In Europe very rare, only known from Finland.

Staurastrum uhtuense is closely allied (if not identic) to *S. columbetoides* W. et G.S. West 1902, in particular its var. *intermedium* Willi Krieger 1932, a species widely distributed in tropical regions.

Staurastrum ungeri Reinsch   Plate 45: 10-13
Reinsch 1867, p. 132, pl. 24B: I: 1-6
Cells (including spines) somewhat broader than long, deeply constricted. Sinus open from an acute-angled apex. Semicells elliptical (dorsal and ventral margins about equally convex), the lat-eral sides ending in a stout spine. Lateral spines of the semicells slightly converging. Cell wall furthermore provided with numerous much shorter, conical spines, irregularly scatterered. Semi-cells in apical view 3-4-angular with straight or slightly concave sides and obtusely rounded an-gles, each of them tipped with a long spine. Zygospore unknown. Cell length (17-)27-28 µm, cell breadth 32-34 µm, length of angular spines ca 8 µm.
Ecology: in benthos of acidic, oligotrophic water bodies.
Distribution: Europe, Africa. Within Europe, *S. ungeri* is known from Germany and Scottland.

The only reliable record of *S. ungeri* is that by Reinsch (1867). Other records are very scarse and seem to refer to *S. echinatum* rather than to *S. ungeri*.

Taxon excluded:
S. ungeri var. *vallesiacum* Viret 1909 (cf *S. echinatum*)

Staurastrum verticillatum W. Archer   Plate 115: 6-7
Archer 1869a, p. 196
First illustration: Cooke 1887, p. 177, pl. 61: 3
Cells about as broad as long, deeply constricted. Sinus widely open with subacute apex. Semicells vase-shaped (subcylindric) with convex apex, the apical angles produced to form slender, diverg-ing arm-like processes. Cell wall of the processes ornamented with many (at least 10) concentric series of denticulations and some 2-3 teeth at the tip. Cell wall of the semicell body with a ring of papillae or acute granules at the base, for the remaining part smooth. Semicells in apical view circular with a whorl of 8-10 radiating processes. Zygospore unknown. Cell length 135-145 µm, cell breadth 125-160 µm.
Ecology: in plankton of acidic, oligotrophic water bodies.
Distribution: Only known with certainty from Scotland and Ireland.

S. verticillatum differs from *S. ophiura* only by a lacking central apical ornamentation (no distinct ring of conical granules or verrucae) and by its diverging processes. The figures in Comère (1901) are a reproduction of those in

163

Cooke (1887) which represent only an apical and an isthmial view and, according to West & West (1903), are 'entirely imaginary'. The figure in Grönblad (1938) is intermediate between *S. verticillatum* and *S. ophiura*. If *S. verticillatum* and *S. ophiura* would be conspecific (possibly only to be distinguished at variety level) the name of *S. verticillatum* should get priority.

Staurastrum vestitum Ralfs

Cells broader than long, deeply constricted. Sinus widely open and acute-angled. Semicells (sub) fusiform, the lateral angles produced to form rather short, stout processes curving downwards in line with the apex of the semicell body. Processes tipped with 3-4 stout spines and towards the semicell body furnished with emarginate verrucae or (bifurcate) denticulations being arranged in indistinct concentric series. Semicell body furnished with two transversal series of emarginate verrucae or (bifurcate) spines: one series on the apex and one series in the median part. Semicells in apical view 3-4-radiate with about straight sides, the angles produced to form processes which may be slightly bent (usually clockwise). Along the sides a marginal and an intramarginal series of emarginate verrucae or (bifurcate) spines, those in the middle of the marginal series (much) more prominent than the others. Zygospore globose, furnished with long spines which are bifurcate at the apex. Cell length 30-45 µm, cell breadth 45-70(-90?) µm.

Ecology: in benthos and tychoplankton of acidic, oligotrophic water bodies.

Distribution: Eurasia, North America. In Europe, both varieties of *S. vestitum* are widely distributed, but usually not common.

In luxuriousness of its ornamentation *S. vestitum* var. *splendidum* represents a transitional form to *S. aculeatum*.

var. ***vestitum***   Plate 82: 1-5

Ralfs 1848, p. 143, pl. 23: 1

Semicells in apical view at each of the sides with two median spines that are much longer than those next to them.

Accounted *Staurastrum vestitum* var. *vestitum*:

Staurastrum vestitum var. *abundans* Korshikov 1941

S. vestitum var. *cedercreutzii* Croasdale, in Croasdale & Grönblad 1964

S. vestitum var. *tortum* W. et G.S. West 1898

var. ***splendidum*** Grönblad   Plate 82: 6-9

Grönblad 1920, p. 81, pl. 3: 100-102

Semicells in apical view at each of the sides with median spines that are but little longer than those next to them.

Taxa excluded:

Staurastrum vestitum var. *denudatum* Nordstedt 1869

S. vestitum var. *distortum* Wolle 1883 (cf *S. controversum*)

S. vestitum var. *gymnocephalum* A.M. Scott et Prescott 1961

S. vestitum var. *incurvum* Hegde 1986

S. vestitum var. *koreana* Skvortsov 1932

S. vestitum var. *montanum* Lenzenweger 1986 (cf *S. borgeanum*)

S. vestitum var. *parvum* Nygaard 1949 (cf *S. anatinum* var. *subanatinum*)

S. vestitum var. *persplendidum* Messikommer 1942 (cf *S. boreale*)

S. vestitum var. *semivestitum* W. West 1892 (= *S. controversum* var. *semivestitum*)

S. vestitum var. *subanatinum* W. et G.S. West 1902 (= *S. anatinum* var. *subanatinum*)

Staurastrum wildemanii Gutwiński   Plate 42: 1-3

Gutwiński 1902, p. 605, pl. 40: 61.

Cells exclusive of spines about as broad as long or slightly broader than long, deeply constricted. Sinus widely open from an acute-angled apex. Semicells in outline elliptic-hexagonal, with (1-)2-3(-4) long, diverging, often curved spines at the upper lateral sides. Cell wall smooth (but distinctly punctate). Semicells in apical view 3-angular with slightly concave sides and broadly rounded (rarely truncate) angles. Zygospore globose, furnished with long, acute, simple spines. Cell length excluding spines 45-70 µm, cell breadth excluding spines 45-75, length of spines 15-40 µm.

Ecology: in oligotrophic water bodies.

Distribution: Europe, Asia, Africa, Australia, particularly in tropical regions. In Europe, *S. wildemanii* is extremely rare and only known from Finland.

Relevant literature: Scott & Prescott (1956).

Actually, the pictures of *S. wildemanii* given by Grönblad (1920, pl. 3: 1-3) from Finland (see our Pl. 42: 3) might refer to a triradiate form of *Xanthidium antilopaeum*. The picture given in Scott et Prescott (1956) also referring to Finnish material, however, is more convincing (our Pl. 42: 1).

Accounted *Staurastrum wildemanii* (var. *wildemanii*):

S. wildemanii var. *horizontale* A.M. Scott et Prescott 1956

S. wildemanii var. *majus* (W. et G.S. West) A.M. Scott et Prescott 1956

S. wildemanii var. *unispiniferum* A.M. Scott et Prescott 1956

Staurodesmus and *Staurastrum* species excluded.

(Synonyms, questionable taxa or questionable European records of extra-European taxa not dealt with in the present flora)

Staurodesmus

Staurodesmus angulatus (W. West) Teiling 1948 = *Staurastrum angulatum*

Std. aversus (P. Lundell) Lillieroth 1950 = *S. aversum*

Std. bieneanus (Rabenhorst) Florin 1957 = *S. bieneanum*

Std. boldtianus (Grönbład) Croasdale 1962 = *S. boldtianum*

Std. boergesenii (Messikommer) Croasdale in Croasdale & Grönbład 1964: cf *Std. omearae*

Std. brevispina (Ralfs) Croasdale 1957 = *S. brevispina*

Std. clepsydra (Nordstedt) Teiling 1948 = *S. clepsydra*

Std. conspicuus (W. et G.S. West) Teiling 1967 = *S. conspicuum*

Std grandis (Bulnheim) Teiling 1967 = *S. grande*

Std. groenbladii (Skuja) Teiling 1967 = *S. groenbladii*

Std. inelegans (W. et G.S. West) Teiling 1948 = *S. inelegans*

Std. insignis (P. Lundell) Teiling 1967 = *S. insigne*

Std. lanceolatus (Archer) Croasdale 1957 = *S. lanceolatum*

Std. mamillatus (Nordstedt) Teiling 1967: cf *Std. cuspidatus*

Std. mimutissimus (Reinsch) Teiling 1967 = *S. minutissimum*

Std. nudus (W.B. Turner) Teiling 1967 = *Cosmarium taxichondrum* var. *nudum* W.B. Turner 1892

Std. obsoletus (Hantzsch)'Teiling 1967 = *Cosmarium obsoletum* (Hantzsch) Reinsch 1867

Std. orientalis (A.M. Scott et Prescott) Coesel 1993, in Coesel & Meesters 2007: cf *Std. triangularis*

Std. pachyrhynchus (Nordstedt) Teiling 1967 = *S. pachyrhynchum*

Std. quiriferus (W. et G.S West) Teiling 1967: cf. *Std. triangularis*

Std. sellatus (Teiling) Teiling 1948: cf *Std. incus*

Std. sibiricus (Borge) Croasdale 1962 = *S. sibiricum*

Std. smolandicus (P. Lundell) Teiling 1967 = *Cosmarium smolandicum* P. Lundell 1871

Std. spencerianus (Maskell) Teiling 1967: cf *Std. omearae*

Std. spetsbergensis (Nordstedt) Teiling 1967: cf *S. bieneanum*

Std. subpygmaeus (W. West) Croasdale 1962 = *S. subpygmaeum*

Std. tortus (Grönbład) Teiling 1967: cf twisted form of *Std. phimus*

Std. tumidus (Ralfs) Teiling 1967 = *S. tumidum*

Std. vulgaris (B. Eichler et Raciborski) Croasdale 1962: cf *Std. extensus* var. *rectus*

Staurastrum

Staurastrum absconditum Grönbład 1938: cf *S. tetracerum*

S. acanthoides Delponte 1878: cf *S. quadrangulare*

S. acestrophorum var. *glabrius* Grönbład 1920: cf *S. diacanthum*

S. acestrophorum var. *subgenuinum* Grönbład 1920: cf S. *diacanthum*

S. adornatum Grönbład 1920: cf *S. arcuatum*

S. affine W. et G.S. West 1905a: possibly a form of *S. paradoxum*

S. affiniforme Grönbład 1920: possibly a reduction form of *S. cingulum*

S. alandicum Teiling 1946: invalid homonym of *S. alandicum* Cedercreutz et Grönbład 1936.

S. alpicolum Schmidle 1900 (= *S. circulare* Schmidle 1896): probably a 3-radiate *Cosmarium* species

S. amoenum Hilse 1865: cf *S. capitulum* (West & West 1912: 124)

S. amoenum var. *italicum* Nordstedt et Wittrock 1876: cf *S. bifasciatum* var. *subkaiseri*

S. amphidoxon W. et G.S. West 1894: unclear picture, cf *S. arcuatum*

S. amphidoxon var. *alpinum* Schmidle 1896: cf *S. suchlandtianum*

S. amphidoxon var. *tripunctatum* Grönblad 1938: cf *S. suchlandtianum*

S. angulare W.B. Turner forma *polonica* B. Eichler 1896: cf *S. quadrangulare*

S. angulosum Schmidt 1903: cf *S. margaritaceum* (West, West & Carter 1923: 131)

S. arachnoides W. West 1892: cf *S. arachne*

S. arcticum G.S. West 1899: cf *S. aculeatum*

S. armigerum Brébisson 1856: cf *S. furcigerum*

S. arnellii var. *spiniferum* W. et G.S. West 1902: cf *S. simonyi* var. *semicirculare*

S. arthrodesmiforme Behre 1956: dubious desmid

S. articulatum (Corda) ex Pritchard 1861: cf *S. furcigerum* (West, West & Carter 1923: 189)

S. asperatum Grönblad 1920: possibly a 3-radiate form of *S. reductum*

S. aspinosum Wolle 1884, in Beijerick 1926: possibly a giant (polyploid?) form of *S. tetracerum*

S. asteroideum W. et G.S. West 1896: described from N. America with sketchy picture of a single semicell

S. asteroideum var. *munitum* Růžička 1972: cf *S. pentasterias*

S. asteroideum var. *nanum* (Wille) Grönblad 1948: cf *S. pentasterias*

S. asteroideum var. *ornatum* Grönblad 1920: cf *S. pentasterias*

S. asteroideum var. *salebrosum* Lenzenweger 1987: cf *S. sexcostatum*

S. aviculoides Grönblad 1938: possibly an anomalous form of *Std. leptodermus* (Grönblad 1947: 26)

S. barbulae Nygaard 1949: cf *S. pingue* var. *planctonicum*

S. basichondroides Gutwiński 1909: poor figure, cf *S. acutum*

S. benkoei Schaarschmidt 1883: poor figure, cf *S. muticum* or *S. pokljukense*

S. bergii Nygaard 1949: cf *S. bohlinianum*

S. bicoronatum var. *alpinum* (Schmidle) Lütkemüller 1900 = *S. amphidoxon* var. *alpinum* Schmidle 1896 (cf *S. suchlandtianum*)

S. bicoronatum var. *simplicius* W. et G.S. West 1896, in Grönblad 1938 and Grönblad 1948: cf *S. suchlandtianum* and *S. vestitum*

S. bidentatum Wittrock 1869: cf *S. longispinum*

S. bioculatum Taylor forma, in Skuja 1948: cf *S. dimazum*

S. blandum Raciborski 1884: possibly some biradiate *Staurodesmus* species (Teiling 1967: 487)

S. bolbothrix Beck-Mannagetta 1926, poor figure, cf. *S. forficulatum*

S. boldtianum Grönblad 1942: cf *S. sibiricum*

S. brachioprominens Børgesen 1890, in Kaiser 1933: cf *S. dimazum*

S. brachioprominens var. *archerianum* Bohlin 1901, in Cedercreutz & Grönblad 1936: cf *S. multinodulosum*

S. breviaculeatum G.M. Smith 1924, in Capdevielle 1978: only an insignificant apical view

S. bullosum Bennet 1886: cf *S. polytrichum*

S. cambricum W. West 1890: cf *S. polytrichum*

S. candianum Delponte 1878: cf *S. furcatum*

S. cedercreutzii Grönblad 1921: cf *S. borgeanum*

S. circulare Schmidle 1896: homonym of *S. circulare* Meyen 1835, see *S. alpicolum*

S. complanatum Delponte 1878: cf *S. vestitum*

S. concinnum W. et G.S. West 1898 forma Grönblad 1947: cf *S. oxyacanthum*

S. congruum Raciborski 1889: cf *S. hantzschii*

S. coniforme Grönblad 1920: cf *Std. mucronatus* var. *subtriangularis*

S. contectum W.B. Turner 1892: cf *S. quadrangulare*

S. contortum Delponte 1878: figures refer to different species (e.g. cf *S. vestitum*)

S. contortum G.M. Smith 1924 = *S. smithii* Teiling 1946

S. cordatum F. Gay 1884: cf *S. ralfsii*

S. cornigerum J. Roy 1893: cf *S. cornutum*

S. cornubiense Bennett 1887: cf *S. furcatum* (West, West & Carter 1923: 174)

S. cosmospinosum (Børgesen) W. et G.S. West 1900: cf *S. echinatum*

S. cracoviense Raciborski 1884: cf *S. pungens*

S. croaticum Pevalek 1929: cf *S. tetracerum*

S. cruciatum Heimerl 1891 = *S. heimerlianum*

S. csorbae Gutwiński 1909: poor, little informative figure

S. cumbricum W. West 1890: cf *S. polytrichum*

S. cyathoides Joshua 1886, in Schulz 1922: cf *S. pseudopelagicum*

S. cyathoides var. *gracile* Schulz 1922: cf *S. pseudopelagicum*

S. cyathoides var. *keuruense* Grönblad 1920: cf *S. pseudopelagicum*

S. cyclacanthum var. *africanum* Croasdale 1971, in Capdevielle 1978: cf *S. manfeldtii*

S. cyclacanthum var. *depressum* A.M.Scott et Grönblad 1957, in Lenzenweger 1986: cf *S. such-landtianum*

S. daaei Huitfeldt-Kaas 1906: cf *Std. cuspidatus*

S. danicum Nygaard 1949: cf *S. paradoxum*

S. decipiens Raciborski 1885: poor, little informative figure

S. decipiens var. *orthobrachium* Schmidle 1898: cf *S. micron*

S. denticulatum (Nägeli) W. Archer in Pritchard 1861: cf *S. avicula.*

S. dentiphorum Uherkovich 1959: cf *S. arcuatum* var. *subavicula*

S. depressiforme Cedercreutz et Grönblad 1936: presumably a 3-radiate *Cosmarium* species.

S. depressum (Nägeli) Turner 1892: cf *S. muticum*

S. detonii B. Eichler et Gutwiński 1895: cf 4-radiate *S. furcatum*

S. diademum Viret 1909: poor, little informative figure

S. diaphoron Beck-Mannagetta 1926: cf *S. furcatum* var. *aciculiferum*

S. diplacanthum De Notaris 1867: possibly an atypical form of *S. monticulosum* (or *S. forficulatum* ?)

S. disputatum W. et G.S. West 1912 = *S. dilatatum* var. *insignis* Raciborski 1892: European records hard to relate to Raciborski's original description from Brasil

S. dorsidentiferum W. et G.S West 1906: cf *S. pingue* var. *planctonicum*

S. dubium W. West 1890: possibly a form of *S. gracile*

S. dubium B. Eichler et Gutwiński 1895 (invalid homonym of *S. dubium* W. West 1890): cf *S. quadrispinatum*

S. dziewulskii B. Eichler et Raciborki 1893: possibly a papillate form of *S. sublaevispinum.*

S. eboracense W.B. Turner 1893: poor, little informative figure; possibly a form of *S. cyrtocerum*

S. ehrenbergianum (Nägeli) Pritchard 1861: cf *S. furcatum* (West, West & Carter 1923: 173)

S. ehrenbergii (Corda) ex Pritchard 1861: cf *S. furcatum* (West, West & Carter 1923: 173)

S. engleri Schmidle 1899: cf *S. polymorphum*

S. enontekiense Grönblad 1942: possibly a 3-radiate *Cosmarium* species

S. erinaceum Viret 1909: poor, little informative figure

S. erlangense Reinsch 1867: refers to a number of different *Staurodesmus* species

S. eurycerum Skuja 1948: cf *S. dybowskii*

S. eustephanum (Ehrenberg) ex Ralfs 1848: cf *S. furcigerum* (West, West & Carter 1923: 190)

S. farquharsonii J. Roy 1893: cf *S. orbiculare* (West & West 1912: 155)

S. felkaensis Kol 1927: poor, little informative figure

S. fennicum Grönblad 1920: cf *S. pingue*

S. fennicum (Grönblad 1948) Kouwets 1901: invalid homonym of *S. fennicum* Grönblad 1920, cf *S. tetracerum* var. *cameloides*

S. fissum var. *perfissum* W. et G.S. West 1895, in Grönblad 1920: cf *S. laeve*

S. franconicum Reinsch 1866: poor figures, presumably referring to different species (West, West & Carter 1923: 86)

S. furcatostellatum Reinsch 1875: cf *S. sexangulare* (West, West & Carter 1923: 194))

S. globulatum Brébisson ex Ralfs 1848: cf *S. bacillare* var. *obesum* (with granule-like mucous extrusions at the lobes?)

S. granulatum Reinsch 1875: poor figure, cf *S. punctulatum*?

S. granulosum Ehrenberg ex Ralfs 1848: presumably a reduction form of *S. lunatum*

S. granulosum var. *acutum* W. et G.S. West 1902: cf *S. acutum*

S. griffithianum (Nägeli) Archer 1866: cf *S. spongiosum* (West, West & Carter 1923: 78)

S. gurgeliense Schmidle 1896: little informative figure, presumably a form of *S. simonyi* or *S. echinatum*

S. hainesii Woodhead et Tweed 1948 : poor, little informative figure

S. hambergii K. Strøm 1923: cf *S. anatinum*

S. haynaldii Schaarschmidt 1883: cf *S. inconspicuum*

S. hexacanthum F. Gay 1884: cf *Std. dejectus*

S. hexagonum Raciborski 1885: cf *S. meriani*

S. inaequale forma *polonica* Raciborski 1884: cf *S. hantzschii*

S. incisum forma *convergens* Gutwiński 1892: poor figure, cf *S. inflexum*?

S. incurvatum W. et G.S. West 1895, in Grönblad 1938: cf *S. inflexum*

S. inflatum W. et G.S. West 1912 = *S. punctulatoides*

S. intricatum Delponte 1878: refers to a number of different species (Nordstedt 1896: 149)

S. iotanum Wolle 1884: original figure very poor. Concept in West & West 1898 reminds of 3-radiate *S. tetracerum*

S. javanicum (Nordstedt) W.B. Turner 1892, in Capdevielle 1978: cf *S. manfeldtii*

S. kaiseri Pevalek 1925: cf *S. capitulum*

S. kaiseri Růžička 1972 = *S. crassangulatum*

S. kebnekaisense Thomasson 1952: cf *S. petsamoense*

S. kitchelii var. *inflatum* Schmidle 1898: cf *S. clevei*

S. kjellmanii Wille 1879: cf *S. punctulatum*

S. krkense Pevalek 1929: cf *S. cyrtocerum* var. *inflexum*

S. lacustre G.M. Smith 1922, in Skuja 1964: cf *S. subboergesenii*

S. laevigatum Messikommer 1960: cf *S. bieneanum*

S. lagerheimii Schmidle 1898: cf *S. anatinum*

S. landmarkii Huitfeldt-Kaas 1906: poor, little informative figure

S. laniatum Delponte 1878: refers to a number of different *Staurodesmus* species

S. lewisianum W.B. Turner 1893: cf *S. pileolatum* var. *cristatum* (West & West 1912: 129)

S. libeltii Raciborski 1889: cf *S. quadrangulare*

S. longiradiatum W. et G.S. West 1896, in Skuja 1948, Nygaard 1949, Scharf 1995: cf *S. pingue* (var. *planctonicum*), in Cobelas & al. 1988: cf 3-radiate *S. johnsonii*

S. longiradiatum var. *breviradiatum* Grönblad 1920: cf *S. manfeldtii*

S. longirostratum Grönblad 1920: cf *S. setigerum*

S. lusitanicum Nauwerck 1962: cf *S. controversum* var. *semivestitum*

S. luetkemuelleri Ruttner et Donat (manuscript) in Messikommer 1942: cf *S. pingue*

S. macedonicum Petkov 1910: cf *S. dispar*

S. messikommeri Lundberg 1931: cf *S. manfeldtii* var. *splendidum*

S. minus Ralfs 1848: poor, little informative figure

S. montanum Raciborski 1885: cf *S. furcigerum*

S. morettii Grönblad 1960: cf *S. avicula*

S. munitum Wood 1873: cf *S. arctiscon*

S. muricatiforme Schmidle 1895: cf *S. lapponicum*

S. muricatiforme var. *waldense* Dick 1926: possibly a 3-radiate *Cosmarium* species

S. navigiolum Grönblad 1920: cf *S. cristatum* var. *cuneatum*

S. nigrae–silvae Schmidle 1892: cf *S. simonyi*

S. nitidum W. Archer 1860: cf *S. cristatum* (West, West & Carter 1923: 47)

S. nodosum W. et G.S. West 1897: likely an anomalous form of *S. inconspicuum*.

S. noduliferum Grönblad 1945, in Förster 1970: cf *S. chaetoceras*

S. nordstedtii Gutwiński 1890 (1892): cf *S. cristatum* (West, West & Carter 1923: 47)

S. notarisii Delponte 1878: cf *S. polytrichum*

S. novae-terrae W.R. Taylor 1935, in Grönblad 1938: cf *S. furcatum* var. *aciculiferum*

S. obliquum (Nordstedt) J. Sampaio 1949 = *Cosmarium obliquum* Nordstedt 1873

S. oblongum Delponte 1878: cf *S. gracile*

S. octoverrucosum var. *simplicius* Scott et Grönblad 1957, in Coesel 1997 (as *S. simplicius*): cf *S. reductum*

S. onegense Grönblad 1948: possibly a reduction form of *S. furcatum* var. *aciculiferum*

S. orchridense Petkov 1910: cf *S. cristatum* var. *oligacanthum*

S. ornatum (Boldt) Turner 1892 : cf *S. pinnatum*

S. ornatum var. *asperum* (Perty) Schmidle 1896 = *S. pertyanum*

S. ornatum var. *morzinense* Laporte 1931: cf *S. pertyanum* (reduction form?)

S. ornithopodon var. *bifurcatum* Borge 1913: cf *S. subboergesenii*

S. osceolense (Georgevitch) Grönblad 1945 (= *S. paradoxum* var. *osceolense* Wolle forma *biradiata* Georgevitch 1919) = *S. tetracerum* var. *cameloides*

S. osceolense var. *fennicum* Grönblad 1948: cf *S. smithii*

S. osteonum W. West 1890: species excluded in West, West & Carter 1923: 197

S. papillosum Kirchner 1878: cf *S. avicula* (West, West & Carter 1923: 41)

S. paucidentatum A. Lemaire 1890: possibly identic to *S. informe* (Grönblad 1920: 67)

S. paxilliferum G.S. West 1899: cf *S. acutum*

S. pecten Perty 1852 : poor figure, cf *S. teliferum* var. *gladiosum?*

S. perinii Auclair 1910: cf *S. vestitum*

S. perundulatum Grönblad 1920: cf *S. tetracerum* var. *irregulare*

S. picum W. et G.S. West 1896: cf *S. echinatum*

S. pileatum Delponte 1878: cf *S. avicula*

S. pilosellum W. et G.S. West 1912: cf *S. punctulatum*

S. pilosum (Nägeli) W. Archer 1861: nomen dubium (West, West & Carter 1923: 64-65); most records refer to *S. brebissonii* or *S. kouwetsii*

S. pinguescens Grönblad 1942: little informative figure possibly refers to *S. micron*

S. pinnatum Turner 1892 var. *subpinnatum* (Schmidle) W. et G.S. West 1902, in Grönblad 1920: cf *S. pertyanum*

S. poikilomazon Beck-Mannagetta 1926: cf *S. borgei*

S. pringsheimii Reinsch 1867: cf *S. polytrichum* (West, West & Carter 1923: 53)

S. protectum W. et G.S. West 1908, in Baïer 1978: cf *S. avicula*

S. pseudincus Reinsch 1867: refers to one or two unidentifiable *Staurodesmus* species

S. pseudo-cosmarium Reinsch 1875: probably a 3-radiate *Cosmarium* species (*C. tetraophthalmum?*)

S. pseudocrenatum P. Lundell 1871: cf *S. maamense*

S. pseudocuspidatum J. Roy et Bisset 1886: cf *Std. cuspidatus*

S. pseudofurcigerum Reinsch 1867: cf *S. furcigerum*

S. pseudoiotanum Grönblad 1921: probably a reduction form of *S. tetracerum* var. *irregulare*

S. pseudolagerheimii Thomasson 1952: poor figure, cf *S. proboscideum?*

S. pseudonanum Grönblad 1920: cf *S. furcatum* var. *aciculiferum*

S. pygmaeum Ralfs 1848: cf *S. punctulatum*

S. quadridentatum Scharf 1995: cf *S. pingue* var. *planctonicum*

S. ravenelii Wood 1873: obscure ornamentation, cf *S. teliferum?*

S. refractum Delponte 1878: cf *S. inconspicuum* (West, West & Carter 1923: 86)

S. reinschii Reinsch ex J. Roy 1883: confusing diagnosis (figure does not match with text)

S. renardii Reinsch 1867: cf *S. furcatum*

S. repandum (Perty) Rabenhorst 1868: unclear figure, cf *S. avicula*

S. reynouardii Auclair 1910: cf *S. manfeldtii*

S. riesengebirgense Grönblad 1926: cf *S. coarctatum*

S. riklii Huber–Pestalozzi 1928: presumably some triradiate *Cosmarium* species

S. robustum Delponte 1878: cf *S. polytrichum*

S. rostafinskii Gutwiński 1890 (for figure, see Gutwiński 1892): cf *S. hirsutum* var. *muricatum*

S. rostellum J. Roy 1893: cf *S. echinatum*

S. rostratum Raciborski 1885: cf *S. sebaldi*

S. rugulosum Ralfs 1848: possibly a form of *S. alternans* (West & West 1912: 179)

S. ruzickae Kouwets 1972: cf *S. boreale*

S. saltans Joshua var. *belgicum* De Wildeman 1897: cf *S. miedzyrzecense*

S. santillanae F. Caballero 1946: cf *S. quadrangulare*

S. sarsii Huitfeldt-Kaas 1906: cf *Std. controversus* var. *crassus* (West & West 1912: 102)

S. saxonicum Rabenhorst 1863 (referring to *Staurastrum* spec. in Bulnheim 1859, p. 22, pl. 2: 7): unclear figure, cf *S. polytrichum*

S. saxonicum Reinsch 1867: invalid homonym of *S. saxonicum* Rabenhorst 1863, cf *S. aculeatum* (Lütkemüller 1900: 76)

S. schmidlei Prescott, in Prescott, Bicudo & Vinyard 1982: invalid synonym of *S. circulare* Schmidle 1896.

S. scorpioideum Delponte 1878: cf *S. oxyacanthum*

S. senarium (Ehrenberg) ex Ralfs 1848: deficient diagnosis (frontal view in Ehrenberg 1843 is wanting). Figures in later literature refer to a series of different taxa, often to *S. forficulatum* var. *verrucosum*

S. senticosum Delponte 1878: cf *S. teliferum*

S. simplicius (A.M. Scott et Grönblad) Coesel 1996 in Coesel 1997 and in Coesel & Meesters 2007: cf *S. reductum*

S. sonthalianum W.B. Turner 1893, in Thomasson 1952: likely refers to some other (not well identifiable) species

S. sphagnicolum P. Magdeburg 1926: cf *S. simonyi*

S. spicatum W. et G.S. West 1896: cf *S. quadrispinatum*

S. spiniferum W. West 1890: poor figure, might refer to *S. simonyi*

S. spiniferum var. *quadratum* Irenee-Marie 1952, in Lenzenweger 1997: cf *S. simonyi*

S. spinosum Brébisson ex Ralfs 1848 = *S. furcatum*

S. spinuliferum Messikommer 1949: invalid homonym of *S. spinuliferum* Maskell 1889, cf *S. simonyi* or *S. echinatum*

S. stellatum Reinsch 1867: scetchy, little informative figure

S. subarcuatum Wolle 1884: cf *S. avicula*

S. subfennicum Grönblad 1920: poor, little informative figure, cf *S. manfeldtii* var. *productum*

S. subgracillimum W. et G.S. West 1896, in West & West 1902 and Williamson 1998: cf *S. subexcavatum*

S. subgrande Borge var. *minor* G.M. Smith 1924: cf *S. myrdalense*

S. sublongipes G.M. Smith, in Grönblad 1938: cf *S. chaetoceras*

S. subnanum Grönblad 1920: cf *S. furcatum* var. *aciculiferum*

S. subpunctulatum Gay 1884: cf *S. punctulatum* (West & West 1912: 182)

S. subrefractum A. Lemaire 1883: cf *S. inconspicuum*

S. subscabrum Nordstedt 1878: cf *S. scabrum*

S. subteliferum Roy et Bisset 1886: European records refer to *S. teliferum*

S. subtile Nordstedt 1878: deficient figure, cf *S. micron* var. *spinulosum?*

S. subtrifurcatum W. et G.S. West 1896, in Grönblad 1920: cf *S. wildemanii*

S. suecicum Cedergren 1938: cf *S. duacense*

S. tectum var. *ayayense* forma *nana* Tell 1980, in Capdevielle 1982: cf *S. levanderi*

S. tenuissimum var. *spinosa* Bock 1969 and var. *anomalum* Lemmermann 1898: obscure and wanting figures, respectively

S. terebrans Nordstedt 1873: cf *S. elongatum*

S. thunmarkii Teiling1946: cf biradiate *S. cingulum*

S. trachygonum W. West 1892: deficient figure, possibly a reduction form of *S. hirsutum*

S. trachynotum W. West 1892 : cf *S. aculeatum* (Lütkemüller 1900: 77)

S. trapezicum Boldt 1888, in Růžička 1956: cf *S. hirsutum* var. *muricatum*

S. trapezicum Boldt var. *campylospinum* Schmidle 1895: cf *S. pyramidatum*

S. trapeziforme Lundberg 1931: cf *S. polytrichum*

S. trelleckense W.B. Turner 1893: cf *S. scabrum*

S. triaculeatum Gutwiński 1890 (for figure, see Gutwiński 1892): deficient description

S. tricorne (Brébisson) ex Ralfs 1848: cf *S. hexacerum* (West, West & Carter 1923: 138).

S. trifidum Nordstedt 1887 forma Eichler 1896: presumably refers to an anomalous form of *S. bifidum*

S. trifolium Gutwiński 1895: obscure, cf *Tetraedron?*

S. trigonum (Boldt) Kossinskaja 1936: cf triradiate *Xanthidium bifidum*

S. triumvirum Skuja, in Thomasson 1952: invalid, for never described by Skuja (nor adequately by Thomasson 1952), cf *S. petsamoense*

S. truncatulum Reinsch 1875 = *S. renardii* Reinsch 1867

S. tuberculatum Bennett 1886: cf *S. sexcostatum* (West, West & Carter 1923: 148)

S. tumidulum F. Gay 1884: unclear for figure does not match with diagnosis, cf *S. punctulatum*

S. tunguscanum Boldt 1885: cf *S. lunatum*

S. tvaerminnense Grönblad 1945: likely based on a single specimen, no later records, possibly a reduction form of some other species

S. uniseriatum Nygaard 1949: cf *S. pingue*

S. upplandicum Teiling var. *italicum* Grönblad 1960: cf *S. multinodulosum* (nota bene: *S. upplandicum* does not exist, Grönblad means *S. alandicum* Teiling 1946!)

S. vastum Schmidle 1896: cf *S. arcuatum* var. *subavicula*

S. ventricosum Delponte 1878: unclear figure, possibly biradiate *S. vestitum?*

S. vulgaris Thomasson 1957: cf *S. pingue* (var. *planctonicum*)

S. waldense (Dick) Růžička 1955: cf *S. hexacerum*

S. westii Turner 1893: possibly a small, anomalous form *of S. forficulatum* var. *verrucosum*

S. wolleanum var. *kissimense* Wolle 1885, in Grönblad 1920: cf *S. gemelliparum*

S. zachariasii Schröder 1898: cf *Std. controversus* var. *crassus*

S. zoniferum Grönblad 1920: cf *S. anatinum*

Original identifications

Plate numbers	Author		Original identification
Pl. 1: 1, 2, 4	Ralfs 1848	pl. 20: 3d, 3b, 3c, 3a	Arthrodesmus convergens
Pl. 1: 3	West & West 1912	pl. 116: 9	Arthrodesmus convergens
Pl. 1: 5, 6, 7, 8, 10,11	Coesel 1994	pl. 11: 3, 1, 6, 2, 5, 4	Staurodesmus convergens
Pl. 1: 9	Lenzenweger 1997	pl. 21: 19	Staurodesmus convergens
Pl. 1: 12	Deflandre 1926	fig. 8	Arthrodesmus convergens forma deplanatum
Pl. 1: 13	West & West 1912	pl. 116: 5	Arthrodesmus convergens
Pl. 1: 14, 15	Woloszynska 1921	fig. 13, 14	Arthrodesmus covergens var. depressum
Pl. 1: 16	Grönblad 1921	pl. 3: 51	Arthrodesmus curvatus var. imatrensis
Pl. 2: 1	Bailey 1841	pl. 1: 12	Euastrum
Pl. 2: 2, 4	Grönblad 1960	pl. 7: 147, 146	Arthrodesmus subulatus
Pl. 2: 3, 6	West & West 1912	pl. 117: 2, 3	Arthrodesmus subulatus var. subaequalis
Pl. 2: 5	Deflandre 1924	fig. 9	Arthrodesmus subulatus var. subaequalis
Pl. 2: 7, 9	Smith 1924	pl. 85: 2, 1-3	Arthrodesmus subulatus var. nordstedtii
Pl. 2: 8	Børgesen 1890	pl. 5: 57	Arthrodesmus subulatus forma major
Pl. 2: 10	Lind & Brook 1980	fig. 117	Staurodesmus subulatus
Pl. 2: 11	West & West 1912	pl. 116: 14	Arthrodesmus subulatus
Pl. 3: 1	Raciborski 1889	pl. 6: 17	Arthrodesmus bulnheimii
Pl. 3: 2	West & West 1912	pl. 116: 1	Arthrodesmus bulnheimii
Pl. 3: 3	Skuja 1964	pl. 41: 1	Arthrodesmus bulnheimii
Pl. 3: 4	Cedergren 1932	pl. 3: 45	Arthrodesmus bulnheimii
Pl. 3: 5	Donat 1926	pl. 1: 7	Arthrodesmus bulnheimii
Pl. 3: 6, 7	West & West 1912	pl. 114: 7a/b, 7a'	Arthrodesmus incus var. subquadratus
Pl. 3: 8, 9	Kossinskaja 1949	fig. 8	Arthrodesmus incus var. subquadratus
Pl. 3: 10	Tomaszewicz & Kowalski 1993	fig. 5: 53	Arthrodesmus incus forma minor
Pl. 3: 11	West & West 1898	pl. 17: 16	Arthrodesmus incus var. validus
Pl. 3: 12	West & West 1912	pl. 114: 9	Arthrodesmus incus var. validus
Pl. 3: 13	West & West 1912	pl. 116: 3	Arthrodesmus incus var. validus
Pl. 3: 14	Lind & Brook 1980	fig. 111	Staurodesmus validus
Pl. 3: 15, 16	Coesel 1994	pl. 11: 8, 7	Staurodesmus bulnheimii var. subincus
Pl. 3: 17	Heimerl 1891	pl. 5: 17	Arthrodesmus incus forma typica
Pl. 4: 1, 2	West & West 1912	pl. 113: 13, 15	Arthrodesmus incus
Pl. 4: 3	West & West 1912	pl. 114: 1	Arthrodesmus incus forma perforata
Pl. 4: 4, 5, 6	Florin 1957	fig. 30: 4, 3, 2	Staurodesmus sellatus
Pl. 4: 7	Nygaard 1979	fig. 290	Staurodesmus sellatus
Pl. 4: 8	Nygaard 1979	fig. 333	Staurodesmus megacanthus var. scoticus
Pl. 4: 9	Teiling 1946	fig. 21	Staurodesmus incus var. sellatus
Pl. 4: 10, 11	West & West 1912	pl. 113: 20, 22	Arthrodesmus incus var. indentatus
Pl. 4: 12	Nygaard 1979	fig. 330	Staurodesmus extensus var. joshuae
Pl. 5: 1	West W. 1892a	pl. 22: 14a/b	Staurastrum jaculiferum
Pl. 5: 2, 3, 4	West & West 1903	pl. 17: 4, 2, 1	Staurastrum jaculiferum

173

Plate numbers	Author		Original identification
Pl. 5: 5, 7	Børgesen 1901	pl. 8: 1a, 1b	Staurastrum jaculiferum
Pl. 5: 6	Teiling 1948	fig. 57	Staurodesmus sellatus
Pl. 5: 8	West & West 1905	pl. 7: 22	Arthrodesmus incus var. longispinum
Pl. 5: 9	Nygaard 1979	fig. 315	Arthrodesmus incus var. longispinum
Pl. 5: 10	Teiling 1967	pl. 5: 20	Staurodesmus extensus var. longispinus
Pl. 6: 1	Ralfs 1848	pl. 20: 4g	Arthrodesmus incus
Pl. 6: 2, 3	Ralfs 1848	pl. 20: 4l, 4e	Arthrodesmus incus
Pl. 6: 4	West & West 1912	pl. 114: 4	Arthrodesmus incus var. ralfsii
Pl. 6: 5	Frémy 1930	fig. 65	Arthrodesmus incus var. ralfsii
Pl. 6: 6	Strøm 1920	pl. 5: 4	Arthrodesmus incus forma perforata
Pl. 6: 7, 8	Borge 1897	pl. 3: 4a,c,4a'	Arthrodesmus incus var. subtriangularis
Pl. 6: 9, 10, 11, 12	West & West 1912	pl. 115: 1,2,4,5	Arthrodesmus triangularis var. subtriangularis
Pl. 6: 13	West W. 1892a	pl. 24: 19a	Arthrodesmus triangularis, forma
Pl. 6: 14	West & West 1912	pl. 114: 16a	Arthrodesmus triangularis var. inflatus forma robusta
Pl. 6: 15, 16	West & West 1912	pl. 114: 15a/b,14	Arthrodesmus triangularis var. inflatus
Pl. 7: 1	Lagerheim 1885	pl. 27: 22a,22c	Arthrodesmus triangularis
Pl. 7: 2, 3	West & West 1912	pl. 114: 12,17	Arthrodesmus triangularis
Pl. 7: 4, 5	Nygaard 1991	pl. 2: 48,57	Staurodesmus triangularis
Pl. 7: 6	Nygaard 1979	fig. 322	Staurodesmus triangularis
Pl. 7: 7, 8	Allorge & Allorge 1931	pl. 10: 42, 43-44	Arthrodesmus triangularis var. brevispinus
Pl. 7: 9	Coesel 1994	pl. 11: 9	Staurodesmus triangularis var. subparellelus
Pl. 7: 10	Lenzenweger 1994	pl. 5: 7	Staurodesmus triangularis
Pl. 7: 11	Coesel 1994	pl. 11: 11	Staurodesmus extensus var. malaccensis
Pl. 7: 12	Borge 1913	pl. 2: 22	Arthrodesmus incus
Pl. 7: 13	Borge 1930	pl. 2: 37	Arthrodesmus incus, forma Borge 1913
Pl. 7: 14	Coesel 1994	pl. 11: 10	Staurodesmus triangularis var. subparallelus
Pl. 7: 15, 16	Förster 1970	pl. 25: 20,18	Staurodesmus indentatus
Pl. 7: 17	Kouwets 1987	pl. 15: 29	Staurodesmus triangularis
Pl. 8: 1, 2	Lundell 1871	pl. 4: 1a/b,1a'/1b'	Staurastrum megacanthum
Pl. 8: 3	Skuja 1964	pl. 50: 3	Staurastrum megacanthum var. tornense
Pl. 8: 4	Lind & Brook 1980	fig. 112	Staurodesmus megacanthus
Pl. 8: 5, 6	West & al. 1923	pl. 131: 8-7	Staurastrum megacanthum
Pl. 9: 1, 2	West & West 1903	pl. 16: 8a-8b	Staurastrum megacanthum var. scoticum
Pl. 9: 3	Skuja 1964	pl. 50: 11	Staurastrum megacanthum var. scoticum
Pl. 9: 4	West & West 1896	pl. 16: 8	Staurastrum glabrum
Pl. 9: 5, 6, 7	West & al. 1923	pl. 129: 4,5,3	Staurastrum glabrum
Pl. 9: 8	West & West 1912	pl. 114: 5a'	Arthrodesmus incus var. ralfsii forma latiuscula
Pl. 9: 9, 13	Coesel 1994	pl. 13: 5,4	Staurodesmus glaber var. debaryanus
Pl. 9: 10, 11	Coesel 1994	pl. 13: 10,11	Staurodesmus glaber var. glaber
Pl. 9: 12, 14	Kouwets 1987	pl.16: 21,20	Staurodesmus glaber var. glaber
Pl. 10: 1	Messikommer 1949	pl. 1: 14	Staurastrum glabrum var. hirundinella
Pl. 10: 2	Kouwets 1987	pl. 16: 22	Staurodesmus glaber var. hirundinella

Plate numbers	Author		Original identification
Pl. 10: 3	Coesel 1994	pl. 13: 3	Staurodesmus glaber var. hirundinella
Pl. 10: 4, 5, 6	Teiling 1948	fig. 14,25,18	Staurodesmus glabrus subsp. brebissonii forma limnophilus
Pl. 10: 7	Williamson 1992	fig. 13: 9	Staurodesmus glaber var. glaber
Pl. 10: 8	Coesel & Meesters 2007	pl. 87: 22	Staurodesmus glaber var. limnophilus
Pl. 10: 9	Skuja 1964	pl. 50: 9	Staurastrum glabrum var. incurvum
Pl. 10: 10	Ralfs 1848	pl. 21: 2	Staurastrum aristiferum
Pl. 10: 11	West & al. 1923	pl. 132: 11	Staurastrum aristiferum
Pl. 10: 12, 13	Coesel 1994	pl. 16: 9,10	Staurodesmus aristiferus
Pl. 10: 14	West & al. 1923	pl. 132: 12	Staurastrum aristiferum var. protuberans
Pl. 10: 15	Skuja 1964	pl. 51: 1	Staurastrum aristiferum
Pl. 11: 1, 2, 3	Ralfs 1848	pl. 21: 1a,1b,1c	Staurastrum cuspidatum var. divergens
Pl. 11: 4, 5, 6, 7, 8, 11	Coesel 1994	pl. 15: 1,2,3,8,12,7	Staurodesmus cuspidatus
Pl. 11: 9, 14, 15	Coesel & Meesters 2007	pl. 89: 13,14,12	Staurodesmus cuspidatus var. divergens
Pl. 11: 10	Coesel & Meesters 2007	pl. 89: 10	Staurodesmus cuspidatus var. cuspidatus
Pl. 11: 12, 13	Lind & Brook 1980	fig. 118,115	Staurodesmus mamillatus
Pl. 11: 16, 17	Teiling 1948	fig. 62,61	Staurodesmus cuspidatus subsp. tricuspidatus
Pl. 11: 18	West & al. 1923	pl. 32: 19	Staurastrum cuspidatum var. maximum
Pl. 11: 19, 20	West & al. 1923	pl. 32: 17,16	Staurastrum cuspidatum var. divergens
Pl. 12: 1, 2	West W. 1892a	pl. 22: 13a'/b,13a	Staurastrum curvatum
Pl. 12: 3	West & West 1903	pl. 17: 12a/b	Staurastrum curvatum
Pl. 12: 4	Skuja 1964	pl. 50: 12	Staurastrum curvatum
Pl. 12: 5	Florin 1957	fig. 30: 10	Staurodesmus megacanthus var. scoticus
Pl. 12: 6	Coesel & Wardenaar 1990	fig. 1b	Staurodesmus cuspidatus var. curvatus
Pl. 12: 7	Teiling 1967	pl. 10: 4	Staurodesmus cuspidatus var. curvatus
Pl. 13: 1	Borge 1913	pl. 2: 23	Arthrodesmus incus var. extensus
Pl. 13: 2	Tomaszewicz 1988	pl. 13: 3	Staurodesmus quiriferus, morpha
Pl. 13: 3, 4	Coesel 1994	pl. 12: 1, 2	Staurodesmus extensus var. extensus
Pl. 13: 5	Lenzenweger 1997	pl. 21: 15	Staurodesmus extensus var. extensus
Pl. 13: 6, 7	Eichler & Raciborski 1893	pl. 3: 22, 24	Arthrodesmus incus var. vulgaris forma recta
Pl. 13: 8, 9	Tomaszewicz 1988	pl. 13: 9, 10	Staurodesmus extensus, morpha
Pl. 13: 10, 11, 12	Coesel 1994	pl. 12: 3, 6, 4	Staurodesmus extensus var. vulgaris
Pl. 13: 13	Coesel & Meesters 2007	pl. 86: 17	Staurodesmus extensus var. vulgaris
Pl. 13: 14, 15	Heimerl 1891	pl. 5: 18	Arthrodesmus incus forma isthmosa
Pl. 13: 16, 17, 18	Coesel 1994	pl. 12: 11, 12, 13	Staurodesmus extensus var. isthmosus
Pl. 13: 19	Tomaszewicz 1988	pl. 13: 15	Staurodesmus isthmosus
Pl. 13: 20	Gutwiński 1892	pl. 3: 6	Arthrodesmus incus forma joshuae
Pl. 13: 21, 22, 23, 24	Coesel & Meesters 2007	pl. 86: 21, 22, 23, 24	Staurodesmus extensus var. joshuae
Pl. 13: 25	West & West 1905	pl. 7: 10	Arthrodesmus incus var. ralfsii, forma spinis brevissimus
Pl. 13: 26, 27	Coesel 1994	pl. 12: 15, 16	Staurodesmus subhexagonus
Pl. 13: 28	Borge 1913	pl. 1: 20	Arthrodesmus incus var. ralfsii
Pl. 13: 29	Borge 1906	pl. 3: 36	Staurastrum dejectum var. debaryanum
Pl. 13: 30	Skuja 1964	pl. 40: 13	Arthrodesmus ralfsii forma subhexagonum
Pl. 14: 1, 2	Raciborski 1889	pl. 3: 9a'/b, 9a	Staurastrum wandae

Plate numbers   Author   Original identification

Plate numbers	Author		Original identification
Pl. 14: 3, 4	Grönblad 1938	fig. 4:1a/b, 1d	Staurastrum wandae var. brevispinum
Pl. 14: 5, 6, 7, 8	Archer 1858	pl. 21: 10/11, 13, 8/9, 12	Staurastrum o'mearii
Pl. 14: 9, 10, 11	Borge 1936	pl. 2: 45a', 45a, 45a"/c'	Staurastrum o'mearii
Pl. 14: 12-19	Coesel 1994	pl. 12: 26-32	Staurodesmus omearii
Pl. 14: 20, 21	West & al. 1923	pl. 132: 7a, 8	Staurastrum o'mearii
Pl. 14: 22	West & al. 1923	pl. 132: 9	Staurastrum o'mearii var. minutum
Pl. 15: 1, 2	Lundell 1871	pl. 3: 29a/b, 29c	Staurastrum pterosporum
Pl. 15: 3, 4	Bourrelly 1966	pl. 101: 7, 8	Staurodesmus pterospermum
Pl. 15: 5, 6, 7	Coesel 1994	pl. 16: 1, 3, 5	Staurodesmus pterosporus
Pl. 15: 8	Coesel & Meesters 2007	pl. 90: 5	Staurodesmus pterosporus
Pl. 15: 9	Turner 1892	pl. 12: 9	Arthrodesmus phimus
Pl. 15: 10, 11	West & West 1912	pl. 115: 17a/b, 17a'	Arthrodesmus phimus var. occidentalis
Pl. 15: 12	Kouwets 1987	pl. 16: 5	Staurodesmus phimus
Pl. 15: 13	Lenzenweger 1991	pl. 3: 33	Staurodesmus phimus
Pl. 15: 14	Lenzenweger 1988	pl. 2: 36	Staurodesmus phimus forma minor
Pl. 15: 15	Schmidle 1896	pl. 16: 9	Arthrodesmus incus forma semilunaris
Pl. 15: 16	Krieger 1930	pl. 5: 10a	Arthrodesmus incus var. semilunaris
Pl. 15: 17	West & West 1912	pl. 117: 22	Arthrodesmus phimus var. hebridarum
Pl. 15: 18	Skuja 1964	pl. 41: 2	Arthrodesmus phimus var. hebridarum
Pl. 15: 19, 20	Turner 1892	pl. 18a/b, 18c	Staurastrum unguiferum
Pl. 15: 21	Grönblad 1938	pl. 2: 10a/b	Staurastrum unguiferum var. extensum
Pl. 15: 22	Grönblad 1947	pl. 1: 22	Staurastrum unguiferum, forma
Pl. 15: 23	Grönblad 1920	pl. 3: 74	Staurastrum connatum var. pseudoamericanum
Pl. 16: 1	Lundell 1871	pl. 3: 23	Staurastrum corniculatum
Pl. 16: 2, 3, 4	West & West 1912	pl. 125: 18b, 17a, 17b	Staurastrum corniculatum
Pl. 16: 5, 6, 7	West & West 1912	pl. 125: 20, 21, 19	Staurastrum corniculatum var. spinigerum
Pl. 16: 8, 9	Coesel 1994	pl. 15: 7, 8	Staurodesmus corniculatus
Pl. 16: 10	Lundell 1871	pl. 3: 26	Staurastrum leptodermum
Pl. 16: 11	Brook 1958	fig. 1	Staurastrum leptodermum
Pl. 16: 12	Lundberg 1931	fig. 12	Staurastrum leptodermum var. inerme
Pl. 16: 13	Grönblad 1942	pl. 4: 27	Staurastrum leptodermum, forma
Pl. 16: 14	Woloszynska 1921	fig. 8/9	Staurastrum andrzejowskii
Pl. 16: 15	Lundell 1871	pl. 3: 28	Staurastrum dejectum var. connatum
Pl. 16: 16, 17	Coesel 1994	pl. 14: 1, 2	Staurodesmus connatus
Pl. 16: 18	Margalef 1956	fig. 30y	Staurastrum connatum
Pl. 17: 1, 2	Teiling 1954	fig. 1, 1	Staurastrum dejectum
Pl. 17: 3-6, 10	Teiling 1967	pl. 9:1-4, 5	Staurodesmus dejectus
Pl. 17: 7, 8	Coesel 1994	pl. 14: 7, 8	Staurodesmus dejectus var. dejectus
Pl. 17: 9	Coesel 1994	pl. 14: 13	Staurodesmus dejectus var. brevispinus
Pl. 17: 11	Brébisson 1856	pl. 1: 23	Staurastrum apiculatum
Pl. 17: 12	Förster 1970	pl. 26: 6	Staurodesmus dejectus var. apiculatus
Pl. 17: 13	Lenzenweger 1997	pl. 22: 9	Staurodesmus dejectus var. apiculatus
Pl. 17: 14, 15	Coesel 1994	pl. 14: 16, 19	Staurodesmus dejectus var. apiculatus
Pl. 17: 16	Messikommer 1928	pl. 8:11	Staurastrum cuspidatum var. robustum
Pl. 17: 17, 18	Coesel 1994	pl. 14: 9, 10	Staurodesmus dejectus var. robustus
Pl. 17: 19, 20	Nygaard 1991	pl. 5: 97, 99	Staurodesmus dejectus var. borealis

Plate numbers	Author		Original identification
Pl. 18: 1	Nordstedt 1888	pl. 4: 16	Staurastrum dejectum var. patens
Pl. 18: 2, 3	West & al. 1923	pl. 129: 6, 8	Staurastrum dejectum var. patens
Pl. 18: 4	Lenzenweger 1981	pl. 13: 7	Staurodesmus dejectum var. patens
Pl. 18: 5	Růžička 1972	pl. 63: 9	Staurastrum dejectum
Pl. 18: 6, 7	Coesel 1994	pl. 3,4	Staurodesmus patens
Pl. 18: 8	Lenzenweger 1986	pl. 4: 14	Staurodesmus brevispina var. kossinskajae
Pl. 18: 9	West W. 1892a	pl. 22: 11	Staurastrum dejectum var. inflatum
Pl. 18: 10	West & al. 1923	pl. 130: 4/5	Staurastrum dejectum var. inflatum
Pl. 18: 11	Skuja 1964	pl. 50: 4	Staurastrum dejectum var. inflatum
Pl. 18: 12	Borge 1930	pl. 2: 40	Staurastrum dejectum var. inflatum
Pl. 19: 1, 2	Ralfs 1848	pl. 20: 5c/d, 5l	Staurastrum dejectum
Pl. 19: 3	West & al. 1923	pl. 130: 11	Staurastrum mucronatum
Pl. 19: 4, 5	Coesel 1994	pl. 13: 10, 9	Staurodesmus mucronatus var. mucronatus
Pl. 19: 6	West G.S. 1909	pl. 5: 5	Staurastrum mucronatum var. delicatulum
Pl. 19: 7	Messikommer 1960	fig. 2: 18	Staurastrum mucronatum var. subtriangulare
Pl. 19: 8	Cedercreutz & Grönblad 1936	pl. 1: 17	Staurastrum mucronatum, forma
Pl. 19: 9	West & West 1903	pl. 17: 11	Staurastrum mucronatum var. subtriangulare
Pl. 19: 10	West & al. 1923	pl. 130: 13	Staurastrum mucronatum var. subtriangulare
Pl. 19: 11, 12, 13	Coesel 1994	pl. 13: 12, 11, 13	Staurodesmus mucronatus var. subtriangularis
Pl. 19: 14	West W. 1892a	pl. 22: 10	Arthrodesmus glaucescens forma convexa
Pl. 19: 15, 16	West & West 1912	pl. 115: 12a, 14	Arthrodesmus controversus
Pl. 19: 17	Allorge & Allorge 1931	pl. 10: 38/39/40	Arthrodesmus controversus
Pl. 19: 18, 19	Borge 1930	pl. 2: 36	Arthrodesmus crassus
Pl. 19: 20	Grönblad 1942	pl. 3: 28	Arthrodesmus crassus
Pl. 19: 21	West & West 1903	pl. 14: 9	Arthrodesmus crassus
Pl. 19: 22-26	Schröder 1898	pl. 2: 4	Staurastrum zachariasii
Pl. 20: 1	Ralfs 1848	pl. 21: 3	Staurastrum dickiei
Pl. 20: 2	Lind & Brook 1980	fig. 121	Staurodesmus dickiei
Pl. 20: 3, 4	Dick 1923	pl. 5: 10, 11	Staurastrum dickiei, forma
Pl. 20: 5, 6, 7	Coesel 1994	pl. 13: 15, 14, 16	Staurodesmus dickiei var. dickiei
Pl. 20: 8	Lenzenweger 1981	pl. 13: 3	Staurodesmus dickiei
Pl. 20: 9	Messikommer 1942	pl. 12: 3	Staurastrum dickiei
Pl. 20: 10	Messikommer 1942	pl. 12: 4	Staurastrum dickiei, forma
Pl. 20: 11	Messikommer 1942	pl. 12: 5	Staurastrum dickiei forma punctata
Pl. 20: 12, 13	Kouwets 1987	Pl. 16: 13, 12	Staurodesmus dickiei var. dickiei
Pl. 21: 1	Turner 1892	pl. 16: 5	Staurodesmus dickiei var. circulare
Pl. 21: 2, 4	Coesel 1994	pl. 13: 19, 17	Staurodesmus dickiei var. circularis
Pl. 21: 3	Tarnogradsky 1960	pl. 2: 23/24	Staurastrum dickiei var. circulare
Pl. 21: 5	West & West 1903	pl. 16: 9	Staurastrum dickiei var. rhomboideum
Pl. 21: 6	Borge 1911	fig. 15	Staurastrum dickiei
Pl. 22: 1	Lundell 1871 in West & West 1912	pl. 12: 9	Staurastrum aversum
Pl. 22: 2	Lenzenweger 1985	pl. 1: 13	Staurodesmus aversus

Plate numbers	Author		Original identification
Pl. 22: 3, 4	West & West 1912	pl. 120: 10, 11	Staurastrum aversum
Pl. 22: 5	Ralfs 1848	pl. 34: 7a/b	Staurastrum brevispina
Pl. 22: 6	West & West 1912	pl. 123: 4	Staurastrum brevispina forma major
Pl. 22: 7	Coesel 1994	pl. 16: 15	Staurodesmus brevispina
Pl. 22: 8	Lenzenweger 1986	pl. 4: 10	Staurodesmus brevispina
Pl. 22: 9, 10	West & West 1912	pl. 123: 7a, 7a'/b	Staurastrum brevispina var. obversum
Pl. 22: 11	Lenzenweger 1986	pl. 4: 11	Staurodesmus brevispina
Pl. 23: 1	Boldt 1885	pl. 5: 30	Staurastrum brevispinum, forma
Pl. 23: 2	Williamson 1992	fig. 12: 7	Staurodesmus brevispina var. brevispinus
Pl. 23: 3	Borge 1894	pl. 3: 42	Staurastrum brevispina var. retusum
Pl. 23: 4	West & West 1912	pl. 123: 5	Staurastrum brevispinum var. altum
Pl. 23: 5	Ralfs 1848	pl. 21: 6a/b	Staurastrum tumidum
Pl. 23: 6	Lundell 1871	pl. 4: 10	Staurastrum tumidum
Pl. 24: 1	West & West 1912	pl. 122: 3, 4	Staurastrum tumidum
Pl. 24: 2	Coesel 1994	pl. 16: 16	Staurodesmus tumidus
Pl. 24: 3	Pevalek 1925	p. 84, fig. 20	Staurastrum julicum
Pl. 24: 4, 5, 6	Archer 1862	pl. 12: 16, 17/18, 22	Staurastrum lanceolatum
Pl. 24: 7	Messikommer 1951	pl. 3: 26a2/b	Staurastrum lanceolatum var. rotundatum
Pl. 24: 8, 9	West & West 1912	pl. 121: 7a', 7a/b'	Staurastrum lanceolatum var. compressum
Pl. 24: 10, 11	Coesel 1997	pl. 1: 11, 12	Staurastrum lanceolatum var. compressum
Pl. 25: 1, 2	West & West 1912	pl. 121: 1, 2	Staurastrum conspicuum
Pl. 25: 3, 4	West & West 1912	pl. 124: 2, 3	Staurastrum inelegans
Pl. 25: 5	Lundberg 1931	p. 289, fig. 11	Staurastrum inelegans var. obtusum
Pl. 26: 1, 2	Nordstedt 1869	pl. 4: 48, 47	Staurastrum clepsydra
Pl. 26: 3	Messikommer 1942	pl. 11: 6	Staurastrum clepsydra, forma
Pl. 26: 4	Boldt 1888	pl. 2: 39	Staurastrum pachyrhynchum forma
Pl. 26: 5, 6	Borge 1891	fig. 4	Staurastrum sibiricum, forma ovalis
Pl. 26: 7	Boldt 1888	pl. 2: 51a/c	Staurastrum sp.
Pl. 26: 8	Børgesen 1894	pl. 2: 22	Staurastrum sibiricum var. crassangulata
Pl. 26: 9	West & West 1912	pl. 122: 8	Staurastrum clepsydra var. sibiricum forma ovalis
Pl. 26: 10	West & West 1912	pl. 122: 10	Staurastrum clepsydra var. sibiricum forma trigona
Pl. 26: 11	Skuja 1964	pl. 48: 10	Staurastrum clepsydra var. sibiricum
Pl. 26: 12, 13	Wittrock & al. 1903	p. 17, fig. 1/8, 5	Cosmarium tortum
Pl. 26: 14, 15	Bijkerk	archives	Staurastrum tortum
Pl. 26: 16	West & West 1912	pl. 123: 9/8	Staurastrum angulatum var. angulatum
Pl. 26: 17	West & West 1903	pl. 16: 10	Staurastrum angulatum var. planctonicum
Pl. 27: 1, 2	West & West 1912	pl. 120: 4, 5	Staurastrum bieneanum
Pl. 27: 3	Nordstedt 1875	pl. 8: 35a/c	Staurastrum bieneanum forma spetsbergensis
Pl. 27: 4	Capdevielle 1985	pl. 9: 3	Staurastrum bieneanum var. angulatum
Pl. 27: 5	Nygaard 1949	fig. 37 bis c	Staurastrum bieneanum var. angulatum
Pl. 27: 6, 7, 8	Coesel 1997	pl. 1: 17, 18, 16	Staurastrum bieneanum
Pl. 27: 9	Florin 1957	fig. 35: 4a	Staurodesmus bieneanus
Pl. 27: 10	Kaiser 1919	fig. 32	Staurastrum orbiculare var. angulatum

Plate numbers	Author		Original identification
Pl. 27: 11	Růžička 1972	pl. 63: 4	Staurastrum kaiseri
Pl. 27: 12, 13	Coesel 1997	pl. 1: 13, 14	Staurastrum kaiseri
Pl. 27: 14	Lenzenweger 1989	fig. 8	Staurastrum kaiseri
Pl. 28: 1, 2, 3	Nordstedt 1875	fig. 31	Staurastrum pachyrhynchum
Pl. 28: 4	West & West 1912	pl. 121: 9	Staurastrum pachyrhynchum
Pl. 28: 5	Lenzenweger 1989b	pl. 5: 6	Staurodesmus pachyrhynchus
Pl. 28: 6	Skuja 1964	pl. 48: 13	Staurastrum pachyrhynchum
Pl. 28: 7	Skuja 1964	pl. 49: 2	Staurastrum pachyrhynchum var. ellipticum
Pl. 28: 8	Raciborski 1889	pl. 7: 14	Staurastrum pachyrhynchum var. convergens
Pl. 28: 9	Lenzenweger 1989	fig. 57	Staurastrum pachyrhynchum var. convergens
Pl. 28: 10	Eichler & Gutwiński 1895	pl. 5: 46	Staurastrum pachyrhynchum var. polonicum
Pl. 28: 11, 12	West, W. 1992	Pl. 23: 8	Staurastrum subpygmaeum var. subpygmaeum
Pl. 28: 13	West & West 1912	pl. 124: 1	Staurastrum subpygmaeum var. subangulatum
Pl. 28: 14, 15, 16	John & Williamson 2009	pl. 29: Ab, Aa; Ac	Staurastrum subpygmaeum var. undulatum
Pl. 28: 17	Dick 1919	pl. 16: 6	Staurastrum spec. nov ?
Pl. 29: 1	Grönblad 1926	pl. 1: 42/43	Cosmarium obliquum var. triquetrum
Pl. 29: 2	Lenzenweger 1997	pl. 23: 20	Staurodesmus groenbladii
Pl. 29: 3, 4	Skuja 1931	pl. 1: 17, 16	Staurastrum groenbladii
Pl. 29: 5	Grönblad 1926	pl. 2: 93/94	Straurastrum schroederi
Pl. 29: 6	Eichler & Gutwiński 1895	pl. 5: 47	Staurastrum ecorne var. podlachicum
Pl. 29: 7	Reinsch 1867 in West & West 1912	pl. 119: 2	Staurastrum minutissimum
Pl. 29: 8	Boldt 1888	pl. 2: 40	Staurastrum minutissimum, forma 3-gona
Pl. 29: 9	Reinsch 1867 in West & West 1912	pl. 119: 3	Staurastrum minutissimum var. convexum
Pl. 29: 10	Boldt 1888	pl. 2: 41	Staurastrum minutissimum, forma 4-gona
Pl. 29: 11, 12	Grönblad 1921	pl. 3: 41/42, 43	Staurastrum minutissimum var. convexum
Pl. 29: 13	Förster 1967	pl. 8: 8a	Staurastrum minutissimum, ad forma tetragona Nordstedt
Pl. 29: 14	Krieger 1938	pl. 2: 13/14	Staurastrum minutissimum var. convexum forma tetragona
Pl. 29: 15	Förster 1967	pl. 8: 8	Staurastrum minutissimum, ad forma tetragona Nordstedt
Pl. 29: 16, 17, 18, 19	Coesel 1996	fig. 34, 36/42, 38, 40	Staurastrum obscurum
Pl. 29: 20, 21	Lundell 1871	pl. 3: 25	Staurastrum insigne
Pl. 29: 22	Peterfi 1963	fig. 34	Staurastrum insigne
Pl. 29: 23	Heimerl 1891	pl.5: 21	Staurastrum insigne
Pl. 29: 24	Irénéé-Marie 1949a	fig. in text	Staurastrum habeebense
Pl. 29: 25, 26	Brook & Williamson 1990	fig. 1D/1H, 1B	Actinotaenium habeebense
Pl. 30: 1	Bulnheim 1861	pl. 9: 14	Staurastrum grande
Pl. 30: 2, 3	West & West 1912	pl. 119: 13, 12	Staurastrum grande
Pl. 30: 4	Cushman 1905 in West & West 1912	pl. 120: 1	Staurastrum grande

Plate numbers   Author   Original identification

Plate numbers	Author		Original identification
Pl. 30: 5, 6	West & West 1912	pl. 120: 3, 2	Staurastrum grande var. parvum
Pl. 31: 1	Grönblad 1920	pl. 3: 107/108	Straurastrum grande var. angulosum
Pl. 31: 2	Ralfs 1848	pl. 21: 4	Staurastrum muticum
Pl. 31: 3, 4, 5	West & West 1912	pl. 118: 18, 16, 19a"	Staurastrum muticum
Pl. 31: 6, 7, 8	Homfeld 1929	pl. 9: 99 (pro parte)	Staurastrum muticum
Pl. 31: 9, 10, 11	Coesel 1997	pl. 1: 10, 8, 9	Staurastrum muticum forma minor
Pl. 31: 12	Smith 1924	pl. 67: 11/12	SStaurastrum subgrande var. minor
Pl. 31: 13	Coesel 1997	pl. 1: 19	Staurastrum subgrande var. minor
Pl. 31: 14	Strøm 1926	pl. 5: 20	Staurastrum bieneanum var. myrdalense
Pl. 31: 15	Messikommer 1956	pl. 2: 21	Staurastrum bieneanum var. myrdalense
Pl. 31: 17	Hirano 1959	pl. 38: 19	Staurastrum coarctatum
Pl. 31: 18	Nordstedt 1888	pl. 4: 20	Staurastrum coarctatum var. subcurtum
Pl. 31: 19	W. West 1980	pl. 5: 8	Staurastrum coarctatum var. subcurtum
Pl. 32: 1, 2	West & West 1912	pl. 124: 10, 11	Staurastrum orbiculare
Pl. 32: 3	Kouwets 1987	pl. 16: 22	Staurastrum orbiculare var. orbiculare
Pl. 32: 4	Lenzenweger 1997	pl. 24: 6	Staurastrum orbiculare
Pl. 32: 5	Coesel 1997	pl. 1: 2	Staurastrum orbiculare var. orbiculare
Pl. 32: 6, 7, 8, 9	West & West 1912	pl. 124: 14, 13, 15, 16	Staurastrum orbiculare var. ralfsii
Pl. 32: 10	Coesel 1997	pl. 1: 3	Staurastrum orbiculare var. ralfsii
Pl. 32: 11	Roy & Bisset 1886	pl. 268: 14	Staurastrum orbiculare var. depressum
Pl. 32: 12, 13	West & West 1912	pl. 124: 19, 17	Staurastrum orbiculare var. depressum
Pl. 32: 14, 15, 18	Coesel 1997	pl. 1: 4, 5, 7	Staurastrum orbiculare var. depressum
Pl. 32: 16	Lind & Brook 1980	fig. 10	Staurastrum orbiculare var. depressum
Pl. 32: 17	Hegde & Bharati 1983	pl. 3: 18	Staurastrum orbiculare var. depressum
Pl. 32: 19	Roy & Bisset 1894	pl. 4: 7	Staurastrum suborbiculare
Pl. 32: 20	West & West 1912	pl. 125: 3	Staurastrum suborbiculare
Pl. 33: 1, 2	West & West 1912	pl. 124: 5, 6	Staurastrum orbiculare var. hibernicum
Pl. 33: 3	Kossinskaja 1953	pl. 3: 18	Staurastrum orbiculare var. hibernicum
Pl. 33: 4	Turner 1892	pl. 13: 13	Staurastrum retusum
Pl. 33: 5, 6, 7	Eichler & Gutwiński 1895	pl. 28: 44	Staurastrum retusum var. punctulatum
Pl. 33: 8, 9	West & West 1912	pl. 125: 8	Staurastrum retusum var. boreale
Pl. 33: 10, 11	West & West 1912	pl. 125: 7, 6	Staurastrum retusum
Pl. 33: 12	Růžička 1972	pl. 63: 3	Staurastrum reusum forma hians
Pl. 33: 13	Eichler & Gutwiński 1895	pl. 28: 45	Staurastrum retusum var. punctulatum forma hians
Pl. 33: 14	Schmidle 1897	fig. IV. fig. 1	Staurastrum orbiculare var. quadratum
Pl. 33: 15	Schröder 1919	pl. 2: 23	Staurastrum kobelianum
Pl. 33: 16	Pevalek 1925	fig. 19	Staurastrum pokljukense
Pl. 33: 17	Nordstedt 1873	pl. 1: 10	Staurastrum orbiculare var. extensum
Pl. 33: 18	West & West 1912	pl. 125: 2	Staurastrum orbiculare var. extensum
Pl. 33: 19	Růžička 1957	fig. 3: 47	Staurastrum orbiculare var. extensum
Pl. 33: 20	Nordstedt 1870 in West & West 1912	pl. 125: 11	Staurastrum cosmarioides
Pl. 33: 21	West & West 1912	pl. 125: 12	Staurastrum cosmarioides
Pl. 34: 1	W. West 1892	fig. 28a/b	Staurastrum ellipticum
Pl. 34: 2	Børgesen 1894	pl. 2: 18	Staurastrum muticum var. subsphaericum
Pl. 34: 3	Skuja 1964	pl. 48: 5	Staurastrum ellipticum var. minor

Plate numbers	Author		Original identification
Pl. 34: 4-6	Nordstedt 1875	pl. 8: 33	Staurastrum subsphaericum
Pl. 34: 7, 8, 9	Nygaard 1991	pl. 4: 81, 80, 79	Staurastrum thomassonii
Pl. 34: 10, 11	Nygaard 1991	pl. 8: 120b, 120a	Staurastrum thomassonii
Pl. 35: 1	Ralfs 1848 in West & al. 1923	pl. 141: 10	Staurastrum bacillare
Pl. 35: 2	Lundell 1871 in West & al. 1923	pl. 141: 11	Staurastrum bacillare var. obesum
Pl. 35: 3	West & al. 1923	pl. 141: 12	Staurastrum bacillare var. obesum
Pl. 35: 4	Kouwets 1987	pl. 18: 3	Staurastrum bacillare var. obesum
Pl. 35: 5	West & West 1898	pl. 18: 22/20	Staurastrum sublaevispinum
Pl. 35: 6	Grönblad 1948	fig. 48	Staurastrum sublaevispinum
Pl. 35: 7	Bisset 1884 in West & al. 1912	pl. 141: 17	Staurastrum laevispinum
Pl. 35: 8	West & al. 1923	pl. 141: 18	Staurastrum laevispinum
Pl. 35: 9	Lenzenweger 1994	pl. 6: 13	Staurastrum laevispinum
Pl. 35: 10, 11, 12, 13	Grönblad 1942	pl. 4: 5, 7, 8, 9	Staurastrum brachiatum var. compactum
Pl. 35: 14, 15	West & al. 1923	pl. 141: 20	Staurastrum subnudibranchiatum
Pl. 35: 16, 17	Nygaard 1991	pl. 7: 118, 119	Staurastrum brachiatum var. bicorne
Pl. 36: 1, 2, 3	Ralfs 1848	pl. 23: 9a, 9d, 9f	Staurastrum brachiatum
Pl. 36: 4, 5, 7-13	Coesel 1997	pl. 12: 9-10, 11-13, 15-17, 14	Staurastrum brachiatum
Pl. 36: 6, 14, 16	Kouwets 1987	pl. 16: 27, 28, 29	Staurastrum brachiatum
Pl. 36: 15	Coesel 1998	fig. 17	Staurastrum brachiatum
Pl. 36: 17	Williamson 1992	fig. 22: 4	Staurastrum brachiatum
Pl. 36: 18, 19	Meesters 2011	archives	Staurastrum brachiatum
Pl. 37: 1	Wittrock 1869	pl. 1: 9	Staurastrum laeve var. clevei
Pl. 37: 2	Ryppowa 1927	pl. 1: 14/15	Staurastrum clevei var. octocornis
Pl. 37: 3	Coesel 1997	pl. 12: 1	Staurastrum clevei
Pl. 37: 4	Messikommer 1960	pl. 2: 17	Staurastrum clevei var. variabile
Pl. 37: 5	Schmidle 1898	pl. 2: 41	Staurastrum kitchellii var. inflatum
Pl. 37: 6, 7, 10, 11	Ralfs 1848	pl. 13: 10a, 10b, 10e, 10d	Staurastrum laeve
Pl. 37: 8	Lenzenweger 1986	pl. 5: 8	Staurastrum laeve
Pl. 37: 9, 12	West & al. 1923	pl. 141: 1, 3	Staurastrum laeve
Pl. 37: 13, 14	Kouwets 1987	pl. 17: 4, 3	Staurastrum laeve
Pl. 37: 15	Nordstedt 1887	pl. 4: 54	Staurastrum gemelliparum
Pl. 37: 16	Tomaszewicz & Skrzeczkowska 1989	fig. 3: 6	Staurastrum gemelliparum
Pl. 37: 17	Nauwerck 1962	pl. 5: 15/16	Staurastrum gemelliparum
Pl. 38: 1, 2	Delponte 1878	pl. 11: 15/16, 20	Staurastrum intricatum
Pl. 38: 3	Homfeld 1929	pl. 9: 97	Staurastrum hantzschii var. congruum
Pl. 38: 4	Dick 1919	pl. 17: 1	Staurastrum senarium
Pl. 38: 5	Coesel 1997	pl. 10: 13	Staurastrum hantzschii var.distentum
Pl. 38: 6	Lenzenweger 1994	pl. 8: 8	Staurastrum hantzschii var. congruum
Pl. 38: 7, 8, 9	West & al. 1923	Pl. 165: 13, 12, 14	Staurastrum tohopekaligense
Pl. 38: 10	Dick 1923	pl. 2: 23	Staurastrum tohopekaligense
Pl. 39: 1	Lundell 1871	pl. 4: 2	Staurastrum bifidum
Pl. 39: 2	West & al. 1923	pl. 134: 4	Staurastrum bifidum
Pl. 39: 3	Schaarschmidt 1883	fig. 19	Staurastrum bifidum var. hexagonum
Pl. 39: 4	Capdevielle 1978	pl. 14: 9	Staurodesmus subpygmaeus var. spiniferus

Plate numbers	Author		Original identification
Pl. 39: 5	West & al. 1923	pl. 134: 16/19	Staurastrum longispinum
Pl. 40: 1	West & al. 1923	pl. 134: 3	Staurastrum longispinum
Pl. 40: 2	Brook 1958	fig. 6	Staurastrum longispinum
Pl. 40: 3	Woloszynska 1921	figs 18/19	Staurastrum besseri
Pl. 40: 4	Nordstedt 1870 in West & al. 1923	pl. 135: 11	Staurastrum brasiliense
Pl. 40: 5	Donat 1926	pl. 1: 1	Staurastrum brasiliense
Pl. 41: 1	Lutkemüller 1910	pl. 3: 10/11	Staurastrum brasiliense
Pl. 41: 2	Brook 1958	figs 7/9	Staurastrum brasiliense
Pl. 42: 1	Scott & Prescott 1956	fig. 5	Staurastrum wildemanii
Pl. 42: 2	Gutwiński 1902	pl. 40: 61a/c'	Staurastrum wildemanii
Pl. 42: 3	Grönblad 1920	pl. 3: 1/3	Staurastrum wildemanii
Pl. 43: 1	Turner 1886	pl. 1: 4	Staurastrum quadrispinatum
Pl. 43: 2	Peterfi 1973	fig. 70	Staurastrum quadrispinatum
Pl. 43: 3	Lenzenweger 1997	pl. 29: 15	Staurastrum quadrispinatum
Pl. 43: 4	West & al. 1923	pl. 135: 5/7	Staurastrum quadrispinatum
Pl. 43: 5	West & al. 1923	pl. 134: 5	Staurastrum quadrangulare
Pl. 43: 6	Grönblad 1960	pl. 8: 159/161	Staurastrum quadrangulare
Pl. 43: 7	Capdevielle 1978	pl. 22: 1	Staurastrum quadrangulare var. prolificum
Pl. 43: 8	Coesel 1997	pl. 10: 9	Staurastrum quadrangulare var. contectum
Pl. 43: 9	Manguin 1936	pl. 7: 100	Staurastrum quadrangulare
Pl. 43: 10	West & West 1902	pl. 2: 35	Staurastrum gatniense
Pl. 43: 11	West & al. 1923	pl. 139: 8/9	Staurastrum echinodermum
Pl. 43: 12	Schaarschmidt 1883	fig. 16	Staurastrum kanitzii
Pl. 43: 13	Ralfs 1848	pl. 22: 7	Staurastrum hystrix
Pl. 43: 14	Lütkemüller 1900	pl. 1: 52/53	Staurastrum hystrix var. pannonicum
Pl. 43: 15	Allorge V.& P. 1931	pl. 13: 7/8	Staurastrum hystrix
Pl. 43: 16	Lenzenweger 1997	pl. 29: 13	Staurastrum hystrix
Pl. 43: 17, 18	Coesel 1997	pl. 5: 10, 11	Staurastrum hystrix
Pl. 43: 19	Beijerinck 1926 in Coesel 1997	pl. 5: 9	Staurastrum hystrix
Pl. 44: 1, 2	Ralfs 1848	pl. 22: 4, pl. 34: 14	Staurastrum teliferum
Pl. 44: 3, 4, 5, 6	West & al. 1923	pl. 136: 2, 5, 3, 4	Staurastrum teliferum
Pl. 44: 7, 8, 9	Coesel 1997	pl. 5: 4, 5, 6	Staurastrum teliferum
Pl. 44: 10	Turner 1885	pl. 16: 21	Staurastrum gladiosum
Pl. 44: 11	West & al. 1923	pl. 137: 3	Staurastrum gladiosum var. delicatulum
Pl. 44: 12, 13	Coesel 1997	pl. 5: 8, 7	Staurastrum teliferum var. gladiosum
Pl. 45: 1	Nordstedt 1873	fig. 13	Staurastrum geminatum
Pl. 45: 2, 3	Printz 1915	pl. 3: 76, 78	Staurastrum geminatum var. longispinum
Pl. 45: 4	Cleve 1864	pl. 4: 4	Staurastrum setigerum
Pl. 45: 5	Lenzenweger 1997	pl. 27: 8	Staurastrum setigerum
Pl. 45: 6, 7	West & al. 1923	pl. 136: 14, 13	Staurastrum setigerum
Pl. 45: 8, 9	Brook 1958	fig. 10	Staurastrum setigerum
Pl. 45: 10, 11, 12, 13	Reinsch 1867	pl. 24B, I: 1, 3, 4, 5	Staurastrum ungeri
Pl. 46: 1	Roy & Bisset 1894	pl. 3: 8	Staurastrum polytrichum

Plate numbers	Author		Original identification
Pl. 46: 2, 3	Coesel 1997	pl. 5: 1, 2	Staurastrum polytrichum
Pl. 46: 4	Schmidle 1894	pl. 28: 15	Staurastrum subbrebissonii
Pl. 46: 5	Bourrelly 1987	pl. 43: 6/7/8	Staurastrum subbrebissonii
Pl. 47: 1, 2, 3	West & al. 1923	pl. 138: 10/11, 12 pl. 139: 16	Staurastrum pyramidatum
Pl. 47: 4	Dick 1930	pl. 9: 9/10	Staurastrum pyramidatum
Pl. 47: 5	Kouwets 1987	pl. 17: 14	Staurastrum pyramidatum
Pl. 48: 1	Lenzenweger 1984	pl. 8: 11	Staurastrum subbrebissonii
Pl. 48: 2	Kouwets 1987	pl. 17: 16	Staurastrum subbrebissonii ?
Pl. 48: 3, 4, 5	Coesel 1997	pl. 4: 6, 4, 5	Staurastrum kouwetsii
Pl. 48: 6	West & al. after Lütkemüller 1923	pl. 137: 4	Staurastrum brebissonii
Pl. 48: 7, 8	Coesel 1997	pl. 4: 1, 2	Staurastrum brebissonii
Pl. 48: 9	Schmidle 1898	pl. 3: 1	Staurastrum brebissonii var. ordinatum
Pl. 48: 10	West & al. 1923	pl. 137: 10/11	Staurastrum brebissonii var. ordinatum
Pl. 48: 11	Messikommer 1942	pl. 15: 4	Staurastrum erasum
Pl. 48: 12	Coesel 1997	pl. 4: 3	Staurastrum erasum
Pl. 49: 1	Grönblad 1926	pl. 2: 85/86	Staurastrum brebissonii var. truncatum
Pl. 49: 2, 5	Homfeld 1926	pl. 8: 93	Staurastrum brebissonii var. truncatum
Pl. 49: 3, 4	Coesel 1997	pl. 4: 10, 11	Staurastrum brebissonii var. truncatum
Pl. 49: 6	West & West 1897	pl. 6: 18	Staurastrum erostellum
Pl. 49: 7, 8	Ralfs 1848	pl. 22: 3b/e, 3h	Staurastrum hirsutum
Pl. 49: 9, 10	West & al. 1923	pl. 138: 4, 5	Staurastrum hirsutum
Pl. 50: 1, 2	Kossinskaja 1950	pl. 2: 5, 4	Staurastrum hirsutum
Pl. 50: 3, 4	Coesel 1997	pl. 4: 8, 9	Staurastrum hirsutum var. hirsutum
Pl. 50: 5, 6	West & al. 1923	pl. 138: 9b, 9a	Staurastrum muricatum
Pl. 50: 7	Coesel 1997	pl. 4: 7	Staurastrum hirsutum var. muricatum
Pl. 50: 8	Kouwets 1987	pl. 18: 6	Staurastrum muricatum
Pl. 50: 9, 10	West & al. 1923	pl. 139: 11, 13/14	Staurastrum arnellii
Pl. 50: 11, 12	Coesel 1997	pl. 6: 1, 2	Staurastrum arnellii
Pl. 50: 13	Boldt 1885	pl. 5: 21	Staurastrum arnellii
Pl. 50: 14	Lenzenweger 1997	pl. 28: 3	Staurastrum arnellii
Pl. 50: 15	Coesel 2009	archives (after Austrian material)	Staurastrum arnellii
Pl. 51: 1	Roy & Bisset 1894 in West & al. 1923	pl. 139: 3	Staurastrum horametrum
Pl. 51: 2, 3	West & al. 1923	pl. 140: 11/13, 12	Staurastrum asperum
Pl. 51: 4	Williamson 1994	fig. 3: 4/5	Staurastrum asperum
Pl. 51: 5	Ralfs 1848 in West & al. 1923	pl. 141: 21	Staurastrum asperum
Pl. 51: 6	Wittrock & Nordstedt 1893	at nr. 1114	Staurastrum scabrum
Pl. 51: 7	Rybnicek 1960	fig. 104	Staurastrum scabrum
Pl. 51: 8, 9	Coesel 1997	pl. 6: 3, 4	Staurastrum scabrum
Pl. 51: 10	Coesel & Meesters 2007	pl. 96: 3	Staurastrum scabrum
Pl. 52: 1	Wille 1879	pl. 13: 58	Staurastrum novae-semliae
Pl. 52: 2, 3	Grönblad 1942	pl. 5: 9, 8	Staurastrum novae-semliae var. crassangulatum
Pl. 51: 4	Förster 1967	pl. 8: 18	Staurastrum novae-semliae var. crassangulatum
Pl. 52: 5	Huber-Pestalozzi 1928	pl. 13: 10	Staurastrum ricklii

Plate numbers	Author		Original identification
Pl. 52: 6	West & West 1912	pl. 126: 4	Staurastrum botrophilum
Pl. 52: 7, 8, 9	Brook & Williamson 1983	fig. 1	Staurastrum botrophilum
Pl. 52: 10	West & West 1912	pl. 126: 7	Staurastrum donardense
Pl. 52: 11	Raciborski 1889	pl. 7: 6	Staurastrum alpinum
Pl. 52: 12	Skuja 1928	pl. 4: 21	Staurastrum alpinum
Pl. 52: 13, 14	Beck-Mannagetta 1926a	fig. 19	Staurastrum hoplotheca
Pl. 53: 1, 2	West & West 1912	pl. 126: 5, 6a	Staurastrum turgescens
Pl. 53: 3	Růžička 1956	pl. 5: 39	Staurastrum turgescens
Pl. 53: 4	Lenzenweger 1994	pl. 6: 7	Staurastrum turgescens
Pl. 53: 5	Schmidle 1898	pl. 3: 5	Staurastrum punctulatum var. muricatiforme forma lapponica
Pl. 53: 6	Grönblad 1926	pl. 2: 106/107	Staurastrum lapponicum
Pl. 53: 7	Coesel 1997	pl. 3: 7	Staurastrum lapponicum
Pl. 53: 8	Kouwets 1987	pl. 18: 17	Staurastrum lapponicum
Pl. 53: 9, 10	West & West 1912	pl. 126: 11, 12	Staurastrum dilatatum
Pl. 53: 11	Růžička 1972	pl. 63: 13	Staurastrum dilatatum
Pl. 53: 12, 15	Coesel 1997	pl. 2: 15, 16	Staurastrum dilatatum
Pl. 53: 13, 14	Coesel & Delfos 1986	fig. 28, 27	Staurastrum dilatatum
Pl. 54: 1	West & West 1912	pl. 126: 8	Staurastrum alternans
Pl. 54: 2	Förster 1967	pl. 8: 15	Staurastrum alternans
Pl. 54: 3, 4, 5, 6, 7	Coesel 1997	pl. 2: 11, 14, 13, 9, 10	Staurastrum alternans
Pl. 54: 8, 9	West & West 1912	pl. 127: 2, pl. 126: 5	Staurastrum striolatum
Pl. 54: 10	Coesel 1997	pl. 2: 19/20	Staurastrum striolatum
Pl. 54: 11	Lütkemüller 1900a	pl. 6: 39/40	Staurastrum sinense
Pl. 54: 12	Borge 1906	pl. 3: 37	Staurastrum sinense
Pl. 55: 1, 3	West & West 1912	pl. 128: 5, 6	Staurastrum punctulatum var. striatum
Pl. 55: 2	Růžička 1957	pl. 3: 54	Staurastrum striatum
Pl. 55: 4, 5, 6	Coesel 1997	pl. 3: 9, 10, 11	Staurastrum striatum
Pl. 55: 7	West & West 1912	pl. 127: 7	Staurastrum dispar, forma
Pl. 55: 8, 9	Coesel 1997	pl. 3: 14, 13	Staurastrum dispar var. semicirculare
Pl. 55: 10	Kouwets 1987	pl. 18: 9	Staurastrum dispar
Pl. 55: 11	West & West 1912	pl. 128: 14	Staurastrum granulosum var. acutum
Pl. 55: 12, 13, 14, 15	Coesel 1997	pl. 3: 17, 18, 15, 16	Staurastrum acutum var. acutum
Pl. 55: 16	Raciborski 1885	pl. 12: 1a	Staurastrum varians
Pl. 55: 17	Schmidle 1896	pl. 16: 18	Staurastrum varians var. badense
Pl. 55: 18	Schmidle 1896	pl. 16: 19	Staurastrum varians, forma trigona
Pl. 55: 19	Lenzenweger 1989	pl. 3: 41	Staurastrum pygmaeum
Pl. 55: 20	Lenzenweger 1989	pl. 3: 42	Staurastrum varians var. badense, forma Schmidle
Pl. 56: 1	Elfving 1881	pl. 1: 4	Staurastrum tristichum
Pl. 56: 2	Grönblad 1921	pl. 5: 1/ 2	Staurastrum tristichum
Pl. 56: 3, 4	West & West 1912	pl. 126: 1, 2	Staurastrum trachytithophorum
Pl. 56: 5	Ralfs 1848	pl. 22: 1	Staurastrum punctulatum
Pl. 56: 6, 7, 8	Wille 1879	pl. 13: 50a/c, 51a/c, 52a/c	Staurastrum punctulatum var. kjellmanii
Pl. 56: 9	West & West 1912	pl. 127: 22	Staurastrum punctulatum var. kjellmanii
Pl. 56: 10	West & West 1912	pl. 127: 9	Staurastrum punctulatum

Plate numbers	Author		Original identification
Pl. 56: 11	Wille 1879 in West & West 1912	pl. 128: 1	Staurastrum punctulatum var. pygmaeum
Pl. 56: 12, 15	Coesel 1997	pl. 3: 7, 8	Staurastrum punctulatum var. pygmaeum
Pl. 56: 13, 14, 16	Coesel 1998	pl. 3: 1, 2, 3	Staurastrum punctulatum
Pl. 56: 17, 18	Coesel 1999	pl. 3: 4, 5	Staurastrum punctulatum
Pl. 56: 19	Förster 1967	pl. 8: 22	Staurastrum punctulatum var. kjellmanii
Pl. 57: 1	West & West 1912	pl. 127: 23	Staurastrum inflatum
Pl. 57: 2, 3	West & West 1912	pl. 118: 5, 6	Staurastrum meriani
Pl. 57: 4, 5	West & al. 1923	pl. 167: 8, 9	Staurastrum meriani
Pl. 57: 6	Růžička 1956	pl. 5: 36	Staurastrum meriani
Pl. 57: 7, 8	Skuja 1964	pl. 164: 6, 7	Staurastrum meriani
Pl. 57: 9	Raciborski 1884	pl. 1: 10	Staurastrum polonicum
Pl. 57: 10	Dick 1923	pl. 3: 9	Staurastrum polonicum
Pl. 57: 11	Lenzenweger 1997	pl. 42: 8	Staurastrum polonicum
Pl. 57: 12	Lundell 1871	pl. 5: 3	Staurastrum mutilatum
Pl. 58: 1	Ralfs 1848	pl. 35: 22	Staurastrum pileolatum
Pl. 58: 2	West & West 1912	pl. 18: 11	Staurastrum pileolatum
Pl. 58: 3	Kouwets 1987	pl. 18: 1	Staurastrum pileolatum
Pl. 58: 4	Rybnicek 1960	fig. 100	Staurastrum pileolatum
Pl. 58: 5	West & West 1912	pl. 118: 14	Staurastrum pileolatum var. cristatum
Pl. 58: 6	Lütkemüller 1892	pl. 9: 16	Staurastrum pileolatum var. cristatum
Pl. 58: 7	Lenzenweger 1984	pl. 8: 2	Staurastrum pileolatum var. cristatum
Pl. 58: 8	West & West 1912	pl. 118: 7	Staurastrum capitulum
Pl. 58: 9	Peterfi 1963	pl. 2: 29/35	Staurastrum capitulum
Pl. 58: 10	Kouwets 1987	pl. 18: 2	Staurastrum capitulum
Pl. 58: 11	Nordstedt 1872	pl. 7: 25	Staurastrum capitulum var. amoenum forma spetsbergensis
Pl. 58: 12	Krieger 1938	pl. 2: 9/10	Staurastrum capitulum var. spetzbergense
Pl. 58: 13, 14	Förster 1967	pl. 9: 13, 14	Staurastrum capitulum var. spetsbergense
Pl. 58: 15, 16	Lütkemüller 1900	pl. 1: 44/45, 46/47	Staurastrum bifasciatum
Pl. 58: 17	Pevalek 1925	fig. 18 (p. 80)	Staurastrum kaiseri
Pl. 58: 18	Dick 1919	pl. 16: 7	Staurastrum capitulum, nov. var.
Pl. 58: 19	Messikommer 1956	pl. 2: 31	Staurastrum subkaiseri
Pl. 58: 20	Bourrelly 1987	pl. 41: 5/7/8	Staurastrum subkaiseri var. lunzense
Pl. 59: 1	Grönblad 1933	pl. 1: 9	Staurastrum capitulum forma borgei
Pl. 59: 2	Lenzenweger 1977	pl. 42: 10	Staurastrum capitulum forma borgei
Pl. 59: 3	Nordstedt 1875	pl. 8: 40	Staurastrum rhabdophorum
Pl. 59: 4	Grönblad 1963	pl. 5: 34	Staurastrum rhabdophorum
Pl. 59: 5	Nordstedt 1872	pl. 7: 26	Staurastrum acarides
Pl. 59: 6	West & al. 1923	pl. 40: 6	Staurastrum acarides
Pl. 59: 7	Růžička 1956	pl. 5: 40	Staurastrum acarides
Pl. 59: 8	Messikommer 1942	pl. 17: 5	Staurastrum acarides
Pl. 59: 9	Cooke 1887	pl. 53: 3	Staurastrum maamense
Pl. 59: 10	West & al. 1923	pl. 139: 11	Staurastrum maamense
Pl. 60: 1	Ralfs 1848	pl. 23: 4	Staurastrum spongiosum
Pl. 60: 2	West & al. 1923	pl. 140: 14	Staurastrum spongiosum
Pl. 60: 3	W. West 1892a	pl. 23: 3	Staurastrum spongiosum var. perbifidum

Plate numbers | Author | Original identification

Plate numbers	Author		Original identification
Pl. 60: 4	Dick 1923	pl. 7: 1	Staurastrum spongiosum
Pl. 60: 5	Förster 1967	pl. 9: 11	Staurastrum spongiosum var. perbifidum
Pl. 60: 6	Coesel 1997	pl. 9: 2	Staurastrum spongiosum var. perbifidum
Pl. 61: 1	Nägeli 1849	pl. 8: C1c	Phycastrum cristatum
Pl. 61: 2	West & al. 1923	pl. 139: 5	Staurastrum cristatum
Pl. 61: 3	Coesel & Meesters 2007	pl. 99: 9	Staurastrum cristatum var. cristatum
Pl. 61: 4	Coesel 1997	pl. 9: 9	Staurastrum cristatum var. cuneatum
Pl. 61: 5	Coesel 1997	pl. 9: 8	Staurastrum cristatum var. navigiolum
Pl. 61: 6	Nordstedt 1875 in West & al. 1923	pl. 139: 6	Staurastrum oligacanthum
Pl. 61: 7	Förster 1970	pl. 27: 20	Staurastrum oligacanthum var. incisum
Pl. 61: 8	Coesel 1997	pl. 9: 10	Staurastrum oligacanthum
Pl. 61: 9	Roy & Bisset 1886	pl. 268: 6	Staurastrum oxyrhynchum
Pl. 61: 10	Lütkemüller 1910	pl. 3: 19-21	Staurastrum oxyrhynchum subsp. truncatum
Pl. 62: 1	Børgesen 1889	pl. 6: 8	Staurastrum aculeatum subspec. cosmospinosum
Pl. 62: 2, 3	Heimans 1926	pl. 5: 5, 6	Staurastrum echinatum
Pl. 62: 4	Messikommer 1962	pl. 51: 8	Staurastrum echinatum
Pl. 62: 5, 6	Coesel 1997	pl. 6: 14, 15	Staurastrum echinatum
Pl. 62: 7	Heimerl 1891	pl. 5: 23	Staurastrum simonyi
Pl. 62: 8	Coesel 1997	pl. 6: 9	Staurastrum simonyi var. simonyi
Pl. 62: 9, 10	Kouwets 1988	pl. 5: 1, 10	Staurastrum simonyi
Pl. 62: 11	West & al. 1923	pl. 135: 1	Staurastrum simonyi
Pl. 62: 12	Coesel 1997	pl. 6: 8	Staurastrum simonyi var. simonyi
Pl. 62: 13	Schmidle 1896	pl. 16: 20a/c	Staurastrum sparciaculeatum
Pl. 62: 14, 15	Coesel 1997	pl. 6: 13, 12	Staurastrum simonyi var. sparciaculeatum
Pl. 62: 16	West & al. 1923	pl. 135: 2	Staurastrum simoyi
Pl. 62: 17, 18, 19	Coesel 1997	pl. 6: 7, 6, 5	Staurastrum simonyi var. semicirculare
Pl. 62: 20	Kouwets 1988	pl. 5: 7	Staurastrum simonyi
Pl. 63: 1	Ralfs 1848	pl. 34: 10a/b	Staurastrum pungens
Pl. 63: 2	West & al. 1923	pl. 135: 8/9	Staurastrum pungens
Pl. 63: 3	Coesel 1997	pl. 9: 4	Staurastrum pungens
Pl. 63: 4, 5, 6, 7	Růžička 1972	pl. 64: 21, 24, 23, 22	Staurastrum pungens
Pl. 63: 8	Dick 1923	pl. 6: 8	Staurastrum pungens
Pl. 63: 9	Ralfs 1848	pl. 34: 9	Staurastrum monticulosum
Pl. 63: 10	West & al. 1923	pl. 154: 10	Staurastrum monticulosum var. groenlandicum
Pl. 63: 11, 12	West & al. 1924	pl. 154: 8, 9	Staurastrum monticulosum var. bifarium
Pl. 63: 13	Grönblad 1920	fig. p. 90	Staurastrum monticulosum
Pl. 63: 14	Grönblad 1920	pl. 1: 17/18	Staurastrum monticulosum var. groenlandicum
Pl. 63: 15	Coesel 1997	pl. 9: 3	Staurastrum monticulosum var. groenlandicum
Pl. 64: 1	Nordstedt 1875	pl. 8: 38	Staurastrum megalonotum
Pl. 64: 2	Roy & Bisset 1894	pl. 4: 5	Staurastrum cornutum
Pl. 64: 3	John & Williamson 2009	pl. 31: D	Staurastrum cornutum
Pl. 64: 4	Grönblad 1920	pl. 2: 27/28	Staurastrum cornutum

Plate numbers Author   Original identification

Pl. 64: 5	Skuja 1948	pl. 18: 6	Staurastrum maamense var. atypicum
Pl. 64: 6	West & West 1905	pl. 7: 17	Staurastrum forficulatum
Pl. 64: 7	West & al. 1923	pl. 154: 15	Staurastrum forficulatum
Pl. 64: 8	Lundell 1871	pl. 4: 5	Staurastrum forficulatum
Pl. 64: 9	Coesel 1997	pl. 8: 13	Staurastrum forficulatum
Pl. 65: 1	West & al. 1923	pl. 155: 7	Staurastrum furcatum var. subsenarium
Pl. 65: 2	Lenzenweger 1986b	pl. 4: 7	Staurastrum furcatum var. subsenarium
Pl. 65: 3	Sieminska 1967	fig. 50	Staurastrum furcatum var. subsenarium
Pl. 65: 4	Grönblad 1920	pl. 3: 50/51	Staurastrum forficulatum var. verrucosum
Pl. 65: 5	Lenzenweger 1981	pl. 14: 8	Staurastrum senarium var. nigrae-silvae
Pl. 65: 6	Kouwets 1987	pl. 17: 7	Staurastrum senarium
Pl. 65: 7, 8	Coesel 1997	pl. 8: 11, 10	Staurastrum senarium
Pl. 65: 9	Eichler & Gutwiński 1894	pl. 5: 49	Staurastrum podlachicum
Pl. 65: 10	Grönblad 1920	pl. 1: 29	Staurastrum oligacanthum var. podlachicum
Pl. 65: 11, 12	Coesel 1997	pl. 9: 6, 7	Staurastrum podlachicum
Pl. 66: 1, 2	Ralfs 1848	pl. 22: 8a/b, 8c	Staurastrum spinosum
Pl. 66: 3, 5	Kouwets 1988	pl. 4: 6, 8	Staurastrum furcatum
Pl. 66: 4	Kouwets 1988	pl.17: 2	Staurastrum furcatum
Pl. 66: 6	Coesel 1997	pl. 10: 1	Staurastrum furcatum
Pl. 66: 7	Coesel 1998	pl. 8: 12	Staurastrum pseudopisciforme, forma
Pl. 66: 8	W. West 1889	pl. 291: 12a/c	Staurastrum avicula var. aciculiferum
Pl. 66: 9, 10	West & al. 1923	pl. 134: 6a, 6a/b	Staurastrum aciculiferum
Pl. 66: 11	Lenzenweger 1989d	fig. 156	Staurastrum aciculiferum
Pl. 66: 12, 13, 14	Coesel 1997	pl. 10: 8, 5, 7	Staurastrum furcatum var. aciculiferum
Pl. 66: 15	Nordstedt 1873	fig. 18	Staurastrum arcuatum
Pl. 66: 16, 17, 18	Coesel 1997	pl. 8: 6, 7, 9	Staurastrum arcuatum
Pl. 67: 1	W. West 1892	fig. 25	Staurastrum arcuatum subsp. subavicula
Pl. 67: 2	Förster 1970	pl. 28: 1	Staurastrum subavicula
Pl. 67: 3, 4, 5, 6	Coesel 1997	pl. 8: 4, 1, 2, 3	Staurastrum subavicula
Pl. 67. 7	Růžička 1963	fig. 1	Staurastrum dicroceros
Pl. 67: 8	Ralfs 1848	pl. 23: 11	Staurastrum avicula
Pl. 67: 9	West & al. 1923	pl. 133: 11	Staurastrum avicula var. subarcuatum
Pl. 67: 10	West & al. 1923	pl. 133: 10	Staurastrum avicula, forma
Pl. 67: 11, 12, 13	West & al. 1923	pl. 133: 14, 15, 13	Staurastrum avicula var. denticulatum
Pl. 67: 14	Messikommer 1943	pl. 14: 15	Staurastrum avicula
Pl. 67: 15	Lenzenweger 1989	fig. 45	Staurastrum avicula var. exornatum
Pl. 67: 16	Nygaard 1992	pl. 4: 85	Staurastrum avicula
Pl. 67: 17	Manguin 1935	pl. 2: 41	Staurastrum avicula var. subarcuatum
Pl. 67: 18	Lenzenweger 1989	fig. 46	Staurastrum avicula var. subarcuatum
Pl. 67: 19, 20	Coesel 1997	pl. 7: 18, 17	Staurastrum subarcuatum
Pl. 67: 21	Lenzenweger 1988	pl. 3: 48	Staurastrum avicula
Pl. 68: 1, 2	Coesel 1997	pl. 7: 6, 5	Staurastrum avicula var. avicula
Pl. 68: 3	Coesel 1997	pl. 7: 10	Staurastrum avicula var. exornatum
Pl. 68: 4	Messikommer 1929	pl. 1: 15	Staurastrum avicula var. exornatum
Pl. 68: 5	Skuja 1956	pl. 37: 8	Staurastrum avicula var. subarcuatum
Pl. 68: 6	Ralfs 1848	pl. 34: 12	Staurastrum lunatum

Plate numbers	Author		Original identification
Pl. 68: 7	Růžička 1972	pl. 64: 25	Staurastrum lunatum
Pl. 68: 8	Lenzenweger 1989	fig. 43 right	Staurastrum lunatum
Pl. 68: 9, 10	Nygaard 1991	pl. 4: 83, 82	Staurastrum lunatum var. planctonicum
Pl. 68: 11, 12, 13, 14	Coesel 1997	pl. 7: 3, 2, 1, 4	Staurastrum lunatum
Pl. 69: 1, 2	West & al. 1923	pl. 133: 21/22, 19	Staurastrum lunatum var. planctonicum
Pl. 69: 3	Skuja 1956	pl. 37: 9	Staurastrum lunatum
Pl. 69: 4	Coesel 1997	pl. 7: 8	Staurastrum avicula var. avicula
Pl. 69: 5	Lenzenweger 1989	fig. 43 left	Staurastrum lunatum
Pl. 69: 6, 7	Coesel 1997	pl. 7: 7, 9	Staurastrum avicula var. avicula
Pl. 69: 8	West & al. 1923	pl. 146: 6	Staurastrum pelagicum
Pl. 69: 9	Cooke 1887	pl. 51: 3b/d	Staurastrum subcruciatum
Pl. 69: 10	West & al. 1923	pl. 133: 6	Staurastrum subcruciatum
Pl. 69: 11, 12	Coesel 1997	pl. 7: 12, 14	Staurastrum subcruciatum
Pl. 70: 1, 2	West & West 1903	pl. 18: 1, 2	Staurastrum pseudopelagicum
Pl. 70: 3	Grönblad 1938	fig. 3: 2a	Staurastrum pseudopelagicum forma biradiata
Pl. 70: 4, 5, 6	Brook 1957	Figs 2, 1, 4a/c	Staurastrum pseudopelagicum
Pl. 71: 1, 2	Schmidle 1898	pl. 3: 3	Staurastrum bohlinianum
Pl. 71: 3	Van Westen	archives	Staurastrum bohlinianum
Pl. 71: 4	Lenzenweger 1989	Fig. 34	Staurastrum bohlinianum
Pl. 71: 5, 6, 7	Coesel 2008	Figs 14, 15, 16	Staurastrum bohlinianum
Pl. 71: 8, 9	Růžička 1972	pl. 63: 27, 26	Staurastrum bohlinianum var. subpygmaeum
Pl. 71: 10, 11	West & al. 1923	pl. 142: 11, 13	Staurastrum hexacerum
Pl. 71: 12, 13, 14	Förster 1970	pl. 28: 6, 7, 13	Staurastrum hexacerum
Pl. 71: 15	Tomaszewicz 1988	pl. 15: 4	Staurastrum hexacerum
Pl. 71: 16	Skuja 1931	pl. 1: 18	Staurastrum hexacerum
Pl. 71: 17, 18, 19, 20, 21	Ralfs 1848	pl. 22: 9d, 9b, 9h, 9i, 9m	Staurastrum polymorphum
Pl. 71: 22	Grönblad 1921	pl. 5: 34	Staurastrum polymorphum var. pygmaeum
Pl. 71: 23	Grönblad 1921	pl. 5: 17-20	Staurastrum polymorphum
Pl. 72: 1	Lütkemüller in West & al. 1923	pl. 142: 24	Staurastrum polymorphum
Pl. 72: 2	Kossinskaja 1950	pl. 2: 8	Staurastrum polymorphum
Pl. 72: 3	Lenzenweger 1994	pl. 7:2	Staurastrum polymorphum
Pl. 72: 4	Lenzenweger 1986	pl. 6: 9	Staurastrum polymorphum var. pygmaeum
Pl. 72: 5	Coesel 1997	pl. 14: 2	Staurastrum polymorphum
Pl. 72: 6	Wille 1881 in Prescott 1982	pl. 401: 15	Staurastrum haaboeliense
Pl. 72: 7, 8	West & al. 1923	pl. 142: 19, 20	Staurastrum haaboeliense
Pl. 72: 9, 10	Coesel 1997	pl. 15: 9, 10	Staurastrum haaboeliense
Pl. 72: 11	Lenzenweger 1987	pl. 4: 31	Staurastrum haaboeliense
Pl. 72: 12	Woloszynska 1919	pl. 3: 53/54	Staurastrum dybowskii
Pl. 72: 13	Coesel 1997	pl. 15: 1	Staurastrum dybowskii
Pl. 72: 14	Lenzenweger 1989	fig. 114	Staurastrum eurycerum
Pl. 72: 15	Skuja 1948	pl. 18: 7	Staurastrum eurycerum
Pl. 73: 1	Homfeld 1929	pl. 9: 96	Staurastrum gracile
Pl. 73: 2, 3	Brook 1959	textfig. 3A, 3B/C/F	Staurastrum gracile

Plate numbers	Author		Original identification
Pl. 73: 4	Růžička 1972	pl. 64: 6	Staurastrum gracile
Pl. 73: 5	Coesel 1997	pl. 2: 3	Staurastrum gracile
Pl. 73: 6	Nygaard 1949	fig. a/a1	Staurastrum gracile
Pl. 73: 7	Lenzenweger 1989	fig. 99	Staurastrum gracile
Pl. 73: 8	West & al. 1923	pl. 146: 5	Staurastrum boreale
Pl. 73: 9	Brook 1959c	pl. 28: 8/9	Staurastrum boreale
Pl. 73: 10	Lenzenweger 1989e	pl. 10: 1	Staurastrum boreale
Pl. 73: 11	Cedercreutz & Grönblad 1936	pl. 1: 10	Staurastrum boreale
Pl. 73: 12, 15, 16	Coesel 1997	pl. 14: 17, 9, 11	Staurastrum boreale
Pl. 73: 13, 14	Coesel 1997	pl. 14: 19, 20	Staurastrum boreale var. quadriradiatum
Pl. 74: 1, 2	Messikommer 1951	pl. 2: 25	Staurastrum glaronense
Pl. 74: 3	Heimerl 1891	pl. 5: 24	Staurastrum cruciatum
Pl. 74: 4, 5	West & al. 1923	pl. 149: 14, 15	Staurastrum heimerlianum
Pl. 74: 6	Lenzenweger 1986	fig. 15	Staurastrum heimerlianum
Pl. 74: 7	Lenzenweger 1987	pl. 4: 32	Staurastrum heimerlianum var. coronatum
Pl. 74: 8	Lenzenweger 1991a	pl. 2: 21	Staurastrum heimerlianum var. heimerlianum, forma
Pl. 74: 9, 10	Lütkemüller 1893	pl. 9: 17	Staurastrum heimerlianum var. spinulosum
Pl. 74: 11	Lenzenweger 1994	pl. 7: 7	Staurastrum heimerlianum var. heimerlianum
Pl. 74: 12	Nägeli 1849	pl. 8B: h	Staurastrum crenulatum
Pl. 74: 13, 16	West & al. 1923	pl. 143: 10/12, 11	Staurastrum crenulatum
Pl. 74: 14	Dick 1930	pl. 9: 3-4	Staurastrum crenulatum
Pl. 74: 15	Brook 1959	pl. 18: 2	Staurastrum crenulatum
Pl. 74: 17, 18	Coesel 1985	archives	Staurastrum crenulatum
Pl. 74: 19, 20, 21	Coesel 1997	pl. : 14, 20, 11	Staurastrum crenulatum
Pl. 75: 1	Grönblad 1963	pl. 5: 35	Staurastrum pentasterias
Pl. 75: 2	Lenzenweger 1989	fig. 117	Staurastrum pentasterias
Pl. 75: 3	Růžička 1972	pl. 64: 1	Staurastrum pentasterias
Pl. 75: 4	Schmidle 1896	pl. 16: 22	Staurastrum ornatum var. asperum
Pl. 75: 5	Laporte 1931	pl. 16: 206/207	Staurastrum ornatum var. asperum
Pl. 75: 6, 7	Lenzenweger 1988	pl. 4: 69	Staurastrum margaritaceum var. ornatum
Pl. 75: 8	West & West 1902	pl. 1: 23	Staurastrum barbaricum
Pl. 75: 9	Messikommer 1942	pl. 18: 4	Staurastrum suchlandtianum
Pl. 75: 10	Lenzenweger 1986b	pl. 5: 5	Staurastrum cyclacanthum var. depressum
Pl. 75: 11	Messikommer 1942	pl. 14: 10	Staurastrum suchlandtianum
Pl. 75: 12	Ralfs 1848	pl. 23: 5a/d	Staurastrum sexcostatum
Pl. 75: 13	West & al. 1923	pl. 150: 15	Staurastrum sexcostatum var. productum
Pl. 75: 14	West & al. 1923	pl. 150: 14	Staurastrum sexcostatum
Pl. 75: 15, 16	Coesel 1997	pl. 20: 4, 5/6	Staurastrum sexcostatum
Pl. 76: 1	Ralfs 1948	pl. 23: 12b/c	Staurastrum asperum var. proboscideum
Pl. 76: 2	West & al. 1923	pl. 143: 15	Staurastrum proboscideum
Pl. 76: 3, 4	Růžička 1972	pl. 64: 17, 18	Staurastrum proboscideum
Pl. 76: 5, 6	Coesel 1997	pl. 13: 1, 2	Staurastrum proboscideum
Pl. 76: 7	Förster 1967	pl. 9: 1	Staurastrum borgeanum

Plate numbers   Author   Original identification

Plate numbers	Author		Original identification
Pl. 76: 8	Schmidle 1898	pl. 3: 7	Staurastrum borgeanum
Pl. 76: 9	Coesel 1997	pl. 13: 3, 6	Staurastrum borgeanum
Pl. 77: 1	Ralfs 1848	pl. 21: 9a/c	Staurastrum margaritaceum
Pl. 77: 2, 3, 4, 5	West & al. 1923	pl. 150: 8, 6, 7, 9	Staurastrum margaritaceum
Pl. 77: 6	Dick 1930	pl. 9: 20/21	Staurastrum margaritaceum
Pl. 77: 7, 8, 9, 10	Coesel 1997	pl. 13: 14, 13, 12, 11	Staurastrum margaritaceum
Pl. 77: 11	Messikommer 1942	pl. 18: 5	Staurastrum subnivale
Pl. 77: 12-15	Coesel 1997	pl. 13: 7-10	Staurastrum borgeanum forma minor
Pl. 77: 16	Lenzenweger 1997	pl. 33: 13	Staurastrum subnivale
Pl. 78: 1	Ralfs 1848	pl. 22: 10	Staurastrum cyrtocerum
Pl. 78: 2	West & al. 1923	pl. 149: 10	Staurastrum cyrtocerum
Pl. 78: 3	Brebisson 1856	fig. 25	Staurastrum inflexum
Pl. 78: 4	Tomaszewicz 1988	pl. 5: 5	Staurastrum inflexum
Pl. 78: 5	West & al. 1923	pl. 143: 7	Staurastrum inflexum
Pl. 78: 6	West & al. 1923	pl. 167: 7	Staurastrum inflexum
Pl. 78: 7	Villeret 1955	pl. 4: 80	Staurastrum inflexum
Pl. 78: 8	Kouwets 1987	pl. 19: 4	Staurastrum inflexum
Pl. 78: 9	Kouwets 1987	pl. 19: 5	Staurastrum brachycerum
Pl. 78: 10	Coesel 1997	pl. 13: 17	Staurastrum inflexum
Pl. 78: 11	Coesel 1997	pl.13: 19/20	Staurastrum inflexum
Pl. 78: 12	Coesel 1997	pl. 13: 18	Staurastrum inflexum var. brachycerum
Pl. 78: 13	Brebisson 1856	fig. 24	Staurastrum brachycerum
Pl. 78: 14	Grönblad 1934	fig. 4: 34/35	Staurastrum brachycerum
Pl. 78: 15	West & al. 1923	pl. 142: 21	Staurastrum brachycerum
Pl. 79: 1	Ralfs 1848	pl. 23: 6	Staurastrum arachne
Pl. 79: 2	West & al. 1923	pl. 150: 1	Staurastrum arachne
Pl. 79: 3	West & al. 1923	pl. 150: 3	Staurastrum arachne var. arachnoides
Pl. 79: 4	Coesel 1997	pl. 16: 17	Staurastrum arachne
Pl. 79: 5	Coesel 1997	pl. 16: 18	Staurastrum arachne var. basiornatum
Pl. 79: 6	West & al. 1923	pl. 150: 2	Staurastrum arachne var. curvatum
Pl. 79: 7	Johnson 1894	pl. 211: 4	Staurastrum gyrans
Pl. 79: 8	Capdevielle & Couté 1980	pl. 1: 8/5/7	Staurastrum arachne var. basiornatum forma pseudogyrans
Pl. 79: 9	Messikommer 1942	pl. 18: 1	Staurastrum arachne var. incurvatum
Pl. 80: 1	Archer 1860	pl. 7: ½	Staurastrum oxyacanthum
Pl. 80: 2	West & al. 1923	pl. 143: 19	Staurastrum oxyacanthum
Pl. 80: 3	Růžička 1972	pl. 64: 13	Staurastrum oxyacanthum
Pl. 80: 4	Homfeld 1929	pl. 9: 100	Staurastrum oxyacanthum
Pl. 80: 5	Lenzenweger 1993	pl. 7: 90	Staurastrum oxyacanthum
Pl. 80: 6, 7	Boldt 1885	pl. 6: 41, 40	Staurastrum oxyacanthum subsp. sibiricum
Pl. 80: 8, 9	Coesel 1997	pl. 18: 6, 5	Staurastrum oxyacanthum var. sibiricum
Pl. 80: 10	Nordstedt 1885	pl. 7: 9	Staurastrum oxyacanthum var. polyacanthum
Pl. 80: 11	West & al. 1923	pl. 143: 20/21/22	Staurastrum oxyacanthum var. polyacanthum
Pl. 80: 12	Růžička 1972	pl. 64: 15	Staurastrum oxyacanthum var. polyacanthum

Plate numbers	Author		Original identification
Pl. 80: 13	Coesel 1997	pl. 18: 14	Staurastrum oxyacanthum var. polyacanthum
Pl. 81: 1, 2, 3	Ralfs 1848	pl. 23: 3b, 3d, 3f	Staurastrum controversum
Pl. 81: 4	West & al. 1923	pl. 154: 3	Staurastrum controversum
Pl. 81: 5	Lenzenweger 1986	pl. 6: 13	Staurastrum controversum
Pl. 81: 6	Lütkemüller 1900 in West & al. 1923	pl. 154: 4	Staurastrum controversum
Pl. 81: 7, 8, 9	Coesel 1997	pl. 18: 7, 8, 9	Staurastrum controversum
Pl. 81: 10	West W. 1892a	fig. 38	Staurastrum vestitum var. semivestitum
Pl. 81: 11	Dubois-Tylski 1969	pl. 2: 23	Staurastrum vestitum
Pl. 81: 12, 13	Coesel 1997	pl. 18: 10, 11	Staurastrum controversum var. semivestitum
Pl. 82: 1	Ralfs 1848	pl. 23: 1a/d	Staurastrum vestitum
Pl. 82: 2	West & al. 1923	pl. 152: 5	Staurastrum vestitum
Pl. 82: 3	Skuja 1934	fig. 106	Staurastrum vestitum
Pl. 82: 4	Lenzenweger 1989c	pl. 5: 54 right	Staurastrum vestitum
Pl. 82: 5	Coesel 1997	pl. 19: 3	Staurastrum vestitum var. vestitum
Pl. 82: 6	Grönblad 1920	pl. 3: 102	Staurastrum vestitum var. splendidum
Pl. 82: 7	Lenzenweger 1989d	fig. 123	Staurastrum vestitum var. splendidum
Pl. 82: 8	Lenzenweger 1989c	pl. 5: 54 left	Staurastrum vestitum var. splendidum
Pl. 82: 9	Kouwets 1989	pl. 19: 15/16	Staurastrum vestitum
Pl. 83: 1, 2, 3	Ralfs 1848	pl. 23: 2a, 2b, 2c	Staurastrum aculeatum
Pl. 83: 4	West & al. 1923	pl. 153: ¾	Staurastrum aculeatum
Pl. 83: 5	Dick 1923	pl. 3: 5a/c/d	Staurastrum aculeatum var. ornatum
Pl. 83: 6	Brook 1959	pl. 8: 6	Staurastrum anatinum forma hirsutum
Pl. 83: 7	Coesel 1997	pl. 19: 15/16	Staurastrum aculeatum
Pl. 84: 1	Cooke 1880	pl. 139: 6	Staurastrum anatinum
Pl. 84: 2	Lind & Brook 1980	fig. 156C	Staurastrum anatinum var. longibrachiatum
Pl. 84: 3	West & al. 1923	pl. 147: 4	Staurastrum anatinum var. lagerheimii
Pl. 84: 4	West & al. 1923	pl. 146: 8	Staurastrum anatinum var. truncatum
Pl. 85: 1	West & al. 1923	pl. 153: 5	Staurastrum vestitum var. subanatinum
Pl. 85: 2	Kouwets 1987	pl. 19: 14	Staurastrum vestitum
Pl. 85: 3	Coesel 1997	pl. 19: 5	Staurastrum vestitum
Pl. 85: 4	West & al. 1923	pl. 147: 5	Staurastrum anatinum var. longibrachiatum
Pl. 86: 1, 2, 3	Capdevielle 1978	pl. 17: 1/1a, 2/2a; pl.18: 1	Staurastrum anatinum var. subfloriferum
Pl. 87: 1	Smith 1924	pl. 75: 23	Staurastrum anatinum var. denticulatum
Pl. 87: 2	Florin 1957	fig. 32: 1a/b	Staurastrum anatinum
Pl. 87: 3	Scharf 1986	fig. 46	Staurastrum anatinum
Pl. 87: 4	Florin 1957	fig. 28: 2	Staurastrum anatinum var. longibrachiatum
Pl. 87: 5, 6, 7	Florin 1957	fig. 33: 11a, 13, 12a	Staurastrum anatinum var. denticulatum
Pl. 87: 8	West & West 1896	pl. 18: 1	Staurastrum floriferum
Pl. 87: 9	Grönblad 1938	fig. 2:3	Staurastrum floriferum
Pl. 87: 10, 11	Florin 1957	fig. 33: 6a/c, 5	Staurastrum floriferum
Pl. 88: 1, 2	Børgesen & Ostenfeld 1903	fig. 148	Staurastrum magdalenae
Pl. 88: 3	Grönblad 1920	pl. 3: 75/76	Staurastrum informe

Plate numbers	Author		Original identification
Pl. 88: 4	Grönblad 1938	fig. 3: 4	Staurastrum informe
Pl. 88: 5	Grönblad 1920	pl. 3: 103/104	Staurastrum subosceolense
Pl. 88: 6	Borge 1939	fig. 11	Staurastrum bullardii var. suecica
Pl. 88: 7	Grönblad 1938	fig. 2: 4	Staurastrum saltator
Pl. 88: 8, 9	Nygaard 1949	fig. 52a/b	Staurastrum pendulum
Pl. 89: 1	Järnefelt 1934	fig. 18	Staurastrum petsamoense
Pl. 89: 2	Thomasson 1957	fig. b1/2	Staurastrum petsamoense
Pl. 89: 3	Lenzenweger 1989	fig. 147	Staurastrum petsamoense
Pl. 90: 1, 2	Ralfs 1848	pl. 23: b/e, a/d	Staurastrum paradoxum
Pl. 90: 3	West & al. 1923	pl. 145: 3	Staurastrum paradoxum
Pl. 90: 4, 5, 6, 7, 8, 9, 10	Coesel 1997	pl. 16: 4, 7, 2, 8, 9, 3, 10	Staurastrum paradoxum
Pl. 90: 11, 12, 13	Coesel 1997	pl. 16: 12, 13, 14	Staurastrum paradoxum var. reductum
Pl. 91: 1	Nordstedt 1888	pl. 4: 9	Staurastrum contortum var. pseudotetracderum
Pl. 91: 2	West & al. 1923	pl. 149: 11	Staurastrum pseudotetracerum
Pl. 91: 3	Dick 1923	pl. 7: 17	Staurastrum contortum var. pseudotetracderum
Pl. 91: 4, 5, 6	Coesel 1996	fig. 21/23, 20, 22	Staurastrum minimum
Pl. 91: 7	Kouwets 2001	pl. 6: 4	Staurastrum minimum
Pl. 91: 8	Lemaire 1890	fig. 2a/b	Staurastrum diacanthum
Pl. 91: 9	Homfeld 1929	pl. 9: 102	Staurastrum paradoxum var. diacanthum
Pl. 91: 10, 11, 12	Kouwets 1987	pl. 20: 8, 10, 13	Staurastrum diacanthum
Pl. 91: 13, 14	Coesel 1997	pl. 17: 1, 2	Staurastrum diacanthum
Pl. 91: 15	Lenzenweger 2000a	pl. 5: 7	Staurastrum acestrophorum var. subgenuinum
Pl. 92: 1	Joshua 1886	pl. 24: 2	Staurastrum platycerum
Pl. 92: 2, 3	Grönblad 1938	fig. 3: 5a/b, 5c	Staurastrum platycerum var. platycerum
Pl. 92: 4	Grönblad 1938	fig. 3: 6	Staurastrum platycerum var. dentatum
Pl. 92: 5, 6, 7, 8	Coesel 1997	pl. 17: 16, 18, 19, 17	Staurastrum micronoides
Pl. 92: 9	West & al. 1923	pl. 149: 6	Staurastrum micron
Pl. 92: 10	Lenzenweger 1991	pl. 10: 20	Staurastrum micron
Pl. 92: 11, 12, 13, 14	Coesel 1997	pl. 17: 12, 9, 10, 8	Staurastrum micron var. micron
Pl. 92: 15	Grönblad 1942	pl. 4: 25	Staurastrum micron, forma
Pl. 92: 16	Coesel 1997	pl. 17: 13, 14	Staurastrum micron var. spinulosum
Pl. 93: 1	Cedercreutz & Grönblad 1936	pl. 2: 40	Staurastrum alandicum
Pl. 93: 2, 3	West & West 1903	pl. 6, 7	Staurastrum paradoxum var. cingulum
Pl. 93: 4, 5	Brook 1959	pl. 7, 2	Staurastrum cingulum var. cingulum
Pl. 94: 1, 2	Smith 1922	pl. 12: 3, 4	Staurastrum cingulum var. obesum
Pl. 94: 3, 4	Nygaard 1949	fig. 39b, 39a	Staurastrum cingulum var. obesum
Pl. 94: 5	Coesel 1997	pl. 15: 7	Staurastrum cingulum var. obesum
Pl. 94: 6, 8	Brook 1959	pl. 14: 8, 7	Staurastrum cingulum var. obesum
Pl. 94: 7	Teiling 1946	fig. 19	Staurastrum thunmarkii
Pl. 94: 9	Coesel & Wardenaar 1990	fig. 1e	Staurastrum cingulum var. obesum
Pl. 95: 1	G.S. West 1909	pl. 3: 12	Staurastrum neglectum
Pl. 95: 2	West & al. 1923	pl. 142: 16	Staurastrum neglectum
Pl. 95: 3	Grönblad 1947	pl. 2: 28	Staurastrum neglectum, forma brachiis divergentibus

Plate numbers	Author		Original identification
Pl. 95: 4, 5	Delponte 1878	pl. 13: 10, 11/7	Staurastrum manfeldtii
Pl. 95: 6	West & al. 1923	pl. 148: 2	Staurastrum manfeldtii
Pl. 95: 7	Lenzenweger 1989d	fig. 141	Staurastrum sebaldi var. ornatum
Pl. 96: 1	Nordstedt 1873	fig. 15	Staurastrum sebaldi var. ornatum
Pl. 96: 2, 3	Coesel 1997	pl. 23 1, 3	Staurastrum manfeldtii var. manfeldtii
Pl. 96: 4	West & al. 1923	pl. 148: 3	Staurastrum manfeldtii var. annulatum
Pl. 96: 5	Coesel 1997	pl. 23: 8	Staurastrum manfeldtii var. annulatum
Pl. 96: 6	Lenzenweger 1989d	fig. 138	Staurastrum messikommeri
Pl. 96: 7	Coesel 1997	pl. 23: 5	Staurastrum manfeldtii var. splendidum
Pl. 97: 1	Messikommer 1928	pl. 9: 14	Staurastrum gracile var. splendidum
Pl. 97: 2	West & al. 1923	pl. 149: 17	Staurastrum sebaldi var. productum
Pl. 97: 3	Coesel 1997	pl. 21: 3	Staurastrum productum
Pl. 97: 4	Wille 1881	pl. 2: 30	Staurastrum pseudosebaldi
Pl. 98: 1	Messikommer 1957	pl. 2: 27	Staurastrum pseudosebaldi var. elongatum
Pl. 98: 2	Lenzenweger 1989d	fig. 134	Staurastrum manfeldtii var. annulatum
Pl. 98: 3	Grönblad 1947	pl. 2: 31	Staurastrum pseudosebaldi, forma
Pl. 98: 4	West & al. 1923	pl. 149: 13	Staurastrum pseudosebaldi var. simplicius
Pl. 98: 5	West & al. 1923	pl. 148: 5	Staurastrum sebaldi
Pl. 98: 6	Lenzenweger 1994	pl. 7: 6	Staurastrum sebaldi
Pl. 98: 7	Dick 1923	pl. 6: 10/13	Staurastrum sebaldi var. ornatum
Pl. 99: 1	Hustedt 1911	fig. 35	Staurastrum traunsteineri
Pl. 99: 2	Dick 1919	pl. 17: 2	Staurastrum traunsteineri
Pl. 99: 3	Lenzenweger 1989a	pl. 8: 5	Staurastrum traunsteineri
Pl. 99: 4	Dick 1923	pl. 6: 15	Staurastrum traunsteineri
Pl. 100: 1	Lundell 1871	pl. 4: 6	Staurastrum cerastes
Pl. 100: 2, 3	West & al. 1923	pl. 151: 1; pl. 150: 16	Staurastrum cerastes
Pl. 100: 4	Børgesen 1889	pl. 6: 9	Staurastrum bicorne
Pl. 100: 5	Förster 1967	pl. 9: 9	Staurastrum bicorne
Pl. 100: 6	West & al. 1923 after Hauptfleisch 1888	pl. 143: 17	Staurastrum bicorne
Pl. 100: 7	Lenzenweger 1967	pl. 5: 2	Staurastrum bicorne
Pl. 101: 1	West & al. 1923	pl. 148: 1	Staurastrum duacense
Pl. 101: 2	Grönblad 1945	pl. 1: 20-22	Staurastrum duacense, forma plus tumida
Pl. 101: 3	Boldt 1885	pl. 6: 36	Staurastrum pseudosebaldi var. bicorne
Pl. 101: 4	Eichler 1896	pl. 4: 51	Staurastrum grallatorium var. miedzyrzecence
Pl. 101: 5	Kouwets 2002	archives	Staurastrum saltans var. miedzyrzecense
Pl. 101: 6, 7	Cedercreutz & Grönblad 1936	pl. 1: 19, 18	Staurastrum saltans var. miedzyrzecense
Pl. 102: 1	W. West 1892	pl. 23: 11	Staurastrum gracile subsp. bulbosum
Pl. 102: 2	W. & G.S. West 1895	pl. 9: 2	Staurastrum gracile var. cyathiforme
Pl. 102: 3	West & al. 1923	pl. 144: 12	Staurastrum gracile var. cyathiforme
Pl. 102: 4, 5	Coesel 1997	pl. 22: 3, 2	Staurastrum bulbosum var. cyathiforme
Pl. 103: 1	W. & G.S. West 1896	pl. 17: 16	Staurastrum johnsonii
Pl. 103: 2	Strøm 1926	pl. 6: 11	Staurastrum johnsonii var. perpendiculatum
Pl. 103: 3	Skuja 1948	pl. 20: 1-3	Staurastrum johnsonii var. perpendiculatum

Plate numbers	Author		Original identification
Pl. 103: 4	Grönblad 1920	pl. 2: 33	Staurastrum johnsonii var. perpendiculatum
Pl. 103: 5	Teiling 1942	fig. 4	Staurastrum pingue
Pl. 103: 6, 7	Förster 1967a	pl. 10: 1/3, 11	Staurastrum pingue
Pl. 104: 1	Messikommer 1942	p. 175 textfig. 2	Staurastrum luetkemuelleri
Pl. 104: 2, 3	Coesel 1997	pl. 25: 3, 5	Staurastrum pingue
Pl. 104: 4	Teiling 1946	fig. 30	Staurastrum planctonicum
Pl. 104: 5	Teiling 1946	fig. 31	Staurastrum planctonicum var. bullosum
Pl. 104: 6, 7, 8, 9	Coesel 1997	pl. 24: 5, 3, 4, 2	Staurastrum planctonicum
Pl. 105: 1	Coesel 1997	pl. 24: 1	Staurastrum planctonicum
Pl. 105: 2	Grönblad 1938	pl. 3: 3	Staurastrum dorsidentiferum var. ornatum
Pl. 105: 3, 4	Florin 1957	fig. 35: 2, fig. 27: 7	Staurastrum planctonicum var. ornatum
Pl. 105: 5	Messikommer 1927a	pl. 2: 15	Staurastrum dimazum var. reductum
Pl. 105: 6	Lenzenweger 2000	pl. 2: 4	Staurastrum simplicius
Pl. 105: 7	Coesel 1997	pl. 22: 4	Staurastrum simplicius
Pl. 106: 1	Lütkemüller 1910	pl. 3: 16-18	Staurastrum natator subsp. dimazum
Pl. 106: 2	Skuja 1948	pl. 20: 7	Staurastrum bioculatum, forma
Pl. 106: 3	Coesel 1997	pl. 22: 6	Staurastrum dimazum
Pl. 106: 4	Smith 1924	p. 106 fig. 12	Staurastrum natator
Pl. 106: 5	West & al. 1923	pl. 147: 7	Staurastrum natator
Pl. 106: 6	Grönblad 1920	pl. 1: 16	Staurastrum natator
Pl. 106: 7	Nordstedt 1869	pl. 4: 57	Staurastrum leptocladum
Pl. 106: 8	Teiling 1944	fig. 160	Staurastrum leptocladum
Pl. 106: 9	Florin 1957	fig. 34: 1	Staurastrum leptocladum var. insigne
Pl. 107: 1	Wille 1884	pl. 1: 39	Staurastrum leptocladum var. cornutum
Pl. 107: 2	Grönblad 1926	pl. 3: 110	Staurastrum leptocladum var. cornutum
Pl. 107: 3	Skuja 1934	fig. 116	Staurastrum leptocladum var. cornutum
Pl. 107: 4	Nordstedt 1873	fig. 17	Staurastrum paradoxum var. longipes
Pl. 107: 5, 6	West & al. 1923	pl. 146: 2, 3	Staurastrum paradoxum var. longipes
Pl. 108: 1, 2	Brook 1959	pl. 16: 4, 3	Staurastrum longipes
Pl. 108: 3	Brook 1958	fig. 51,	Staurastrum longipes
Pl. 108: 4	Brook 1958	fig. 52	Staurastrum longipes var. contractum
Pl. 108: 5	Teiling 1946	fig. 37	Staurastrum longipes var. contractum
Pl. 108: 6, 7	Brook 1959	pl. 16: 6, 5	Staurastrum longipes var. contractum
Pl. 109: 1	Grönblad 1926	pl. 3: 113/114	Staurastrum multinodulosum
Pl. 109: 2, 3	Coesel & Alfinito 2006	pl. fig. 1/3, 2	Staurastrum multinodulosum
Pl. 109: 4	Smith 1924	pl. 74: 19/20	Staurastrum bullardii
Pl. 109: 5, 6	Skuja 1948	pl. 20: 5, 4	Staurastrum bullardii
Pl. 109: 7, 8	Teiling 1942a	figs 6, 7	Staurastrum bullardii var. alandicum
Pl. 110: 1	Schröder 1898	fig. a/b	Staurastrum polymorphum var. chaetoceras
Pl. 110: 2, 3, 4, 5, 6	Coesel 1997	pl. 26: 6, 2, 4, 5, 7	Staurastrum chaetoceras
Pl. 110: 7, 8, 9	Brook 1959	pl. 15: 4, 5, 6	Staurastrum chaetoceras
Pl. 111: 1	Grönblad 1938	fig. 3: 1	Staurastrum subboergenesii
Pl. 111: 2, 3, 4	Skuja 1964	pl. 53: 13, 9/10, 11/12	Staurastrum subboergenesii
Pl. 111: 5	Borge 1913	pl. 3: 38	Staurastrum ornithopodon var. bifurcatum

Plate numbers	Author		Original identification
Pl. 111: 6, 7	Grönblad 1938	fig. 1: 9	Staurastrum levanderi
Pl. 111: 8	Capdevielle 1982	pl. 4: 6	Staurastrum tectum var. ayayense forma nana
Pl. 111: 9, 10, 11	Coesel & Joosten 1996	figs 9/10, 7, 8	Staurastrum hollandicum
Pl. 111: 12, 13	Nygaard 1949	fig. 49	Staurastrum iversenii
Pl. 112: 1, 2	Coesel & Joosten 1996	figs 3, 1	Staurastrum bloklandiae
Pl. 112: 3	Lenzenweger 2003	pl. 5: 3	Staurastrum bloklandiae
Pl. 112: 4, 5, 6	Lenzenweger 1999	fig. 3	Staurastrum octodontum var. tetrodontum forma torta
Pl. 112: 7, 8, 9	Nygaard 1949	fig. 40a-d	Staurastrum smithii var verrucosum
Pl. 112: 10, 11	Smith 1924	pl. 76: 17, 18/19	Staurastrum contortum
Pl. 112: 12	Florin 1957	fig. 29: 1a	Staurastrum smithii
Pl. 112: 13, 14	Coesel 1997	pl. 27: 8, 6	Staurastrum smithii
Pl. 113: 1, 2, 3	Grönblad & Scott 1955	pl. 1: 1, 3/4, 2	Staurastrum bibrachiatum
Pl. 113: 4-6	Lenzenweger 2001	pl. 3: 22	Staurastrum bibrachiatum
Pl. 113: 7	Ralfs 1948	pl. 23: 7c/7d	Staurastrum tetracerum
Pl. 113: 8, 9, 10	West & al. 1923	pl. 149: 2, 4, 3	Staurastrum tetracerum
Pl. 113: 11, 12	Kouwets 1987	pl. 20: 22, 21	Staurastrum tetracerum var. tetracerum
Pl. 113: 13	Brook 1982	fig. 1a	Staurastrum tetracerum
Pl. 113: 14, 15	Růžička 1972	pl. 60: 47, 48	Staurastrum tetracerum
Pl. 113: 16, 17	Coesel 1997	pl. 27: 20, 21	Staurastrum tetracerum var. tetracerum
Pl. 113: 18	Grönblad 1921	pl. 5: 31/32	Staurastrum biverruciferum
Pl. 113: 19, 20	West & al. 1923	pl. 149: 7	Staurastrum irregulare
Pl. 113: 21, 22	Coesel 1997	pl. 27: 13/14, 12	Staurastrum tetracerum var. irregulare
Pl. 113: 23, 24, 25	Kouwets 1987	pl. 20: 23, 24, 25/26	Staurastrum tetracerum var. irregulare
Pl. 113: 26-27	Florin 1957	fig. 29: 5a, 6/7	Staurastrum tetracerum var. cameloides
Pl. 113: 28	Grönblad 1960	pl. 8: 178	Staurastrum tetracerum var. cameloides
Pl. 113: 29	Coesel 1997	pl. 27: 15	Staurastrum tetracerum var. excavatum
Pl. 113: 30	Georgevitch 1910	text fig. 6	Staurastrum paradoxum var. osceolense forma biradiata
Pl. 114: 1	Allorge & Allorge 1931	pl. 14: 11	Staurastrum excavatum
Pl. 114: 2	Grönblad 1921	pl. 5: 28/29	Staurastrum tetracerum var. subexcavatum
Pl. 114: 3	Bohlin 1901	p. 56 fig. 15	Staurastrum chavesii
Pl. 114: 4	Coesel 1997	pl. 17: 6	Staurastrum chavesii
Pl. 114: 5	West & West 1902	pl. 1: 20	Staurastrum latiusculum
Pl. 114: 6	West & al. 1923	pl. 149: 8	Staurastrum latiusculum
Pl. 114: 7	Krieger 1932	pl. 16: 20	Staurastrum dentatum
Pl. 114: 8	Lenzenweger 1997	pl. 31: 10	Staurastrum dentatum
Pl. 114: 9	Nordstedt 1873	fig. 11	Staurastrum inconspicuum
Pl. 114: 10	Tomaszewicz & Kowalski 1993	fig. 72	Staurastrum inconspicuum
Pl. 114: 11, 12	Coesel 1997	pl. 12: 6, 5	Staurastrum inconspicuum
Pl. 114: 13, 14	Kouwets 1987	pl. 19: 18, 18	Staurastrum inconspicuum
Pl. 114: 15	Lütkemüller 1900	pl. 1: 54	Staurastrum inconspicuum
Pl. 114: 16	Gay 1884	pl. 2: 10	Staurastrum inconspicuum var. crassum
Pl. 114: 17	Børgesen 1901	pl. 8: 4	Staurastrum inconspicuum
Pl. 114: 18	Grönblad 1921	pl. 5: 30	Stauratrum uhtuense

Plate numbers	Author		Original identification
Pl. 114: 19	Grönblad 1938	fig. 2: 1	Stauratrum uhtuense
Pl. 115: 7	John & Williamson 2009	pl. 38: F	Staurastrum verticillatum
Pl. 116: 1	W. West 1892a	pl. 23: 15	Staurastrum archeri
Pl. 116: 2	West & al. 1923	pl. 153: 6	Staurastrum archeri
Pl. 116: 3	West & al. 1923	pl. 152: 3 / 4	Staurastrum ophiura
Pl. 117: 1	Lundell 1871	pl. 4: 7	Staurastrum ophiura
Pl. 117: 2	Donat 1926	pl. 1: 2	Staurastrum ophiura
Pl. 117: 3	Borge 1939	fig. 14	Staurastrum ophiura var. subcylindricum
Pl. 118: 1	Lundell 1871	pl. 4: 9a/b	Staurastrum sexangulare
Pl. 118: 2, 3	West & al. 1923	pl. 157: 4, 2	Staurastrum sexangulare
Pl. 118: 4	Williamson 1992	fig. 24: 2	Staurastrum sexangulare
Pl. 119: 1	Bailey 1841	pl. 1: 15	Staurastrum arctiscon
Pl. 119: 2	West & al. 1923	pl. 157: 5	Staurastrum arctiscon
Pl. 119: 3	Heimans 1960	fig. 8	Staurastrum arctiscon
Pl. 119: 4	Ralfs 1848	pl. 33: 12b/d	Didymocladon furcigerus
Pl. 120: 1	West & al. 1923	pl. 156: 9	Staurastrum furcigerum var. reductum
Pl. 120: 2	West & al. 1923	pl. 156: 8	Staurastrum furcigerum
Pl. 120: 3	Coesel 1997	pl. 11: 3	Staurastrum furcigerum
Pl. 120: 4	Coesel 2002	archives	Staurastrum furcigerum
Pl. 120: 5	Brook 1958	figs 81	Staurastrum furcigerum
Pl. 120: 6	Coesel & Meesters 2002	fig. 2B	Staurastrum furcigerum forma armigera
Pl. 120: 7	Eichler & Gutwiński 1859	pl. 5: 50	Staurastrum pseudopisciforme
Pl. 120: 8	Grönblad 1920	pl. 1: 25/26	Staurastrum pseudopisciforme var. dimidio-minor

References

Allorge, V. & P. Allorge, 1931. Hétérocontes, Euchlorophycées et Conjuguées de Galice. — Rev. Algol. 5: 327-382.

Andersen, S., M. Heldal & G. Knutsen, 1987. Cell wall structure and iron distribution in the green alga *Staurastrum luetkemuelleri* (Desmidiaceae). — J. Phycol. 23: 669-672.

Andersson, O.F., 1890. Bidrag till kännedomen om Sveriges Chlorophyllophyceer. I. Chlorophyllophyceer från Roslagen. — Bih. Kongl. Svenska Vetensk.-Akad. Handl. 16, Afd. III (5): 1-20, 1 pl.

Archer, W., 1858. Supplementary catalogue of Desmidiaceae found in the neighbourhood of Dublin, with description and figures of a proposed new genus and of four new species. — Nat. Hist. Rev. 5: 234-263.

—, 1860. Description of two new species of *Staurastrum*. — Quart. J. Microscop. Sci. 8: 75-79. pl. 7.

—, 1861. Sub-group Desmidieae or Desmidiaceae. — In: A. Pritchard: A history of Infusoria including the Desmidiaceae and Diatomaceae, British and foreign. Ed. 4, p. 715-752. — Whittaker and Co., London, 968 pp.

—, 1862. Description of new species of *Cosmarium* (Corda), of *Staurastrum* (Meyen), of two new species of *Closterium* (Nitzsch), and of *Spirotaenia* (Bréb.). — Quart. J. Microsc. Sci., n.s. 2: 247-255, pl. 12.

—, 1866. *Staurastrum oligacanthum* and *Staurastrum cristatum* exhibited. — Quart. J. Microsc. Sci., n.s. 6: 67, 189.

—, 1869. A new *Staurastrum* from Galway (*S. maamense*). — Quart. J. Microscop. Sci., n.s. 9: 200.

—, 1869a. A fine new *Staurastrum* from Connemara (*S. verticillatum*). — Quart. J. Microsc. Sci., n.s. 9: 196.

—, 1878. Zygospore of *Staurastrum turgescens* De Not. — Quart. J. Microsc. Sci., n.s. 18: 105-106.

Bailey, J.W., 1841. A sketch of the infusoria, of the family Bacillaria, with some account of the most interesting species which have been found in recent or fossil state in the United States. — Amer. J. Sci. Arts 41: 284-305.

—, 1851. Microscopical observations made in South Carolina, Georgia and Florida. — Smithsonian Contr. Knowl. 11 (Art. 8): 1-48, pls 1-3.

Barker, J., 1869. On a supposed new *Staurastrum* (*S. elongatum* Bark.). Quart. J. Microscop. Sci., n.s. 9: 424.

Beck-Mannagetta, G., 1926. Algenfunde im Riesengebirge. — Vestn. Kral. Ceske Spolecn. Nauk. Tr. Mat.-Prir. 1926 (10): 1-18.

—, 1926a. Neue Grünalgen aus Kärnten. — Arch. Protistenk. 55: 173-183.

—, 1931. Die Algen Kärntens. Erste Grundlagen einer Algenflora von Kärnten. — Beih. Bot. Centralbl. 47(2): 211-342.

Bennett, A.W., 1886. Fresh-water algae (including Chlorophyllaceous Protophytes) of the English Lake District; with descriptions of twelve new species (I). — J. Roy. Microsc. Soc. (London) II, 6: 1-15, pls 1, 2.

Bergh, R.S. & C.F. Løtken, 1887. Dijmphna-Togtets zoologisk-botaniske udbytte. — H. Hagerup, Kjøbenhavn, 515 pp, 49 pls.

Bernard, C., 1908. Protococcacées et Desmidiées d'eau douce recoltées à Java et décrites par Ch. Bernard, Dr. és Sciences. — Dept. de l'Agric. aux Indes Néerl., Batavia, 230 pp, 8 pls.

—, 1909. Sur quelques algues unicellulaires d'eau douce récoltées dans le Domaine Malais. — Dept. de l'Agric. aux Indes-Néerl., Buitenzorg, 94 pp, 6 pls.

Bicudo, C.E.M., 1967. Two new varieties of *Staurastrum* (Desmidiaceae) from Sao Paulo. — J. Phycol. 3: 55-56.

— & I. Ungaretti, 1986. Desmids (Zygnemaphyceae) from the Aguas Belas Reservoir, State of Rio Grande do Sul, Brazil. — Revista Brasil. Biol. 46: 285-308.

Bisset, J.P., 1884. List of Desmideae found in gatherings made in the neighbourhood of Lake Windermere during 1883. — J. Roy. Microsc. Soc. (London), ser. 2, 4: 192-197.

Bohlin, K., 1901. Étude sur la flora algologique d'eau douce des Açores. — Bih. Kongl. Svenska Vetensk.-Akad. Handl. 27, afd. 3(4): 1-85, pl. 1.

Boldt, R., 1885. Bidrag till kännedomen om Sibiriens Chlorophyllophycéer. — Öfvers. Förh. Kongl. Svenska Vetensk.-Akad. 1885, no.2: 91-128, pls 5, 6.

—, 1888. Studier öfver sötvattensalger och deras utbredning. II. Desmidieer fran Grönland. — Bih. Kongl. Svenska Vetensk.-Akad. Handl. 13 (5): 1-48, pls 1, 2.

Borge, O., 1891. Ett litet bidrag till Sibiriens Chlorophyllophycé-flora. — Bih. Kongl. Svenska Vetensk.-Akad. Handl. 17 (2): 1-16, pl. 1.

—, 1892. Chlorophyllophyceer från Norska Finmarken. — Bih. Kongl. Svenska Vetensk.-Akad. Handl. 17 (4): 1-15.

—, 1894. Süsswasser-Chlorophyceen gesammelt von Dr. A. Osw. Kihlman im nördlichsten Russland, Gouvernement Archangel. — Bih. Kongl. Svenska Vetensk.-Akad. Handl. 19 (5): 1-41, pls 1-3.

—, 1896. Australische Süsswasserchlorophyceen. — Bih. Kongl. Svenska Vetensk.-Akad. Handl. 22 (9): 1-32, pls 1-4.

—, 1897. Algologiska Notiser. 3-4. — Bot. Not. 1897: 210-215.

—, 1901. Süsswasseralgen aus Süd-Patagonien. — Bih. Kongl. Svenska Vetensk.-Akad. Handl. 27 (10): 1-40, pls 1, 2.

—, 1903. Die Algen der ersten Regnellschen Expedition. II. Desmidiaceen. — Ark. Bot. 1: 71-138, pls 1-5.

—, 1906. Beiträge zur Algenflora von Schweden. — Ark. Bot. 6 (1): 1-88, pls 1-3.

—, 1913. Beiträge zur Algenflora von Schweden. 2. Die Algenflora um den Torne-Träsksee in Schwedisch-Lappland. — Bot. Not. 1913: 1-32, 49-64, 97-110.

—, 1918. Die von Dr. A. Löfgren in Sao Paulo gesammelten Süsswasseralgen. — Ark. Bot. 15(13): 1-108.

—, 1925. Die von Dr. F.C. Hoehne während der Expedition Roosevelt-Rondon gesammelten Süsswasseralgen. — Ark. Bot. 19 (17): 1-56, pls 1-6.

—, 1930. Beiträge zur Algenflora von Schweden. 4. — Ark. Bot. 23 (2): 1-64.

—, 1939. Beiträge zur Algenflora von Schweden. 6. — Ark. Bot. 29A (16): 1-26.

Børgesen, F., 1889. Et lille Bidrag til Bornholms Desmidié-Flora. — Bot. Tidsskr. 17: 141-152, pl. 6.

—, 1890. Desmidiaceae. — In: E. Warming (ed.): Symbolae ad floram Brasiliae centralis cognoscendam. Particula 34. — Vidensk. Meddel. Naturh. Forening Kjøbenhavn 1890: 929-958, pls 2-5.

—, 1894. Ferskvandsalger fra Østgrønland. — Meddel. Grønland 18: 1-41, pls 1, 2.

—, 1901. Freshwater algae of the Faeröes. — In: Botany of the Faeröes. Part I: 198-259 — Det Nordiske Forlag, København, pp 1-338, pls 1-10.

— & C.H. Ostenfeld, 1903. Phytoplankton of lakes in the Faeröes. — In: Botany of the Faeröes. Part II: 613-624 — Det Nordiske Forlag, København, pp 339-681, pls 11-12.

Bourrelly, P., 1966. Les algues d'eau douce. I. Les algues vertes. — Boubée & Cie, Paris, 511 pp.

—, 1987. Algues d'eau douce des mares d'alpage de la région de Lunz am See, Autriche. — Bibliotheca Phycologica 76, J. Cramer, Berlin/Stuttgart, 182 pp.

— & A. Couté, 1991. Desmidiées de Madagascar (Chlorophyta, Zygophyceae). — Bibliotheca Phycologica 86, J. Cramer, Berlin/Stuttgart, 349 pp.

Brébisson, A. de, 1856. Liste des Desmidiées observées en Basse-Normandie. — Mém. Soc. Sci. Nat. Cherbourg 4: 113-166, 301-304, pls 1, 2.

— & P. Godey, 1835. Algues des environs de Falaise décrites et dessinnées par... — Mém. Soc. Acad. Sci. Falaise 1835: 1-62, 266-269.

Brook, A.J., 1957. Notes on desmids of the genus *Staurastrum*. I. — Naturalist (London) 1957: 97-100.

—, 1958. Desmids from the plankton of some Irish loughs. — Proc. Roy. Irish Acad. 59B: 71-91.

—, 1958a. Notes on desmids of the genus *Staurastrum*. II. *Staurastrum leptodermum, S. longispinum, S. brasiliense, S. setigerum, S. clevei* and *S. tohopekaligense* var. *trifurcatum*. — Naturalist (London) 1958: 91-95.

—, 1959. "*Staurastrum paradoxum*" Meyen and "*S. gracile*" Ralfs in the British freshwater plankton and a revision of the "*S. anatinum*"-group of radiate desmids. — Trans. Roy. Soc. Edinb. 63 (part III, no. 26): 589-628.

—, 1959a. Notes on desmids of the genus *Staurastrum*. III. *Staurastrum paradoxum* Meyen in the Jenner herbarium of the British Museum. — Naturalist (London) 1959: 81-83.

—, 1959b. The published figures of the desmid *Staurastrum paradoxum*. — Rev. Algol. 4: 239-255.

—, 1959c. De Brébisson's determinations of *Staurastrum paradoxum* Meyen and *S. gracile* Ralfs. — Nova Hedwigia 1: 163-166.

—, 1959d. *Staurastrum pendulum* var. *pinguiforme* Croasdale, S. *minor* West f. *major* f. nov., fac. quadrata and S. *micron* var. *perpendiculatum* (Grönblad) nov. comb., desmids new to the British freshwater plankton. — Nova Hedwigia 1: 157-162, pls 25-27.

—, 1967. Possible type material of *Staurastrum avicula*. — Nova Hedwigia 14: 107-110.

—, 1981. The Biology of Desmids. — Blackwell Scientific Publications, Oxford, 276 pp.

—, 1982. Desmids of the *Staurastrum tetracerum*-group from a eutrophic lake in mid-Wales. — Br. phycol. J. 17: 259-274.

— & D.B. Williamson, 1983. On *Staurastrum botrophilum* Wolle, a rare and inadequately described desmid. — Br. phycol. J. 18: 69-72.

— & D.B. Williamson, 1990. *Actinotaenium habeebense* (Irénée-Marie) nov. comb., a rare, drought-resistant desmid. — Br. phycol. J. 25: 321-327.

Brummitt, R.K. & C.E. Powell, 1992. Authors of Plant Names. — Royal Botanic Gardens, Kew, 732 pp.

Bulnheim, A., 1861. Beiträge zur Flora der Desmidieen Sachsens. I. — Hedwigia 2: 50-52, pl. 9.

Capdevielle, P., 1978. Recherches écologiques et systématiques sur le phytoplancton du lac de Cazaux-Sanguinet-Biscarosse. — Ann. Station Biologique Besse-en-Chandesse 12: 304 pp, 36 pls.

—, 1982. Algues d'eau douce rares ou nouvelles pour la flora de France. — Cryptogamie, Algol. 3: 211-225.

—, 1985. Observation dans la région des Landes d'algues d'eau douce rares ou nouvelles pour la flore de France. — Cryptogamie, Algol. 6: 141-170.

— & A. Couté, 1980. Quelques *Staurastrum* Meyen (Chlorophycées, Desmidiacées) rares ou nouveaux pour la France. — Nova Hedwigia 33: 859-872.

Cedercreutz, C., 1932. Süsswasseralgen aus Petsamo. II. — Memoranda Soc. Fauna Fl. Fenn. 7: 236-248.

— & R. Grönblad, 1936. Bemerkungen über einige Desmidiaceen von Åland. — Commentat. Biol. 7: 1-9, pls 1, 2.

Cedergren, G.R., 1926. Beiträge zur Kenntnis der Süsswasseralgen in Schweden II. Die Algen aus Bergslagen und Wästerdalarne. — Bot. Not. 1926: 289-319.

—, 1932. Die Algenflora der Provinz Härjedalen. — Ark. Bot. 25A (4): 1-109, pls 1-4.

Claassen, M.I., 1961. A contribution to our knowledge of the freshwater algae of the Transvaal Province. — Bothalia 7: 559-666.

Cleve, P.T., 1864. Bidrag till kännedomenom Sveriges sötvattensalger af familjen Desmidieae. — Öfvers. Förh. Kongl. Svenska Vetensk.-Akad. 20 (10): 481-498, pl. 4.

Coesel, P.F.M., 1987. Taxonomic notes on Colombian desmids. — Cryptogamie, Algol. 8: 127-142.

—, 1992. The *Staurastrum manfeldtii* complex (Chlorophyta, Desmidiaceae): morphological variability and taxonomic implications. — Algol. Studies 67: 69-83.

—, 1993. Taxonomic notes on Dutch desmids II. — Cryptogamie, Algol. 14: 105-114.

—, 1994. De Desmidiaceeën van Nederland. Deel 5. — KNNV Uitgeverij, Utrecht, 52 pp.

—, 1996. Taxonomic notes on Dutch desmids III. — Cryptogamie, Algol. 17: 19-34.

—, 1996a. The Dutch representatives of the *Staurastrum manfeldtii* complex (Desmidiaceae, Chlorophyta): a taxonomic revision. — Nord. J. Bot. 16: 99-106.

—, 1996b. Biogeography of desmids. — Hydrobiologia 336: 41-53.

—, 1997. De Desmidiaceeën van Nederland. Deel 6. — KNNV Uitgeverij, Utrecht, 93 pp.

—, 1998. Desmids from mountain pools in Sierra de Gredos (Central Spain), biogeographical aspects. — Biologia (Bratislava) 53: 437-443.

—, 2004. Some new and otherwise interesting desmid species from Kakadu National Park (Northern Australia). — Quekett J. Microsc. 39: 779-782.

—, 2007. Taxonomic notes on Dutch desmids IV: new species, new names, new combinations. — Syst. Geogr. Pl. 77: 5-14.

— & E.M.G. Hoogendijk, 1975. Bijdragen tot de kennis der Nederlandse Desmidiaceeënflora. 2. — Gorteria 7: 123-128.

— & A.M.T. Joosten, 1996. Three new planktic *Staurastrum* taxa (Chlorophyta, Desmidiaceae) from eutrophic water bodies and the significance of microspecies in desmid taxonomy. — Algol. Studies 80: 9-20.

— & L. Krienitz, 2008. Diversity and geographical distribution of desmids and other coccoid green algae. — Biodivers. Conserv. 17: 381-392.

— & J. Meesters, 2002. Signalen van sieralgen. — Natura 99: 68-70.

— & J. Meesters, 2007. Desmids of the Lowlands. — KNNV Publishing, Zeist, 351 pp.

Comère, J., 1901. Les Desmidiées de France. — Paul Klincksieck, Paris, 224 pp, 16 pls,

Compère, P., 1967. Algues du Sahara et de la région du lac Tchad. — Bull. Jard. Bot. Belg. 37: 109-288.

—, 1976. Observations taxonomiques et nomenclaturales sur quelques Desmidiées (Chlorophycophyta) de la région du lac Tchad (Afrique centrale). — Bull. Jard. Bot. Belg. 46: 455-470.

—, 1976a. The typification of the genus name *Arthrodesmus* (Algae-Chlorophyta). — Taxon 25: 359-360.

—, 1977. *Staurodesmus* Teiling (Desmidiacées). Typification du genre et combinaisons nouvelles. — Bull. Jard. Bot. Belg. 47: 262-265.

—, 1983. Some algae from Kashmir and Ladakh, W. Himalayas. — Bull. Soc. Roy. Bot. Belgique 116: 141-160.

Cooke, M.C., 1880. British species of *Spirulina* and *Staurastrum*. — Grevillea 9: 44-45, pl. 139.

—, 1881. Notes on British desmids. — Grevillea 9: 89-92, pls 140, 141.

—, 1887. British Desmids. A Supplement to British Freshwater Algae. —Williams & Norgate, London, 205 pp, 66 pls.

Couté, A. & G. Rousselin, 1975. Contribution à l'étude des algues d'eau douce du moyen Niger (Mali). — Bull. Mus. Hist. Nat. (Paris), sér. 3, no 277, Bot. 21: 73-175.

Croasdale, H., 1957. Freshwater algae of Alaska 1. Some desmids from the interior. Part 3. Cosmariae concluded. — Trans. Amer. Microscop. Soc. 76: 116-158.

—, 1958. Freshwater algae of Alaska 2. Some new forms from the plankton of Karluk Lake. — Trans. Amer. Microscop. Soc. 77: 31-35.

—, 1962. Freshwater algae of Alaska 3. Desmids from the Cape Thompson Area. — Trans. Amer. Microscop. Soc. 81: 12-42.

—, 1965. Desmids of Devon Island, N.W.T., Canada. — Trans. Amer. Microscop. Soc. 84: 301-335.

— & R. Grönblad, 1964. Desmids of Labrador, 1. Desmids of the southeastern coastal area. — Trans. Amer. Microscop. Soc. 83: 142-212.

— , E.A. Flint & M.M. Racine, 1994. Flora of New Zealand: freshwater algae, Chlorophyta, desmids. Vol. III. — Manaaki Whenua Press, Lincoln, Canterbury, New Zealand, pp 1-218, pls 62-146.

Cushman, J.A., 1904. Desmids from southwestern Colorado. — Bull. Torrey Bot. Club 31: 161-164, pl. 7.

—, 1905. Notes on the zygospores of certain New England desmids with descriptions of a few new species. — Bull. Torrey Bot. Club 32: 223-229, pls 7, 8.

Deflandre, G., 1926. Contributions à la flore algologique de France. I. — Bull. Soc. Bot. France 73: 987-999.

Delponte, J. B., 1878. Specimen Desmidiacearum subalpinarum. — Mem. Reale Accad. Sci. Torino, ser. 2, 30: 1-186, pls 7-23.

De Notaris, G., 1867. Elementi per lo studio delle Desmidiacee Italiche: Desmidiacee della Val Intrasca. — Genova, 84 pp, 9 pls.

De Toni, G.B., 1889. Sylloge algarum omnium hucusque cognitarum. Vol. 1. Sylloge Chlorophycearum. — Patavii, 1315 pp.

De Wildeman, É., 1897. Les algues du Limbourg. — Ann. Soc. Belge Microscop. 21: 42-68.

Dick, J., 1919. Beiträge zur Kenntnis der Desmidiaceenflora von Südbayern. — Kryptog. Forsch. 1: 230-262.

—, 1923. Beiträge zur Kenntnis der Desmidiaceen-Flora von Süd-Bayern. II. — Bot. Arch. 3: 214-236.

—, 1926. Beiträge zur Kenntnis der Desmidiaceen-Flora von Süd-Bayern III. Folge: Oberschwaben (Bayr. Allgäu). — Kryptog. Forsch. 1 (7): 444-454, pls 18-21.

Donat, A., 1926. Zur Kenntnis der Desmidiaceen des norddeutschen Flachlandes. — Pflanzenforschung 5: 1-51.

Dubois-Tylski, T., 1969. Florule algologique d'un marais d'Ardenne. — Rev. Algol. n.s. 9: 316-325.

Ducellier, F., 1918. Contribution à l'étude de la flore desmidiologique de la Suisse, 2. — Bull. Soc. Bot. Genève, II, 10: 85-154.

Ehrenberg, C.G., 1834. Dritter Beitrag zur Erkentniss grosser Organisation in der Richting des kleinsten Raumes. — Abh. Königl. Akad. Wiss. Berlin 1833: 145-336, pls. 1-6.

—, 1838. Die Infusionsthierchen als vollkommene Organismen. — Leopold Voss, Leipzig, 548 pp, 64 pls.

—, 1839. Über das im Jahre 1686 in Curland vom Himmel gefallene Meteorpapier und über dessen Zusammensetzung aus Conferven und Infusorien. — Abh. Königl. Akad. Wiss. Berlin 1838: 45-58, pls 1, 2.

—, 1843. Verbreitung und Einfluss des mikroskopischen Lebens in Süd- und Nordamerika. — Abh. Königl. Akad. Wiss. Berlin 1841: 291-415, pls 1-4.

Eichler, B., 1896. Materyaly do flory wodorostów okolic Miedzyrzecza. —Pamietn. Fizyogr. 16: 119-136, pls 2-4.

— & R. Gutwiński, 1895. De nonnulis specibus algarum novarum. — Rozpr. Akad. Umiejetn. Wydz. Mat.-Przyr. 28: 162-178, pls 4, 5.

— & M. Raciborski, 1893. Nowe gatunki zielenic. — Rozpr. Akad. Umiejetn. Wydz. Mat.-Przyr. 26: 116-126.

Evens, F., 1849. Le plancton du lac Moero et de la région d'Elisabethville. — Rev. Zool. Bot. Africaines 42: 1-64.

Florin, M.-B., 1957. Plankton of fresh and brackish waters in the Södertälje area. — Acta Phytogeogr. Suecic. 37: 1-144.

Förster, K., 1963. Liste der Desmidiaceen der Torne-Lappmark (südl. des Torneträsk) mit Beschreibung neuer Desmidiaceen. — Naturwiss. Mitt. Kempten-Allg. 7: 47-56.

—, 1963a. Desmidiaceen aus Brasilien I. Nord Brasilien. — Rev. Algol. 7: 38-92.

—, 1964. Desmidiaceen aus Brasilien. 2. Teil: Bahia, Piauhy und Nord-Brasilien. — Hydrobiologia 23: 321-505.

—, 1965. Beitrag zur Kenntnis der Desmidiaceen-Flora von Nepal. — Ergebn. Forsch.-Unternehmen Nepal, Khumbu Himal. (Berlin, Springer-Verlag): 25-58.

—, 1967. Beitrag zur Desmidiaceen-Flora der Torne-Lappmark in Schwedisch-Lappland. — Ark. Bot. 6: 109-161, pls 1-12.

—, 1967a. *Staurastrum pingue* Teiling und einige andere *Staurastren* aus dem Titisee (Schwarzwald). — Arch. Hydrobiol. Suppl. 33: 121-126.

—, 1969. Amazonische Desmidieen. 1. Teil: Areal Santarém. — Amazoniana 2: 5-116, pls 1-56.

—, 1970. Beitrag zur Desmidieenflora von Süd-Holstein und der Hansestadt Hamburg. — Nova Hedwigia 20: 253-411.

—, 1974. Amazonische Desmidieen. 2. Teil: Areal Maués - Abacaxis. — Amazoniana 5: 135-242.

—, 1981. Revision und Validierung von Desmidiaceen-Namen aus früheren Publikationen. 2. — Algol. Studies 28: 236-251.

Frémy, P., 1930. Algues provenant des récoltes de M. Henri Gadeau de Kerville dans le canton de Bagnères-de-Luchon (Haute Garonne). — Bull. Soc. Amis Sci. Nat. Rouen 1930: 159-227.

Fritsch, F.E. & F. Rich, F. 1937. Contributions to our knowledge of the freshwater algae of Africa. — Trans. Roy. Soc. South Africa 25: 153-228.

Gauthier-Lièvre, L., 1931. Recherches sur la flore des eaux continentales de l'Algérie et de la Tunisie. — Thesis, Université de Paris, 299 pp, 14 pls.

—, 1958. Algues. — In: P. Quezel (ed.): Mission botanique au Tibesti, p. 27-43. — Université d'Alger, Institut de Recherches Sahariennes, 357 pp.

Gay, F., 1884. Essai d'une monographie locale des Conjuguées. — Thesis, Montpellier, 112 pp, 4 pls.

Georgevitch. P., 1910. Desmidiaceen aus dem Prepasee in Macedonien. —Beih. Bot. Centralbl. 26: 237-246.

Gontcharov, A.A. & M. Melkonian, 2005. Molecular phylogeny of *Staurastrum* Meyen ex Ralfs and related genera (Zygnematophyceae, Streptophyta) based on coding and noncoding rDNA sequence comparisons. — J. Phycol. 41: 887-899.

— & M. Melkonian, 2011. A study of conflict between molecular phylogeny and taxonomy in the Desmidiaceae (Streptophyta, Viridiplantae): analyses of 291 *rbc*L sequences. — Protist 162: 253-267.

Grönblad, R., 1920. Finnländische Desmidiaceen aus Keuru. — Acta Soc. Fauna Fl. Fenn. 47 (4): 1-98, pls1-6.

—, 1921. New desmids from Finland and northern Russia with critical remarks on some known species. — Acta Soc. Fauna Fl. Fenn. 49: 1-78, pls 1-7.

—, 1926. Beitrag zur Kenntnis der Desmidiaceen Schlesiens. — Commentat. Biol. 2 (5): 1-39, pls 1-3.

—, 1933. A contribution to the knowledge of sub-aërial desmids. — Commentat. Biol. 4 (4): 1-7.

—, 1934. A short report of the freshwater-algae recorded from the neighbourhood of the Zoological Station at Twärminne. — Memoranda Soc. Fauna Fl. Fenn. 10: 256-271.

—, 1936. Desmids from North Russia (Karelia) collected 1918 at Uhtua (Ukhtinskaya) and Hirvisalmi. — Commentat. Biol. 5 (6): 1-12, pls 1, 2.

—, 1938. Neue und seltene Desmidiaceen. — Bot. Not. 1938: 49-66.

—, 1942. Algen, hauptsächlich Desmidiaceen, aus dem Finnischen, Norwegischen und Schwedischen Lappland. — Acta Soc. Sci. Fenn., ser. B, Opera Biol. 2 (5): 1-46.

—, 1945. De algis Brasiliensibus, praecipue Desmidiaceis, in regione inferiore fluminis Amazonas a Professore August Ginzberger (Wien) anno 1927 collectis. — Acta Soc. Sci. Fenn., ser. B, Opera Biol. 2 (6): 1-43, pls 1-16.

—, 1947. Desmidiaceen aus Salmi. — Acta Soc. Fauna Fl. Fenn. 66: 1-29.

—, 1948. Freshwater algae from Täcktom träsk. — Bot. Not. 1948: 413-424.

—, 1948a. A list of desmids and plankton-organisms from the surroundings of Velikaja Guba (Suurlahti) in Easts-Carelia (Onega). — Commentat. Biol. 10 (5): 1-12.

—, 1960. Contributions to the knowledge of the freshwater algae of Italy. — Commentat. Biol. 22: 1-85.

—, 1962. Sudanese desmids II. — Acta Bot. Fennica 63: 1-19.

—, 1963. Desmids from Jämtland, Sweden and adjacent Norway. — Commentat. Biol. 26 (1): 1-43.

— & J. Růžička, 1959. Zur Systematik der Desmidiaceen. — Bot. Not. 112: 205-226.

— & A.M. Scott, 1955. On the variation of *Staurastrum bibrachiatum* Reinsch as an example of variability in a desmid species. — Acta Soc. Fauna Fl. Fenn. 72 (6): 1-11.

— , G.A. Prowse & A.M. Scott, 1958. Sudanese desmids. — Acta Bot. Fennica 58: 1-82.

Gutwiński, R., 1890. Zur Wahrung der Priorität, Vorläufige Mitteilungen über enige neue Algen-Species und -Varietäten aus der Umgebung von Lemberg. — Bot. Centralbl. 43: 65-73.

—, 1892. Flora glonów okolic Lwowa. — Spraw. Komis. Fizjogr. 27: 1-124, pls 1-3.

—, 1896. De nonnulis algis novis vel minus cognitis. — Rozpr. Akad. Umiejetn. Wydz. Mat.-Przyr. 33: 32-63, pls 5-7.

—, 1902. De algis a Dre M. Raciborski anno 1899 in insula Java collecties. — Bull. Acad. Sci. Cracovie 9: 575-617.

—, 1909. Flora algarum montium Tatrensium. — Bull. Int. Acad. Sci. Cracovie, Cl. Sci. Math. 4: 416-569, pls 7, 8.

Harvey, F.L., 1892. The freshwater algae of Maine. III. — Bull. Torrey Bot. Club 19: 118-125, pl. 126.

Hauptfleisch, P., 1888. Zellmembran und Hüllgallerte der Desmidiaceen. — Mitth. Naturwiss. Verein Neu-Vorpommern Greifswald 36: 59-136.

Hegde, G.R. & S.G. Bharati, 1980. Zygospore formation in some species of desmids. — Phykos 19: 213-221.

— & S.G. Bharati, 1983. Zygospore formation in some species of desmids, part 2. — Phykos 22: 4-12.

Heimans, J., 1926. A propos du *Staurastrum echinatum* Bréb. — Recueil Trav. Bot. Néerl. 23: 73-93.

—, 1940. Desmidiaceeën van Winterswijk. — Ned. Kruidk. Arch. 50: 206-214.

—, 1960. Desmidiaceeën in de vennen van het natuurreservaat Oisterwijk. —In: Hydrobiologie van de Oisterwijkse Vennen, p. 25-42. —Hydrobiologische Vereniging, Amsterdam, 90 pp.

—, 1969. Ecological, phytogeographical and taxonomic problems with desmids. — Vegetatio 17: 50-82.

Heimerl, A., 1891. Desmidiaceae alpinae. — Verh. Zool.-Bot. Ges. Wien 41: 587-609, pl. 5.

Hilse, L., 1865. Beiträge zur Algenkunde Schlesiens, als Fortsetzung der Beiträge im Jahres-Bericht für 1864. — Ber. Schles. Gesell. Vaterl. Cultur 1865: 109-129.

Hinode, T., 1959. Desmidian flora of the Sandankyo Gorge and the Yawata Highland in Hiroshima Prefecture. — Sci. Res. Sandankyo Gorge and Yawata Highl., Hiroshima, Japan 1959: 276-301, pls 1-14.

—, 1960. On some Japanese desmids (2). — Hikobia 2: 61-64.

—, 1965. Desmid flora of the southern district of Tokushima Prefecture II. — Hikobia 4: 188-208.

—, 1967. Some newly found desmids from the north-eastern areas of Shikoku. — Hikobia 5: 69-82.

—, 1971. A study on the desmids of Kurozo, a Sphagnum-moor in Shikoku. — Hikobia 6: 95-130.

Hirano, M., 1948. Desmidiaceae novae Japonicae (I). — Mem. Coll. Sci. Kyoto Imp. Univ., ser. B, Biol. 19: 65-69.

—, 1949. Some new or noteworthy desmids from Japan. — Acta Phytotax. Geobot. 14: 1-4.

—, 1950. Some new or noteworthy desmids from Japan, II. — Acta Phytotax. Geobot. 14: 35-38.

—, 1951. Some new or noteworthy desmids fom Japan. III. —Acta Phytotax. Geobot. 14: 69-71.

—, 1953. The alpine desmids from the Japanese Alps. 2. — Bot. Mag. (Tokyo) 66: 205-210.

—, 1955. Freshwater Algae. — In: H. Kihara (ed.): Fauna and Flora of Nepal Himalaya, Vol. 1, p. 5-42, pls 1-8. — Kyoto University, Fauna and Flora Research Society, 390 pp.

—, 1959. Flora Desmidiarum Japonicarum, V. — Contr. Biol. Lab. Kyoto Univ. 7: 226-301, pls 31-38.

—, 1959a. Flora Desmidiarum Japonicarum, VI. — Contr. Biol. Lab. Kyoto Univ. 9: 303-386, pls 39-52.

—, 1968. Desmids of Arctic Alaska. — Contr. Biol. Lab. Kyoto Univ. 21: 1-53.

Homfeld, H., 1929. Beitrag zur Kenntnis der Desmidiaceen Nordwestdeutschlands, besonders ihrer Zygoten. — Pflanzenforschung 12: 1-96.

Huitfeldt-Kaas, H.A., 1906. Planktonundersøgelser i norske vande. — Christiania 1906: 1-199, pls 1-9.

Irénée-Marie, F., 1938. Flore Desmidiale de la Région de Montréal. — Laprairie, Canada, 547 pp.

—, 1949. Contribution à la connaissance des Desmidiées de la région des Trois-Rivières. V. — Naturaliste Canad. 76: 99-133.

—, 1949a. Un nouveau *Staurastrum* canadien. — Ann. ACFAS 15: 94-95.

—, 1952. Contribution à la connaissance des Desmidiées de la région du Lac-St-Jean. — Hydrobiologia 4: 1-208.

—, 1957. Les *Staurastrum* de la région des Trois-Rivières. — Hydrobiologia 9: 145-209.

—, 1959. Expédition algologique dans le Haute Mauricie. — Hydrobiologia 13: 319-381.

— & D.K. Hilliard, 1963. Desmids from southcentral Alaska. — Hydrobiologia 21: 90-124.

Jao, C.C., 1949. Studies on the freshwater algae of China. XIX. Desmidiaceae from Kwangsi. — Bot. Bull. Acad. Sin. 3: 37-95.

Järnefelt, H., 1934. Zur Limnologie einiger Gewässer Finlands. XI. — Ann. Soc. Zool. Bot. Fenn. Vanamo 1934: 172-347.

John, D.M. & D.B. Williamson, 2009. — A practical guide to the desmids of the West of Ireland. — Martin Ryan Institute, Galway, 196 pp.

Johnson, L.N., 1894. Some new and rare desmids of the United States I. — Bull. Torrey Bot. Club 21: 285-291, pl. 211.

Joshua, W., 1886. Burmese Desmidieae, with descriptions of new species occurring in the neighbourhood of Rangoon. — J. Linn. Soc. Bot. 21: 634-654.

Kaiser, P.E., 1916. Beiträge zur Kenntnis der Algenflora von Traunstein und dem Chiemgau, III. — Kryptog. Forsch. 1: 30-38.

—, 1919. Desmidiaceen des Berchtesgadener Landes. — Kryptog. Forsch. 4: 216-230.

—, 1926. Beiträge zur Kenntnis der Algenflora von Traunstein und dem Chiemgau, V. — Kryptog. Forsch. 7: 419-443.

Kanetsuna, Y., 2002. New and interesting desmids (Zygnematales, Chlorophyceae) collected from Asia. — Phycological Research 50: 101-113.

Korshikov, A.A., 1941. A contribution to the algal flora of the Kola peninsula. — Proc. Inst. Kharkov 4: 53-76.

Kossinskaja, E.K., 1936. Desmidien der Arktis. — Trudy Bot. Inst. Akad. Nauk SSSR, ser. 2, Sporov. Rast. 3: 401-449, pls 1-5.

—, 1949. Desmidiaceae rariores et novae in Valdaj inventae. — Bot. Mat. Otd. Sporov. Rast. Bot. Inst. Akad. Nauk SSSR 6: 47-50.

—, 1953. Desmidievye mesotenievye i gonatozigovye vodorosli okrestnostej g. Valdaja. — Acta Inst. bot. nom. V. Komarovii Acad. Sci. URSS, ser. 2, 8: 5-37.

Kouwets, F.A.C., 1987. Desmids from the Auvergne (France). — Hydrobiologia 146: 193-263.

—, 1988. Remarkable forms in the desmid flora of a small mountain bog in the French Jura. — Cryptogamie, Algol. 9: 289-309.

—, 1991. Notes on the morphology and taxonomy of some rare or remarkable desmids (Chlorophyta, Zygnemaphyceae) from South-West France. — Nova Hedwigia 53: 383-408.

—, 2008. The species concept in desmids: the problem of variability, infraspecific taxa and the monothetic species definition. — Biologia, Bratislava 63: 877-883.

Krieger, W., 1932. Die Desmidiaceen der Deutschen Limnologischen Sunda-Expedition. — Arch. Hydrobiol. Suppl. 11: 129-230, pls 3-26.

—, 1938. Süsswasseralgen aus Spitzbergen. (Conjugatae und Chlorophyceae). — Ber. Deutsch. Bot. Ges. 56: 55-72.

— & P. Bourrelly, 1956. Desmidiacées des Andes du Venezuela. — Ergebn. d. Deutsch. limnol. Venezuela-Expedition 1: 141-195.

Kützing, F.T., 1833. Synopsis Diatomacearum oder Versuch einer systematischen Zusammenstellung der Diatomeen. — Linnaea 8: 529-620, pls 13-19.

—, 1849. Species algarum auctore Friderico Traug. Kützing. — F.A. Brockhaus, Leipzig, 922 pp.

Lagerheim, G., 1885. Bidrag till Amerikas Desmidiéflora. — Öfvers. Förh. Kongl. Svenska Vetensk.-Akad. 42: 225-255.

—, 1893. Chlorophyceen aus Abessinien und Kordofan. — Nuova Notarisia 4: 153-165.

Laporte, L.J., 1931. Recherches sur la biologie et la systématique des Desmidiées. — Encyclop. Biol. 9: 1-147, pls 1-22.

Larsen, E., 1907. Ferskvandsalger fra Vest-Grønland. — Meddel. Grønland 33: 305-364, pls 7, 8.

Lemaire, A., 1883. Liste des Desmidiées observées dans les Vosges jusqu'en 1882 precedée d'une introduction contenant des indications sur la récolte et la preparation de ces algues. — Bull. Soc. Sci. Nancy, sér. 2, 6: 1-28, pl 1.

—, 1890. Liste des Desmidiées observées dans quelques lacs des Vosges et aux environs d'Étival. — Bull. Soc. Sci. Nancy, sér. 2, 10, sep.: 1-10.

Lemmermann, E., 1896. Zur Algenflora des Riesengebirges. — Forschungsber. Biol. Stat. Plön 4: 88-133.

Lenzenweger, R., 1980. Algenalarm am Badesee. — Mikrokosmos 10: 319-320.

—, 1981. Zieralgen aus dem Hornspitzgebiet bei Gosau. Teil 1. — Naturk. Jahrb. Stadt Linz 27: 25-82.

—, 1984. Beitrag zur Kenntnis der Zieralgen der Nördlichen Kalkalpen Österreichs (Steiermark und Oberösterreich). — Algol. Studies 36: 251-281.

—, 1985. Zieralgen aus dem Plankton und Sublitoral einiger Oberösterreichischen Seen. — Jahrb. Oberösterr. Musealvereins 130: 193-208.

—, 1986. Beitrag zur Kenntnis der Zieralgen der Nördlichen Kalkalpen Österreichs (Steiermark). — Algol. Studies 42: 93-122.

—, 1986a. Zur Zieralgenflora der Schwarzen Lacken am Gerzkopf bei Eben/Pongau (Salzburg, Österreich). — Linzer Biol. Beitr. 18/1: 101-115.

—, 1986b. Interessante Zieralgen-Funde im Bergland nördlich von Gröbming und Stainach (Steiermark, Österreich). — Mitt. Abt. Bot. Landesmus. "Joanneum" Graz 13/14: 29-43.

—, 1987. Beitrag zur Kenntnis der Zieralgenflora des Salzburger Lungaues. —Algol. Studies 46: 47-64.

—, 1988. Zur Zieralgenflora einiger Moore und Seeuferzonen in Kärnten. — Carinthia II 178/98: 537-559.

—, 1989. Die Staurastren (Desmidiaceae) Österreichs und ihre bislang bekannte Verbreitung. — Stapfia (Linz) 22: 1-44.

—, 1989a. Beitrag zur Kenntnis der Zieralgenflora des Egelsees bei Abtenau (Salzburg, Österreich). — Stapfia (Linz) 22: 45-79.

—, 1989b. Zieralgen von Süd-Grönland. — Stapfia (Linz) 22: 81-140.

—, 1989c. Beitrag zur Zieralgenflora von Kärnten. — Carinthia II 179/99: 509-536.

—, 1989d. Zieralgen von Süd-Grönland. — Stapfia (Linz) 22: 81-140.

—, 1991. Beitrag zur Desmidiaceenflora im Nationalpark Hohe Tauern (Mölltal, Kärnten). — Carinthia II 181/101: 367-385.

—, 1991a. Die Zieralgenflora der Seewiesenalm bei Lienz/Osttirol. — Veröff. Mus. Ferdinandeum 71: 117-134.

—, 1994. Die Desmidiaceenflora des Rosanin-Sees in den Nockbergen (Salzburg, Österreich). — Nova Hedwigia 59: 163-187.

—, 1997. Desmidiaceenflora von Österreich. Teil 2. — Bibliotheca Phycologica 102, J. Cramer, Berlin/Stuttgart, 216 pp.

—, 2000. Neue bemerkenswerte Zieralgenfunde aus Österreich. — Algol. Studies 98: 27-41.

—, 2000a. Vorläufiges Ergebnis der Untersuchungen zur Zieralgenflora der Schwemm bei Walchsee in Nordtirol. — Ber. Naturwiss.-Med. Vereins Innsbruck 87: 41-66.

Lhotsky, O., 1949. Poznámka k flore Desmidiacei Hrubého Jeseniku. — Zvláštní otisk z Cas. Vlast. spolku Mus. Olomouci 58: 149-155.

Lillieroth, S., 1950. Über Folgen kulturbedingter Wasserstandsenkungen für Makrophyten und Planktongemeinschaften in seichten Seen des südschwedischen Oligotrophiegebietes. — Acta Limnol. 3: 1-288.

Lind, E. & A.J. Brook, 1980. — A key to the commoner desmids of the English Lake District. — Freshwater Biological Association, Ambleside, 123 pp.

— & W. Pearsall, 1945. Plankton algae from north-western Ireland. — Proc. Roy. Irish Acad. 50 (B): 311-320.

Ling, H.U. & P.A. Tyler, 1995. The *Staurastrum tohopekaligense* species cluster. — Algol. Studies 76: 27-60.

— & P.A. Tyler, 2000. — Australian freshwater algae (exclusive of diatoms). —Bibliotheca Phycologica 105, J. Cramer, Berlin/Stuttgart, 643 pp.

Lundberg, F., 1931. Beiträge zur Kenntnis der Algenflora von Schweden. I. Über das Phytoplankton einiger Seen in Dalarne. — Bot. Not. 1931: 269-296.

Lundell, P.M., 1871. De Desmidiaceis, quae in Suecia inventae sunt, observationes criticae. — Nova Acta Regiae Soc. Sci. Upsal., ser. III, 8 (2): 1-100, pls 1-5.

Lütkemüller, J., 1893. Desmidiaceen aus der Umgebung des Attersees in Oberösterreich. — Verh. Zool.-Bot. Ges. Wien 42: 537-570, pls 8, 9.

—, 1900. Desmidiaceen aus der Umgebung des Millstättersees in Kärnten. — Verh. Zool.-Bot. Ges. Wien 50: 60-84.

—, 1900a. Desmidiaceen aus den Ningpo-Mountains in Centralchina. — Ann. K. K. Naturhist. Hofmus. 15: 115-126, pl. 6.

—, 1910. Zur Kenntnis der Desmidiaceen Böhmens. — Verh. Zool.-Bot. Ges. Wien 60: 478-503, pls 2, 3.

Manguin, M., 1935. Catalogue des algues d'eau douce du Canton de Fresnay-sur-Sarthe. 2. — Bull. Soc. Agric. Sarthe 54: 57-95.

—, 1939. Florule algologique des cuvettes tourbeuses de la Foret de Sillé (Sarthe). — Rev. Algol. 8: 298-310, pls 40-42.

Margalef, R., 1956. Algas de agua dulce del noroeste de España. — Publ. Inst. Biol. Aplicada 22: 43-152.

Maskell, W.M., 1883. On the New Zealand Desmidieae. Additions to catalogue and notes on various species of New Zealand Desmidieae. — Trans. & Proc. New Zealand Inst. 15: 237-259.

—, 1889. Further notes on the Desmidieae of New Zealand, with descriptions of new species. — Trans. & Proc. New Zealand Inst. 21: 3-32.

McNeill, J., 2006. International Code of Botanical Nomenclature (Vienna Code) adopted by the Seventeenth International Botanical Congress Vienna, Austria, July 2005. — Regnum Veg. 146.

Meneghini, G., 1840. Synopsis Desmidiearum hucusque cognitarum. — Linnaea, Halle a.d. S. 14: 201-240.

Messikommer, E., 1927. Biologische Studien im Torfmoor von Robenhausen, unter besonderer Berücksichtigung der Algenvegetation. — Thesis, Universität Zürich, 171 pp.

—, 1927a. Beiträge zur Kenntnis der Algenflora des Kantons Zürich. II. Die Algenvegetation des Böndlerstück. — Vierteljahrsschr. Naturf. Ges. Zürich 72: 332-351, pls 1, 2.

—, 1928. Beiträge zur Kenntnis der Algenflora des Kantons Zürich. III: Die Algenvegetation des Hinwiler- und Oberhöflerriedes. — Vierteljahrsschr. Naturf. Ges. Zürich 73: 195-213, pls 8, 9.

—, 1929. Beiträge zur Kenntnis der Algenflora des Kantons Zürich. IV: Die Algenvegetation der Moore am Pfäffikersee. — Vierteljahrsschr. Naturf. Ges. Zürich 74: 139-162, pl. 1.

—, 1935. Algen aus dem Obertoggenburg. — Mitt. Bot. Mus. Univ. Zürich 148: 95-130.

—, 1942. Beitrag zur Kenntnis der Algenflora und Algenvegetation des Hochgebirges um Davos. — Beitr. Geobot. Landesaufn. Schweiz 24: 1-452.

—, 1949. Algologische Erhebungen im St.-Gallischen Abschnitt der NW-Sardonagruppe II. — Vierteljahrsschr. Naturf. Ges. Zürich 94: 231-251.

—, 1951. Grundlagen zu einer Algenflora des Kantons Glarus. — Mitt. Naturf. Ges. Kantons Glarus 8: 1-122.

—, 1956. Alte und neuere Untersuchungen über die Algenflora des östlichen Berner Oberlandes. — Mitth. Naturf. Ges. Bern 13: 81-149, 3 pls.

—, 1957. Beitrag zur Kenntnis der Algenflora der Dombes. — Rev. Algol. n.s. 3: 71-93.

—, 1960. Algenflora der Gewässer des St.-Gotthard-Gebietes. — Schweiz. Z. Hydrol. 22: 177-224.

—, 1962. Algen aus dem Hinterrheingebiete. — Nova Hedwigia 4: 131-164, pl. 51.

—, 1966. Tessiner Algen. — Nova Hedwigia 11: 353-386.

—, 1971. Zur Kenntnis der Algenflora in der Transfluenzlandschaft zwischen Hombrechtikon und Bubikon. — Schweiz. Z. Hydrol. 33: 138-170.

Meyen, F.J.F., 1828. Beobachtungen über einige niedere Algenformen. — Nova Acta Phys.-Med. Acad. Caes. Leop.-Carol. Nat. Cur. 14: 768-778.

Mix, M., 1973. *Staurastrum tumidum* Bréb. ex Ralfs var. *quadrangulatum* M.Mix, var. nov. — Mitt. Staatsinst. Allg. Bot. Hamburg 14: 37-41.

Mollenhauer, D., 1988. *Staurastrum* — procrustean bed or natural form? — Arch. Protistenk. 135: 35-40.

Nägeli, C., 1849. Gattungen einzelliger Algen physiologisch und systematisch bearbeitet. — Schulthess, Zürich, 139 pp, 8 pls.

Nauwerck, A. 1962. Zur Systematik und Ökologie Portugiesischer Planktonalgen. — Mem. Soc. Brot. 15: 5-55, pls 1-7.

Noda, M. & B. Skvortsov, 1969. New and rare desmids from Genho River of Great Chingan Mountains, autonomous region of Inner Mongolia, China. — Sci. Rep. Niigata Univ., ser. D 6: 65-86.

Nordstedt, O., 1870. Desmidiaceae. — In: E. Warming (ed.): Symbolae ad floram Brasiliae centralis cognoscendam. — Vidensk. Meddel. Naturh. Forening Kjøbenhavn 1869: 195-234, pls 2-4.

—, 1872. Desmidiaceae ex insulis Spetsbergensibus et Beeren Eiland in expeditionibus annorum 1868 et 1870 suecanis collectae. — Öfvers. Förh. Kongl. Svenska Vetensk.-Akad. 1872 (6): 23-42, pls 6, 7.

—, 1873. Bidrag till kännedomen om sydligare Norges Desmideer. — Lunds Univ. Årsskrift 9: 1-51.

—, 1875. Desmidieae arctoae. — Öfvers. Förh. Kongl. Svenska Vetensk.-Akad. 1875 (6): 13-43.

—, 1878. De algis aquae dulcis et de Characeis ex insulis Sandvicensibus a Sv. Berggren 1875 reportatis. — Minneskr. Kongl. Fysiogr. Sällsk. Lund, Hundraarsfest 1878: 1-24, pls 1, 2.

—, 1880. Desmidieae. — In: Points-förteckning öfver Skandinaviens växter. (*Enumerantur plantae Scandinaviae*) 4. Characeer, Alger och lafvar. — Lund.

—, 1887. Algologiska småsaker. 4. Utdrag ur ett arbete öfver de af Dr. S. Berggren på Nya Seland och in Australien samlade sötvattensalgerna. — Bot. Not. 1887: 153-164.

—, 1888. Desmidieer från Bornholm, samlade och delvis bestämda af R.T. Hoff, granskade af O. Nordstedt. — Vidensk. Meddel. Naturh. Forening Kjøbenhavn 1888: 188-213, pl. 6.

—, 1888a. Freshwater algae collected by Dr. S. Berggren in New Zealand and Australia. — Kongl. Svensk. Vet.-Akad. Handl. 22 (8): 1-98, pls 1-7.

— & V. Wittrock, 1876. Desmidieae et Oedogonieae ab O. Nordstedt in Italia et Tyrolia collectae, quas determinaverunt. — Öfvers. Förh. Kongl. Svenska Vetensk.-Akad. 1876 (6): 25-56, pls 12, 13.

Nygaard, G., 1926. Plankton from two lakes of the Malayan region. — Vidensk. Meddel. Naturh. Forening Kjøbenhavn 82: 197-240, pls 1-8.

—, 1949. Hydrobiological studies in some ponds and lakes. Part II: The quotient hypothesis and some new or little known phytoplankton organisms. — Biol. Skr. 7 (1): 1-293.

—, 1979. Freshwater phytoplankton from the Narssaq Area, South Greenland. — Bot. Tidsskr. 73: 191-238.

—, 1991. Seasonal periodicity of planktonic desmids in oligotrophic lake Grane Langsø, Denmark. — Hydrobiologia 211: 195-226.

Palamar-Mordvintseva, G.M., 1961. New representatives of the genus *Staurastrum* Meyen in the Ukraine. — Ukrajins'k Bot. Zurn. 18: 81-86.

—, 1964. Nova flora *Staurastrum leptocladum* Nordst. z ozer zakhidnoukrains'kogo Pollissya. — Ukrajins'k Bot. Zurn. 21: 87-89.

—, 1976. A taxonomic analysis of the genus *Staurastrum*. — Ukrajins'k Bot. Zurn. 33: 31-38.

—, 1976a. New genera of Desmidiales. — Ukrajins'k Bot. Zurn. 33: 396-398.

—, 1981. Novije vidij Desmidievijkh (Desmidiales). — Novosti Sist. Vyssh. Nizsh. Rast. 1980: 226-232.

—, 1982. Opredelitel presnovodnyh vodorosley SSSR. Vol.11 (2). Zelenye vodorosli. Klass Konjugaty. Poriadok Desmidievye (2). —Nauka, Leningrad, 624 pp.

Patel, R.J. & C.K.A. Kumar, 1980. Contributions to the desmid flora of India: genus *Staurastrum* Meyen from Gujarat. — Phykos 19: 177-185.

Perty, M., 1849. Mikroskopische Organismen der Alpen und der Italienischen Schweiz. — Mitth. Naturf. Ges. Bern 1849: 153-176.

—, 1852. Zur Kenntnis kleinster Lebensformen nach Bau, Funktionen, Systematik, mit Spezialverzeichnis der in der Schweiz beobachteten. —Jent & Reinert, Bern, 228 pp, 17 pls.

Péterfi, L.S., 1963. Alge din Bazinul Superior al Riului Sebes. — Studia Cluj Universitatis Babes-Bolyai, ser. Biologia 8: 13-30.

—, 1972. Variability of Staurastra in natural populations with remarks on its taxonomic and nomenclatural implications. — Rev. Roumaine Biol. Sér. Bot. 17: 19-28.

—, 1973. Studies on Romanian Staurastra. I. Variability and taxonomy of *Staurastrum spinosum* (Brébisson) Ralfs. — Nova Hedwigia 24: 121-144.

Péterfi, S. (1943). Über einige *Staurastrum*-Arten des Gyaluer-Gebirges. — Múz. Füz. 1: 183-203.

Petkoff, S., 1910. La flore aquatique et algologique de la Macedoine du S. O. Philippopoli. — Philippopoli Acad. Bulg. Sci. 1910: 1-189, pls 1-4.

Pevalek, I., 1925. Geobotanicka i algoloska istrazivanja cretova u Hrvatskoj i Sloveniji. — Rad Jugoslav. Akad. Znan. 230: 29-117.

Playfair, G.J., 1908. Some Sydney desmids. — Proc. Linn. Soc. New South Wales 33: 603-628, pls 11-13.

—, 1910. Polymorphism and life-history in the Desmidiaceae. — Proc. Linn. Soc. New South Wales 35: 459-495.

Prescott, G.W., 1936. Notes on alpine and subalpine desmids from western United States. — Pap. Michigan Acad. Sci. 21: 135-146, pls 15-17.

—, 1966. Algae of the Panama Canal and its tributaries. II. Conjugales. — Phykos 5: 1-49.

— & A. Magnotta, 1935. Notes on Michigan desmids, with descriptions of some species and varieties new to science. — Pap. Michigan Acad. Sci. 20: 157-170.

— & A.M. Scott, 1942. The freshwater algae of southern United States. I. Desmids from Mississipi, with descriptions of some new species and varieties. — Trans. Amer. Microscop. Soc. 61: 1-29.

—, C.E. de M. Bicudo & W.C. Vinyard, 1982. A Synopsis of North American Desmids, Part 2, Section 4. — University of Nebraska Press, Lincoln, 700 pp.

Printz, H., 1915. Beiträge zur Kenntnis der Chlorophyceen und ihrer Verbreitung in Norwegen. — Kongel. Norske Vidensk. Selsk. Skr. 2: 1-76.

Pritchard, A., 1861. A history of Infusoria, including the Desmidiaceae and Diatomaceae, British and foreign. Ed. 4. — Whittaker and Co., London, 968 pp.

Rabenhorst, L., 1862. Die Algen Europa's [Coll. exsicc]. — Dresden.

—, 1863. Kryptogamen-Flora von Sachsen, der Ober-Lausitz, Thüringen und Nordböhmen, mit Berücksichtigung der benachbarten Länder. 1. Abth.: Algen im weitesten Sinne, Leber- und Laubmoose. — Eduard Kummer, Leipzig, 653 pp.

—, 1868. Flora europaea algarum aquae dulcis et submarinae. Sect. 3. — E. Kummer, Leipzig.

Raciborski, M., 1884. Desmidyje okoloc Krakowa. — Spraw. Komis. Fizjogr. 19: 3-24.

—, 1885. De nonnulis Desmidiaceis novis vel minus cognitis, quae in Polonia inventae sunt. — Pamietn. Akad. Umiejetn. w Krakowie, Wydz. Mat.-Przyr. 10: 57-100.

—, 1888. Materyaly do flory glonów Polski. — Spraw. Komis. Fizjogr. 22: 80-122.

—, 1889. Nowe Desmidyje. — Pamietn. Akad. Umiejetn. w Krakowie, Wydz. Mat.-Przyr. 17: 73-113, pls 5-7.

—, 1889a. Su alcune Desmidiaceae Lituanae. — Notarisia 4: 659-663.

—, 1892. Desmidyja zebrane przez Dr. E. Ciastonia w podrozy na okolo ziemi. — Rozpr. Akad. Umiejetn. Wydz. Mat.-Przyr. Ser. 2, 22: 361-392, pls 6, 7.

—, 1895. Die Desmidieenflora des Tapakoomasees. — Flora 81: 30-35.

Ralfs, J., 1845. On the British Desmidieae. — Ann. Mag. Nat. Hist. 15: 149-160, 401-406.

—, 1848. The British Desmidieae. — Reeve, Benham & Reeve, London, 226 pp, 35 pls.

Reinsch, P.F., 1867. De speciebus generibusque nonnulis novis ex algarum et fungorum classe. — Acta Societ. Senckenb. 6: 111-144.

—, 1867a. Die Algenflora des mittleren Teiles von Franken, enthaltend die vom Autor bis jetzt in diesem Gebiete beobachteten Süsswasseralgen. — Abh. Naturhist. Ges. Nürnberg 3 (2): 1-238.

—, 1875. Contributiones ad algologiam et fungologiam. Vol. I. — T.O. Weichel, Leipzig, 103 pp, 124 pls.

Ricci, S., 1990. *Staurastrum dilatatum* var. *thomassonii* var. nov. — Hydrobiologia 194: 115-118.

Rich, F., 1925. Further notes on the algae of Leicestershire. — J. Bot. 63: 71-78.

—, 1932. Contributions to our knowledge of the freshwater algae of Africa. — Trans. Roy. Soc. South Africa 20: 149-188.

—, 1935. Contributions to our knowledge of the freshwater algae of Africa. 11. Algae from a Pan in Southern Rhodesia. — Trans. Roy. Soc. South Africa 23: 107-160.

Round, F.E. & A.J. Brook, 1959. The phytoplankton of some Irish loughs and an assessment of their trophic status. — Proc. Roy. Irish Acad. 60: 168-191.

Roy, J., 1893. On Scottish Desmidieae.— Ann. Scott. Nat. Hist. 1893: 106-111, 170-180, 237-245.

—, 1894. On Scottish Desmidieae. — Ann. Scott. Nat. Hist. 1894: 40-46.

— & J.P. Bisset, 1886. Notes on Japanese desmids - 1. — J. Bot. 24: 193-196, 237-242.

— & J.P. Bisset, 1894. On Scottish Desmidieae. — Ann. Scott. Nat. Hist. 1894: 100-105, 167-178, 241-256.

Růžička, J., 1956. Die Desmidiaceen der Moravice-Quellen (Grosser Kessel, Gesenke). — Acta Rerum Nat. Distr. Ostrav. 17: 38-58.

—, 1957. Die Desmidiaceen der oberen Moldau (Böhmerwald). — Preslia 29: 132-154.

—, 1957a. Desmidiaceen uas dem Quellgebiete auf dem 'Maly Ded' (Gesenke). — Cas. Slez. Mus., ser. A, Hist. Nat. 6: 108-121.

—, 1962. *Cosmarium bulliferum* spec. nova. — Preslia 34: 415-416.

—, 1963. *Staurastrum dicroceros* spec. nova. — Preslia 35: 73.

—, 1972. Die Zieralgen der Insel Hiddensee. — Arch. Protistenk. 114: 453-485.

—, 1973. Die Zieralgen des Naturschutzgebietes 'Rezabinec' (Südböhmen). — Preslia 45: 193-241.

—, 1977. Die Desmidiaceeen Mitteleuropas. Band 1, Lieferung 1. — E. Schweizerbart'sche Verlagsbuchhandlung, Stuttgart, pp 1-292, pls. 1-44.

—, 1981. Die Desmidiaceeen Mitteleuropas. Band 1, Lieferung 2. — E. Schweizerbart'sche Verlagsbuchhandlung, Stuttgart, pp 293-736, pls. 45-117.

Rybnicek, K., 1960. Mesotaeniales and Desmidiales of the Moravskoslezké Beskydy Mountains. 1. Desmidiales in the peat bogs at Súlov and Jancik. — Práce Brnenské Zákl. Ceskoslov. Akad. Ved 32: 125-156.

Ryppowa, H., 1927. Les algues de petits lacs tourbeux nommés 'Suchary' dans les environs du Lac de Wigry. — Arch. Hydrobiol. Rybactwa 2: 41-66.

Salisbury, R.K., 1936. The desmids of Florida. — Ohio J. Sci. 36: 55-61.

Sampaio, J., 1944. Desmídias Portuguesas. — Bol. Soc. Brot., ser. 2, 18: 1-563.

Schaarschmidt, J., 1883. Tanulmányok a Magyarhoni Desmidiaceákról. — Magyar Tud. Akad. Értes. 18: 259-280, 1 pl.

Schmidle, W., 1893. Beiträge zur Algenflora des Schwarzwaldes und der Rheinebene. — Ber. Naturf. Ges. Freiburg 7: 68-112, pls 2-6.

—, 1894. Algen aus dem Gebiete des Oberrheins. — Ber. Deutsch. Bot. Ges. 11: 544-555.

—, 1894a. Einzellige Algen aus den Berner-Alpen. — Hedwigia 33: 86-96.

—, 1895. Weitere Beiträge zur Algenflora der Rheinebene und des Schwarzwaldes. — Hedwigia 34: 66-83, pl. 1.

—, 1895a. Einige Algen aus Sumatra. — Hedwigia 34: 293-307.

—, 1896. Beiträge zur alpinen Algenflora. — Oesterr. Bot. Z. 7: 1-40, pls 14-17.

—, 1896a. Süsswasseralgen aus Australien. — Flora 82: 297-313, pl. 9.

—, 1898. Über einige von Knut Bohlin in Pite Lappmark und Vesterbotten gesammelte Süsswasseralgen. — Bih. Kongl. Svenska Vetensk.-Akad. Handl. 24, Afd. III (8): 1-71.

—, 1898a. Die von Professor Dr. Volkens und Dr. Stuhlmann in Ost-Afrika gesammelten Desmidiaceen. — Bot. Jahrb. Syst. 26: 1-59, pls 1-4.

Schmidt, M., 1903. Grundlagen einer Algenflora der Lüneburger Heide. —Thesis, Georg-August-Universität, Göttingen.

Schröder, B., 1897. Die Algen der Versuchsteiche des Schlesischen Fischereivereins zu Trachenberg. — Forschungsber. Biol. Stat. Plön 5: 29-66, pls 2-4.

—, 1898. Neue Beiträge zur Kenntnis der Algen des Riesengebirges. — Forschungsber. Biol. Stat. Plön 6: 9-46.

—, 1919. Beiträge zur Kenntnis der Algenvegetation des Moores von Groß-Iser. — Ber. Deutsch. Bot. Ges. 37: 250-261.

Schulz, P., 1922. Desmidiaceen aus dem Gebiete der freien Stadt Danzig und dem benachbarten Pomerellen. — Bot. Arch. 2: 113-173.

Schwarz, A. & R. Lenzenweger, 1999. Ein bemerkenswerter Fund von *Staurastrum octodontum* Skuja var. *tetrodontum* Scott & Grönblad im Dobbertiner See (Norddeutschland). — Algol. Studies 95: 73-79.

Scott, A.M., 1950. New varieties of *Staurastrum ophiura* Lund. — Trans. American Microscop. Soc. 69: 248-253.

— & R. Grönblad, 1957. New and interesting desmids from the southeastern United States. — Acta Soc. Sci. Fenn., ser. B, Opera Biol. 2 (8): 1-62, pls 1-37.

— & G.W. Prescott, 1956. Notes on Indonesian freshwater algae - I. *Staurastrum wildemani* Gutw. (Desmidiaceae). — Reinwardtia 3: 351-362.

— & G.W. Prescott, 1958. Some freshwater algae from Arnhem Land in the Northern Territory of Australia. — Rec. American-Australian Sci. Exped. Arnhem Land 3: 8-136.

—, R. Grönblad & H. Croasdale, H., 1965. Desmids from the Amazon basin, Brazil. — Acta Bot. Fennica 69: 1-94.

Shirshov, P.P., 1935. Ecological-geographical essay on the freshwater algae of Novaya Zemlya and Franz-Josef Land. — Trudy Leningrad Vsesojuzn. Arkticheskii Institut 14: 73-162.

Skuja, H., 1928. Vorarbeiten zu einer Algenflora von Lettland. IV. — Acta Horti Bot. Univ. Latv. 3: 103-218, pls 1-4.

—, 1931. Die Algenflora der Insel Moritzholm im Usmaitensee. — Arbeiten Naturf. Vereins Riga n.s. 19: 1-20, pl. 1.

—, 1934. Beitrag zur Algenflora Letlands I. — Acta Horti Bot. Univ. Latv. 7: 3-85.

—, 1948. Taxonomie des Phytoplanktons einiger Seen in Uppland, Schweden. — Symbol. Bot. Upsal. 9 (3): 1-399, pls 1-39.

—, 1956. Taxonomische und biologische Studien über das Phytoplankton Schwedischer Binnengewässer. — Nova Acta Regiae Soc. Sci. Upsal., ser. IV, 16 (3): 1-404, pls 1-63.

—, 1964. Grundzüge der Algenflora und Algenvegetation der Fjeldgegenden um Abisko in Schwedisch-Lappland. — Nova Acta Regiae Soc. Sci. Upsal., ser. IV, 18 (3): 1-465, pls 1-69.

—, 1976. Zur Kenntnis der Algen Neuseeländischer Torfmoore. — Nova Acta Regiae Soc. Sci. Upsal., ser. V:C, 2: 1-159.

Smith, G.M., 1922. The phytoplankton of the Muskoka Region, Ontario, Canada. — Trans. Wisconsin Acad. Sci. 20: 323-364.

—, 1924. Phytoplankton of the inland lakes of Wisconsin. Part II, Desmidiaceae. — Wisconsin Geological and Natural History Survey 57 (2): 1-227.

Strøm, K.M., 1920. Freshwater algae from Tuddal in Telemark. — Nyt Mag. Naturvidensk. 57: 1-53, pls 1-3.

—, 1926. Norwegian mountain algae. — Skr. Norske Vidensk.-Akad. Oslo, Mat.-Naturvidensk. Kl. 6: 1-263, pls 1-25.

Taft, C.E., 1945. The desmids of the west end of Lake Erie. — Ohio J. Sci. 45: 180-205.

Tarnavschi, I.T. & D. Radulescu, 1956. Forme noi Desmidiaceae turficole descrise din Bazinul Dornelor (Regiunea Suceava). — Comun. Acad. Republ. Populare Romine 6: 437-442.

Tarnogradsky, D.A., 1960. Genus *Staurastrum* Meyen (Desmidiaceae) in Lacubus Sakoczavi (Georgia, Caucasus). — Bot. Mater. Otd. Sporov. Rast. Bot. Inst. Komarova Akad. Nauk SSSR 13: 87-100.

Teiling, E., 1942. Schwedische Planktonalgen 3. Neue und wenig bekannte Formen. — Bot. Not. 1942: 63-68.

—, 1942a. Schwedische Planktonalgen 4. Phytoplankton aus Roslagen. — Bot. Not. 1942: 207-217.

—, 1946. Zur Phytoplanktonflora Schwedens. — Bot. Not. 1946: 61-88.

—, 1947. *Staurastrum planctonicum* and *St. pingue*. A study of planktic evolution. — Svensk Bot. Tidskr. 41: 218-234.

—, 1948. *Staurodesmus*, genus novum. — Bot. Not. 1948: 49-83.

—, 1950. Radiation of desmids, its origin and its consequences as regards taxonomy and nomenclature. — Bot. Not. 1950: 299-327.

—, 1954. L'authentique *Staurodesmus dejectus* (Bréb.). — Compte-rendue VIII Congr. Int. Bot., sect. 17: 128-129.

—, 1957. Morphological investigations of asymmetry in desmids. — Bot. Not. 110: 49-82.

—, 1967. The desmid genus *Staurodesmus*. A taxonomic study. — Ark. Bot. 6 (11): 467-629, pls 1-31.

Tell, G., 1980. Le genre *Staurastrum* (Algues Chlorophycées,Desmidiées) dans le nord-est de l'Argentine. — Bull. Mus. Hist. Nat. (Paris), sér. 4, 2, section B 2: 145-207.

Teodoresco, E.C., 1907. Matériaux pour la flora algologique de la Roumaine. — Beih. Bot. Centralbl. 21: 103-219.

Thérézien, Y., 1985). Contribution à l'étude des algues d'eau douce de la Guyane Française, à l'exclusion des diatomées. — Bibliotheca Phycologica 72, J. Cramer, Berlin/Stuttgart.

Thomasson, K., 1957. Contributions to the knowledge of the plankton in Scandinavian mountain lakes. 4. — Bot. Not. 110: 251-264.

—, 1957a. Notes on the plankton of Lake Bangweulu. — Nova Acta Regiae Soc. Sci. Upsal., ser. 4, 17 (3): 3-18.

—, 1959. Nahuel Huapi — plankton of some lakes in an Argentine national park, with notes on terrestrial vegetation. — Acta Phytogeogr. Suecic. 42: 1-83.

—, 1960. Notes on the plankton of Lake Bangweulu. 2. — Nova Acta Regiae Soc. Sci. Upsal., ser. 4, 17 (12): 1-43.

—, 1960a. Some planktic *Staurastra* from New Zealand. — Bot. Not. 113: 225-245.

—, 1963. Araucanian Lakes. — Acta Phytogeogr. Suecic. 47: 1-139.

—, 1965. Notes on algal vegetation of Lake Kariba. — Nova Acta Regiae Soc. Sci. Upsal., ser. 4, 19 (1): 1-34.

—, 1966. Phytoplankton of Lake Shiwa Ngandu. — Expl. hydrobiol. Bangweolo - Luapula 4, 2: 1-91.

—, 1971. Amazonian algae. — Mém. Inst. Roy. Sci. Nat. Belgique, sér. 2, 86: 1-57, pls 1-24.

— & P.A. Tyler, 1971. Taxonomy of Australian freshwater algae. 2. Some planktic Staurastra from Tasmania. — Nova Hedwigia 21: 287-319.

Thunmark, S., 1948. Sjöar och myrar i Lenhovda socken. — In: Lenhovda - En Värendssocken berättar, pp. 665-710. — Lenhovda Hembygdsförening, Moheda, 813 pp.

Tomaszewics, G.H., 1988. Desmids of the transitional bogs of the Middle Mazowsze Lowland. — Monogr. Bot. (Warszawa) 70: 1-86.

— & W.W. Kowalski, 1993. Desmids of some polyhumic lakes in the Wigry National Park, northeastern Poland. — Fragm. Florist. Geobot. 38: 525-548.

— & E. Skrzeczkowska, 1989. Some critical remarks about new and rare to Polish flora desmids from the Suwalki Lakeland. — Acta Soc. Bot. Poloniae 58: 117-126.

Turner, W.B., 1886. Notes on freshwater algae with descriptions of new species. — Naturalist (London) 1886: 33-35.

—, 1892. The freshwater algae (principally Desmidieae) of East India. — Kongl. Svensk. Vetensk. Acad. Handl. 25 (5): 1-187, pls 1-23.

Vijverman, W., 1991. Desmids from Papua New Guinea. — Bibliotheca Phycologica 87, J. Cramer, Berlin/Stuttgart.

Villeret, S., 1955. Contribution à la biologie des algues des tourbières à Sphaignes. — Bull. Soc. Sci. Bretagne 29: 5-246.

Wade, W.E., 1957. Additions to our knowledge of the desmid flora of Michigan. — Rev. Algol., n.s. 2: 249-273.

Watanabe, M., M.H. Watanabe & M. Saitow, 1980. Phytoplankton studies of Lake Kasumigaura. (1) *Staurastrum chaetoceras* var. *tricrenatum* Skuja. — Bull. Natl. Sci. Mus., Tokyo, B. 6: 147-156.

West, G.S., 1899. The alga-flora of Cambridgeshire. — J. Bot. 37: 49-58, 109-116, 216-225, 262-268, 290-299, pls 394-396.

—, 1899a. On variation in the Desmidieae, and its bearings on their classification. — J. Linn. Soc. Bot. 34: 366-416, pls 8-11.

—, 1905. Desmids from Victoria. — J. Bot. 43: 252-254.

—, 1907. Report on the freshwater algae, including phytoplankton, of the Third Tanganyika Expedition conducted by Dr. W.A. Cunnington, 1904-1905. — J. Linn. Soc., Bot. 38: 81-197, pls 2-10.

—, 1909. The algae of the Yan Yean Reservoir, Victoria: a biological and ecological study. — J. Linn. Soc., Bot. 39: 1-88, pls 1-6.

—, 1914. A contribution to our knowledge of the freshwater algae of Colombia. — Mém. Soc. Sci. Nat. Neuchatel 5: 1013-1051.

West, W., 1889. List of desmids from Massachusetts, U.S.A. — J. Roy. Microsc. Soc. (London) 5: 16-21, pls 2, 3.

—, 1889a. The freshwater algae of North Yorkshire. — J. Bot. 27: 289-298.

—, 1890. Contribution to the freshwater algae of North Wales. — J. Roy. Microsc. Soc. (London) 6: 277-306, pls 5, 6.

—, 1891. The freshwater algae of Maine. — J. Bot. 29: 353-357, pl. 315.

—, 1891a. Additions to the freshwater algae of West Yorkshire. — Naturalist (London) 16: 243-252.

—, 1892. Algae of the English Lake District. — J. Roy. Microsc. Soc. (London) 8: 713-748.

—, 1892a. A contribution to the freshwater algae of West Ireland. — J. Linn. Soc., Bot. 29: 103-216, pls 18-24.

— & G.S. West, 1894. New British freshwater algae. — J. Roy. Microsc. Soc. (London) 1894: 1-17.

— & G.S. West, 1895. A contribution to our knowledge of the freshwater algae of Madagascar. — Trans. Linn. Soc. London, Bot. 5 (2): 41-90, pls 5-9.

— & G.S. West, 1895a. Some recently published Desmidieae. — J. Bot. 33: 65-70.

— & G.S. West, 1896. On some North American Desmidiaceae. — Trans. Linn. Soc. London, Bot. 5: 229-274.

— & G.S. West, 1896a. On some new and interesting freshwater algae. — J. Roy. Microsc. Soc. (London) 1896: 149-165, pls 3, 4.

— & G.S. West, 1897. A contribution to the freshwater algae of the south of England. — J. Roy. Microsc. Soc. (London) 1897: 467-511, pls 6, 7.

— & G.S. West, 1897a. Fresh-water algae from Burma, including a few from Bengal and Madras. — Ann. Roy. Bot. Gard. (Calcutta) 6: 175-260, pls 10-16.

— & G.S. West, 1897b. Welwitsch's African freshwater algae. — J. Bot. 35: 1-7, 33-42, 77-89, 113-183, 235-243, 264-272, 297-304, pls 365-370.

— & G.S. West, 1898. On some desmids of the United States. — J. Linn. Soc. Bot. 33: 279-322.

— & G.S. West, 1898a. Notes on freshwater algae. 1. — J. Bot. 36: 330-338.

— & G.S. West, 1900. Notes on freshwater algae. II. — J. Bot. 38: 289-299, pl. 412.

— & G.S. West, 1901. Freshwater Chlorophyceae. — In: J. Schmidt: Flora of Koh Chang, part IV. — Bot. Tidsskr. 24: 157-186.

— & G.S. West, 1901a. The alga-flora of Yorkshire. III. — Trans. Yorkshire Naturalists' Union 5 (25): 101-164.

— & G.S. West, 1902. A contribution to the freshwater algae of the north of Ireland. — Trans. Roy. Irish Acad. 32, B, I: 1-100, pls 1-3.

— & G.S. West, 1902a. A contribution to the freshwater algae of Ceylon. — Trans. Linn. Soc. London, Bot. 2nd ser. 6: 123-215.

— & G.S. West, 1903. Scottish freshwater plankton. No. I. — J. Linn. Soc., Bot. 35: 519-556, pls 14-18.

— & G.S. West, 1903a. Notes on freshwater algae. III. — J. Bot. 41: 33-41, 74-82, pls 446-448.

— & G.S. West, 1905. A further contribution to the freshwater plankton of the Scottish lochs. — Trans. Roy. Soc. Edinburgh 41: 477-518.

— & G.S. West, 1905a. Freshwater algae from the Orkneys and Shetlands. — Trans. & Proc. Bot. Soc. Edinburgh 23: 3-41, pls 1, 2.

— & G.S. West, 1906. A comparative study of the plankton of some Irish lakes. — Trans. Roy. Irish Acad. 33 (B): 77-116, pls 6-11.

— & G.S. West, 1907. Freshwater algae from Burma, including a few from Bengal and Madras. — Ann. Roy. Bot. Gard. (Calcutta) 6: 175-260.
— & G.S. West, 1909. The phytoplankton of the English Lake District. — Naturalist (London) 1909: 115-122, 134-141, 186-193, 260-267, 287-292, 323-331.
— & G.S. West, 1909a. The British freshwater phytoplankton with special reference to the desmid plankton and the distribution of British desmids. — Proc. Roy. Soc. London, ser. B, Biol. Sci. 81: 165-206.
— & G.S. West, 1912. A Monograph of the British Desmidiaceae. Vol. 4. — Ray Society, London, pp 1-194, pls 96-128.
— , G.S. West & N. Carter, 1923. A Monograph of the British Desmidiaceae. Vol. 5. — Ray Society, London, pp 1-300, pls 129-167.
Wille, N., 1879. Ferskvandsalger fra Novaja Semlja samlede af Dr. F. Kjellman paa Nordsenskiölds Expedition 1875. — Öfvers. Förh. Kongl. Svenska Vetensk.-Akad. 1879 (5): 13-74.
—, 1881. Bidrag till Kundsgaben om Norges Ferskvandalger. I. Smaalenenes Chlorophyllophyceer. — Skr. Vidensk.-Selsk. Christiania, Math.-Naturvidensk. Kl. 1880 (11): 1-72, pls 1, 2.
—, 1884. Bidrag til Sydamerikas algflora. I-III. — Bih. Kongl. Svenska Vetensk.-Akad. Handl. 8 (18): 1-64, pls 1-3.
Williamson, D.B., 1992. A contribution to our knowledge of the desmid flora of the Shetland Islands. — Bot. J. Scotl. 46: 233-285.
—, 1994. Observations on desmids from Scotland, Shetland, Cumbria and the English Midlands. — Bot. J. Scotl. 47: 113-122.
—, 1997. Rare desmids from Scotland. — Algol. Studies 84: 53-81.
Wittrock, V.B., 1869. Anteckningar om Skandinaviens Desmidiacéer. — Nova Acta Regiae Soc. Sci. Upsal., ser. 3, 7: 1-28.
—, 1872. Om Gotlands och Ölands sötvattens-alger. — Bih. Kongl. Svenska Vetensk.-Akad. Handl. 1 (1): 1-72.
— & O. Nordstedt, 1883. Algae aquae dulcis exsiccatae praecipue Scandinavicae, quas adjectis algis marinis chlorophyllaceis et phycochromaceis distribuerunt. — Bot. Not. 1883: 145-153.
— & O. Nordstedt, 1886. Algae aquae dulcis exsiccatae praecipue Scandinavicae, quas adjectis algis marinis chlorophyllaceis et phycochromaceis distribuerunt. Fasc. 15-17. Nos. 701-850. — Stockholm.
— & O. Nordstedt, 1893. Algae aquae dulcis exsiccatae. Fasc. 24, nr. 1101-1150. — Bot. Not. 1893: 185-200.
— , O. Nordstedt & G. Lagerheim, 1903. Algae aquae dulcis exsiccatae praecipue Scandinavicae, quas adjectis algis marinis chlorophyllaceis et phycochromaceis distribuerunt. Fasc. 35: 1-42. — Stockholm.
Wolle, F., 1881. American freshwater algae. Species and varieties of desmids new to science. — Bull. Torrey Bot. Club 8: 1-4.
—, 1884. Desmids of the United States and List of American Pediastrums. — Moravian Publication Office, Bethlehem, Pennsylvania,168 pp, 53 pls.
—, 1885. Freshwaster algae X. — Bull. Torrey Bot. Club 12: 125-129, pl. 51.
Woloszynska, J., 1919. Przyczynek do znajomosci glonow Litwy. — Rozpr. Wydz. Mat.-Przyr. Polsk. Akad. Umiejetn., Dzial B, Nauki Biol. 57: 1-65, pl. 3.
—, 1921. Glony okolic Kijowa. — Rozpr. Wydz. Mat.-Przyr. Polsk. Akad. Umiejetn., Dzial A/B, Nauki Mat.-Fiz. Biol. 60: 127-140.
Wood, H.C., 1873. A contribution to the history of the freshwater algae of North America. — Smithsonian Contr. Knowl. 19: 1-262, pls 1-21.
Woodhead, N. & R.D. Tweed, 1960. Additions to the algal flora of Newfoundland. — Hydrobiologia 15: 309-362.

Woronichin, N.N., 1930. Vodorosli Poljarnogo i Severnogo Urala (Algen des Polar- und des Nord-Urals). — Trudy Leningradsk. Obsc. Estestvoisp., Vyp. 3, Otd. Bot. 60 (3): 1-77, 3 pls.

Yacubson, S., 1980. The phytoplankton of some freshwater bodies from Zulia State (Venezuela). — Nova Hedwigia 33: 279-340.

Yamaguchi, H. & M. Hirano, 1953. Plankton desmids from Lake Biwa, 2. — Acta Phytotax. Geobot. 15: 56-60.

Zacharias, O., 1898. Untersuchungen über das Plankton der Teichgewässer. — Forschungsber. Biol. Stat. Plön 6: 89-139, pl. 4.

Plates

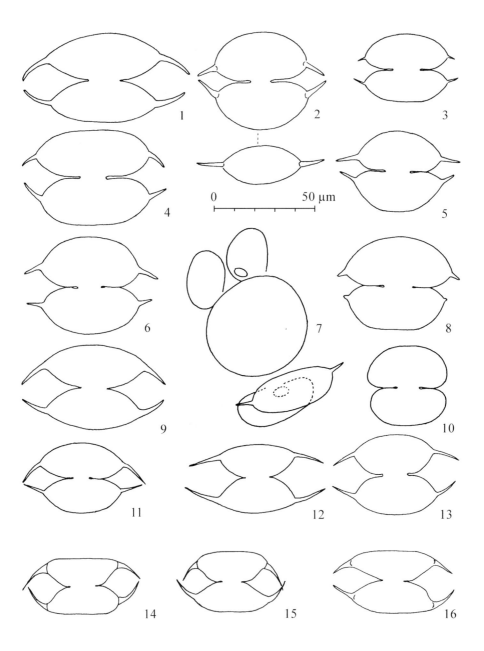

Plate 1. — 1-11. *Staurodesmus convergens* var. *convergens* ; 12-13. *Std. convergens* var. *deplanatus* ; 14-16. *Std. depressus* (1-2, 4. after Ralfs 1848; 3, 13. after West & West 1912; 5-8, 10-11. after Coesel 1994; 9 after Lenzenweger 1997; 12. after Deflandre 1926; 14-15. after Woloszynska 1921; 16. after Grönblad 1921).

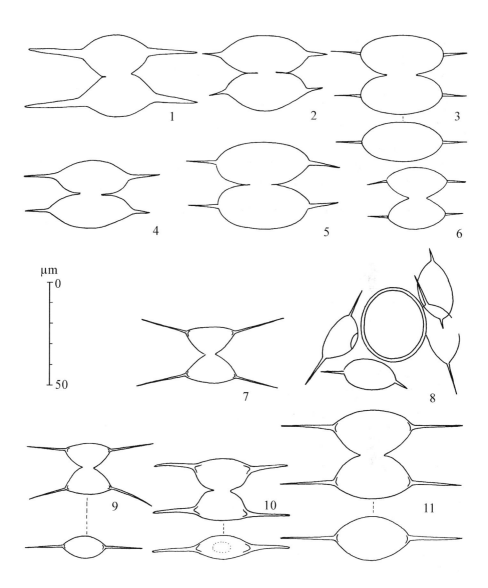

Plate 2. — 1-6. *Staurodesmus subulatus* var. *subulatus* ; 7-11. *Std. subulatus* var. *nordstedtii* (1. after Bailey 1841; 2, 4. after Grönblad 1960; 3, 6, 11. after West & West 1912; 5. after Deflandre 1924; 7, 9. after Smith 1924; 8. after Børgesen 1890; 10. after Lind & Brook 1980).

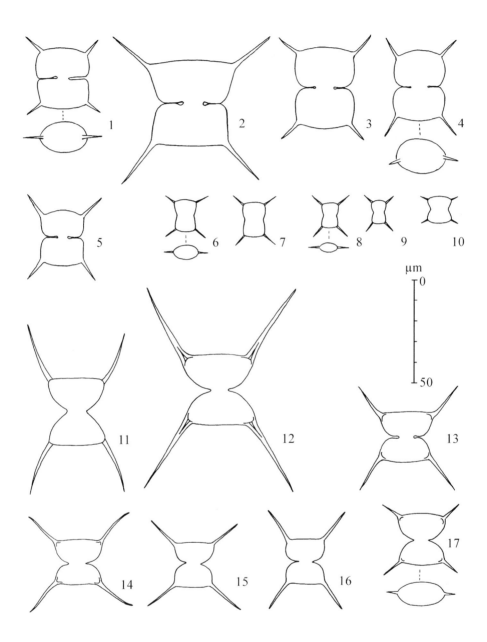

Plate 3. — 1-5. *Staurodesmus bulnheimii* ; 6-10. *Std. subquadratus* ; 11-12. *Std. validus* var. *validus* ; 13-17. *Std. validus* var. *subincus* (1. after Raciborski 1889; 2, 6-7, 11-13. after West & West 1912; 3. after Skuja 1964; 4. after Cedergren 1932; 5. after Donat 1926; 8-9. after Kossinskaja 1949; 10. after Tomaszewicz & Kowalski 1993; 14. after Lind & Brook 1980; 15-16. after Coesel 1994; 17. after Heimerl 1891).

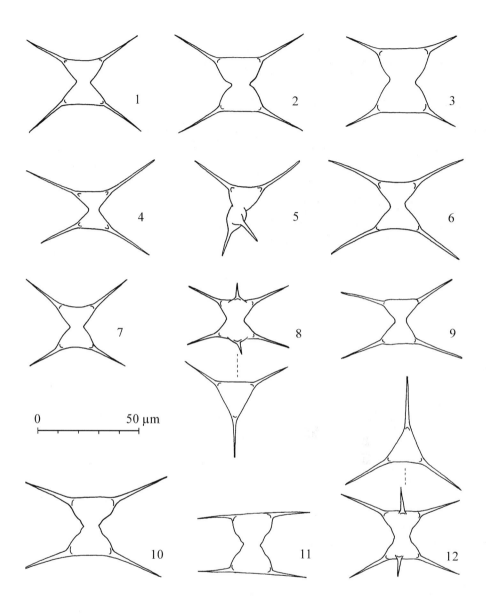

Plate 4. — 1-9. *Staurodesmus incus* var. *incus* ; 10-12. *Std. incus* var. *indentatus* (1-3, 10, 11. after West & West 1912; 4-6. after Florin 1957; 7, 8, 12. after Nygaard 1979; 9. after Teiling 1946).

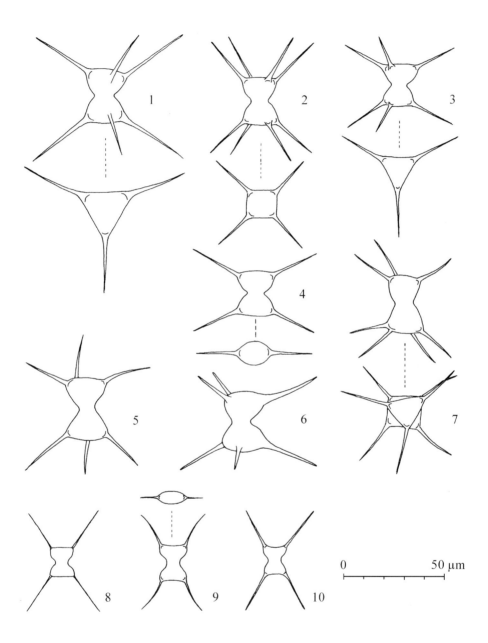

Plate 5. — 1-7. *Staurodesmus incus* var. *jaculiferus* ; 8-10. *Std. longispinus* (1. after W. West 1892a; 2-4. after West & West 1903; 5, 7. after Børgesen 1901; 6. after Teiling 1948; 8. after West & West 1905; 9. after Nygaard 1979; 10. after Teiling 1967).

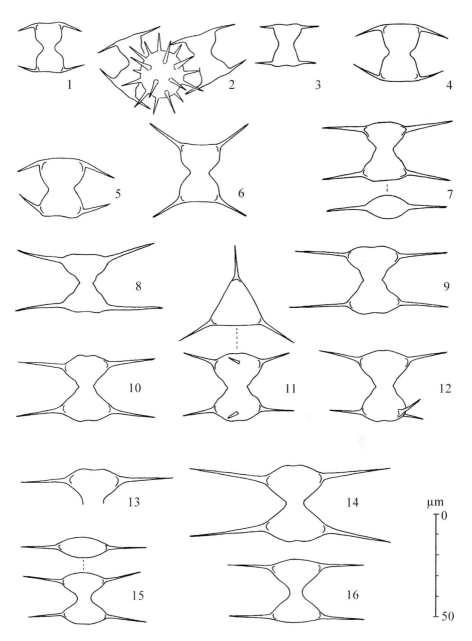

Plate 6. — 1-6. *Staurodesmus ralfsii* ; 7-12. *Std. subtriangularis* var. *subtriangularis* ; 13-16. *Std. subtriangularis* var. *inflatus* (1-3. after Ralfs 1848; 4, 9-12, 14-16. after West & West 1912; 5. after Frémy 1930; 6. after Strøm 1920; 7-8. after Borge 1897; 13. after W. West 1892a).

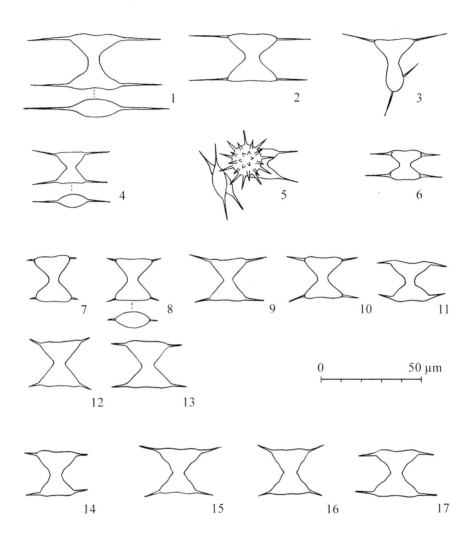

Plate 7. — 1-6. *Staurodesmus triangularis* var. *triangularis* ; 7-13. *Std. triangularis* var. *brevispina* ; 14-17. *Std. triangularis* var. *indentatus* (1. after Lagerheim 1885; 2-3. after West & West 1912; 4-5. after Nygaard 1991; 6. after Nygaard 1979; 7-8. after Allorge & Allorge 1931; 9, 11, 14. after Coesel 1994; 10. after Lenzenweger 1994; 12. after Borge 1913; 13. after Borge 1930; 15-16. after Förster 1970; 17. after Kouwets 1987).

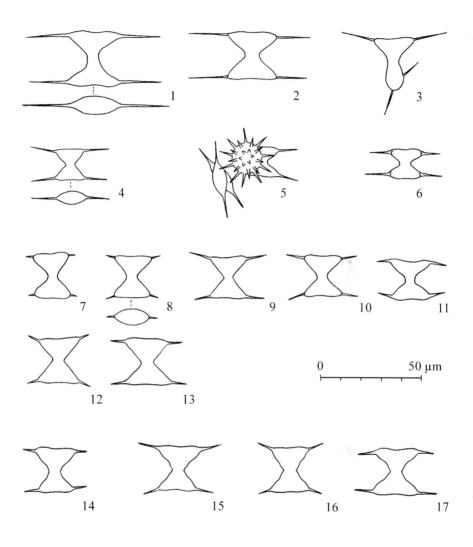

Plate 7. — 1-6. *Staurodesmus triangularis* var. *triangularis* ; 7-13. *Std. triangularis* var. *brevispina* ; 14-17. *Std. triangularis* var. *indentatus* (1. after Lagerheim 1885; 2-3. after West & West 1912; 4-5. after Nygaard 1991; 6. after Nygaard 1979; 7-8. after Allorge & Allorge 1931; 9, 11, 14. after Coesel 1994; 10. after Lenzenweger 1994; 12. after Borge 1913; 13. after Borge 1930; 15-16. after Förster 1970; 17. after Kouwets 1987).

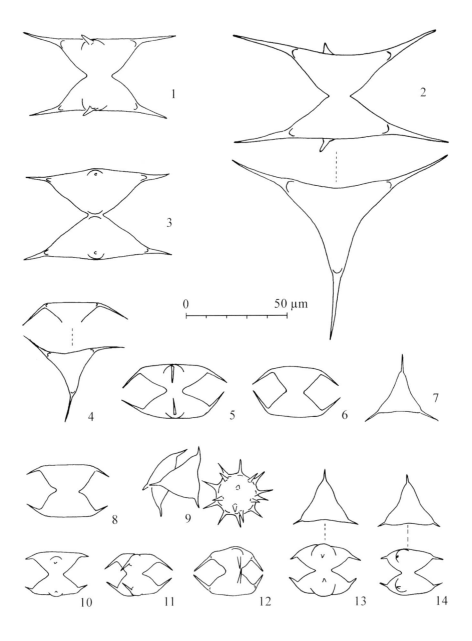

Plate 9. — 1-3. *Staurodesmus megacanthus* var. *scoticus* ; 4-14. *Std. glaber* var. *glaber* (1-2, 5-7. after West & al. 1923; 3. after Skuja 1964; 4. after West & West 1896; 8. after West & West 1912; 9-11, 13. after Coesel 1994; 12, 14. after Kouwets 1987).

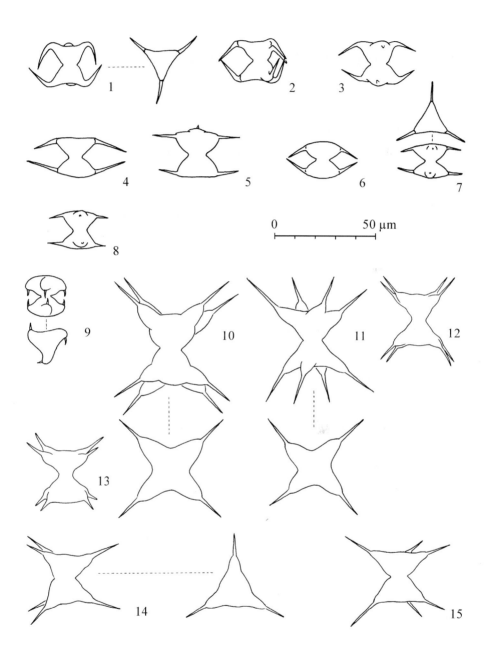

Plate 10. — 1-3. *Staurodesmus glaber* var. *hirundinella* ; 4-8. *Std. glaber* var. *limnophilus* ; 9. *Std. recurvus* ; 10-13. *Std. aristiferus* var. *aristiferus* ; 14-15. *Std. aristiferus* var. *protuberans* (1. after Messikommer 1949; 2. after Kouwets 1987; 3. after Coesel 1994; 4-6. after Teiling 1948; 7. after Williamson 1992; 8. after Coesel 1994; 9, 15. after Skuja 1964; 10. after Ralfs 1848; 11, 14. after West & al. 1923; 12-13. after Coesel 1994).

Plate 11. — 1-20. *Staurodesmus cuspidatus* (1-3. after Ralfs 1848; 4-8, 11. after Coesel 1994; 9-10, 14, 15. after Coesel & Meesters 2007; 12-13. after Lind & Brook 1980; 16-17. after Teiling 1948; 18-20. after West & al. 1923).

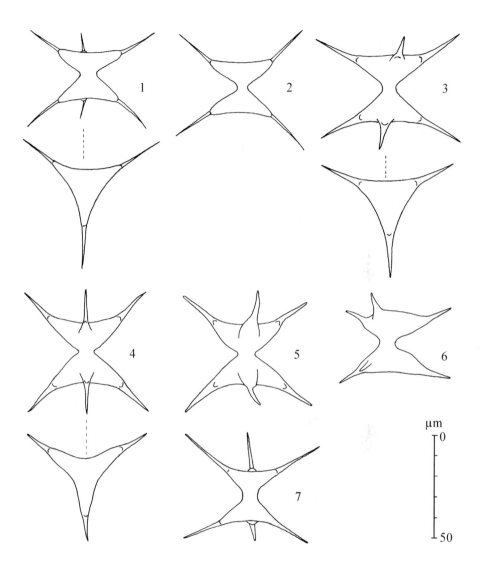

Plate 12. — 1-7. *Staurodesmus cuspidicurvatus* (1-2. after W. West 1892a; 3. after West & West 1903; 4. after Skuja 1964; 5. after Florin 1957; 6. after Coesel & Wardenaar 1990; 7. after Teling 1967).

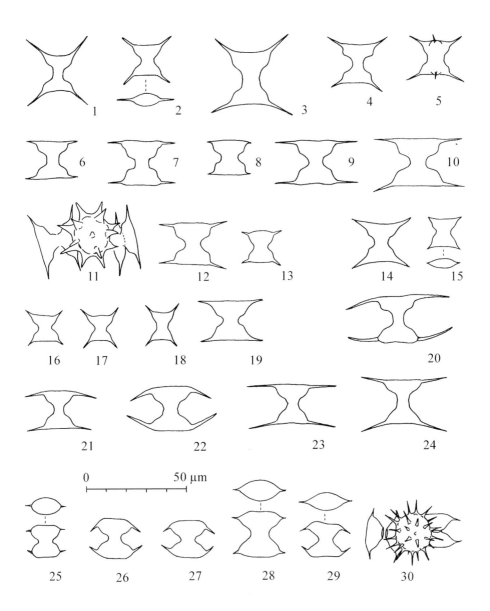

Plate 13. — 1-5. *Staurodesmus extensus* var. *extensus* ; 6-13. *Std. extensus* var. *rectus* ; 14-19. *Std. extensus* var. *isthmosus* ; 20-24. *Std. extensus* var. *joshua* ; 25-30. *Sd. subhexagonus* (1. after Borge 1913; 2, 8-9, 19. after Tomaszewicz 1988; 3-4, 10-12, 16-18. after Coesel 1994; 5. after Lenzenweger 1997; 6-7. after Eichler & Raciborski; 13, 21-24, 26-27. after Coesel & Meesters 2007; 14-15. after Heimerl 1891; 20. after Gutwiński 1892; 25. after West & West 1912; 28. after Borge 1913; 29. after Borge 1906; 30. after Skuja 1964).

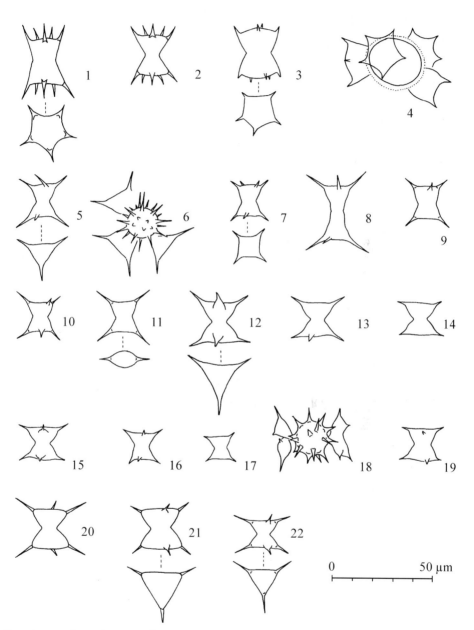

Plate 14. — 1-4. *Staurodesmus wandae* ; 5-22. *Std. omearae* (1-2. after Raciborski 1889; 3-4. after Grönblad 1938; 5-8. after Archer 1858; 9-11. after Borge 1936; 12-19. after Coesel 1994; 20-22. after West & al. 1923).

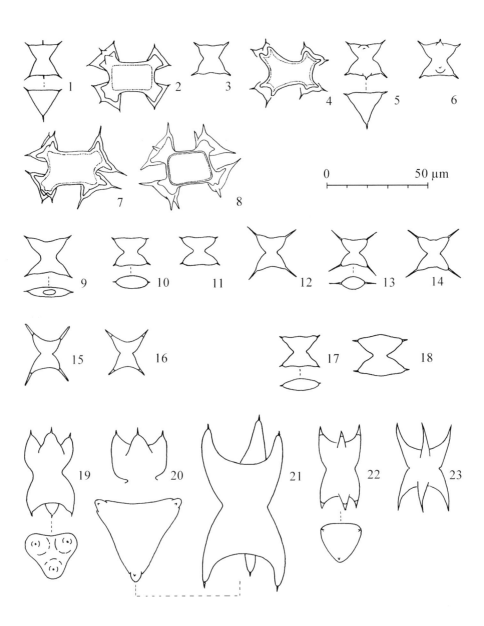

Plate 15. — 1-8. *Staurodesmus pterosporus* ; 9-14. *Std. phimus* ; 15-16. *Std. semilunaris* ; 17-18. *Std. hebridarus*; 19-23. *Std. unguiferus* (1-2. after Lundell 1871; 3-4. after Bourrelly 1966; 5-7. after Coesel 1994; 8. after Coesel & Meesters 2007; 9, 19-20. after Turner 1892; 10-11, 17. after West & West 1912; 12. after Kouwets 1987; 13. after Lenzenweger 1991; 14. after Lenzenweger 1988; 15. after Schmidle 1896; 16. after Krieger 1930; 18. after Skuja 1964; 21. after Grönblad 1938; 22. after Grönblad 1920; 23. after Grönblad 1947).

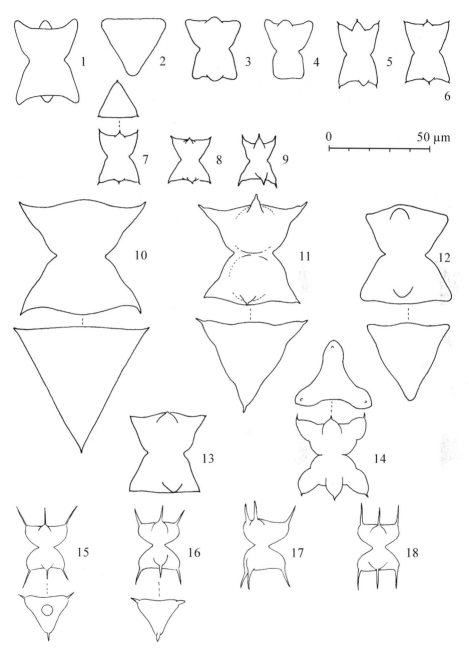

Plate 16. — 1-9. *Staurodesmus corniculatus* ; 10-13. *Std. leptodermus* ; 14. *Std. andrzejowskii* ; 15-18. *Std. connatus* (1, 10, 15. after Lundell 1871; 2-7. after West & West 1912; 8-9. after Heimans 1940; 11. after Brook 1958; 12. after Lundberg 1931; 13. after Grönblad 1942; 14. after Woloszynska 1921; 16-17. after Coesel 1994; 18. after Margalef 1956).

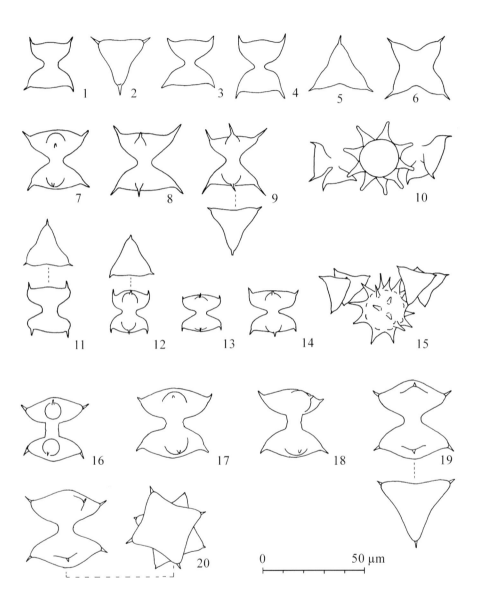

Plate 17. — 1-10. *Staurodesmus dejectus* var. *dejectus* ; 11-15. *Std. dejectus* var. *apiculatus* ; 16-20. *Std. dejectus* var. *robustus* (1-2. after Teiling 1954; 3-6, 10. after Teiling 1967; 7-9. after Coesel 1994; 11. after Brébisson 1856; 12. after Förster 1970; 13. after Lenzenweger 1997; 14-15, 17-18. after Coesel 1994; 16. after Messikommer 1928; 19-20. after Nygaard 1991).

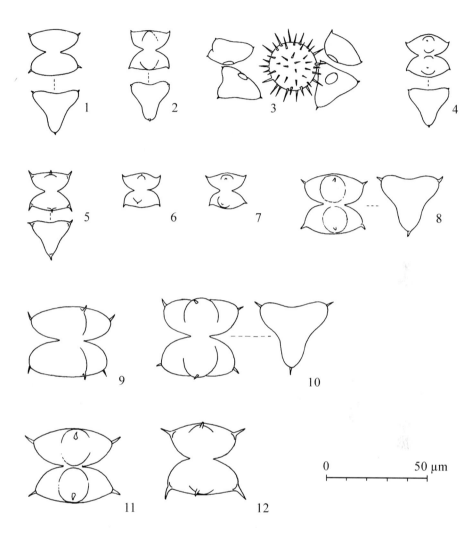

Plate 18. — 1-8. *Staurodesmus patens* var. *patens* ; 9-12. *Std. patens* var. *inflatus* (1. after Nordstedt 1888, 2-3, 10. after West & al. 1923, 4. after Lenzenweger 1981, 5. after Růžička 1972, 6-7. after Coesel 1994; 8. after Lenzenweger 1986; 9. after W. West 1892a; 11. after Skuja 1964, 12. after Borge 1930).

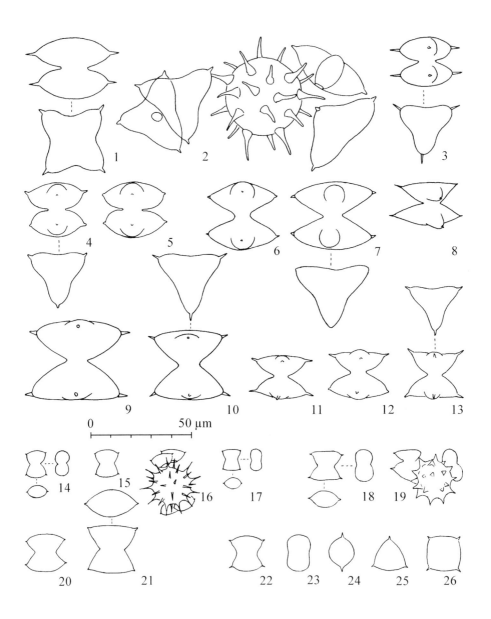

Plate 19. — 1-5. *Staurodesmus mucronatus* var. *mucronatus* ; 6-8. *Std. mucronatus* var. *delicatulus* ; 9-13. *Std. mucronatus* var. *subtriangularis* ; 14-17. *Std. controversus* var. *controversus* ; 18-26. Std. *controversus* var. *crassus* (1-2. after Ralfs 1848; 3, 10. after West & Carter 1923; 4-5, 11-13. after Coesel 1994; 6. after G.S. West 1909; 7. after Messikommer 1960; 8. after Cedercreutz & Grönblad 1936; 9, 21. after West & West 1903; 14. after W. West 1892a; 15-16. after West & West 1912; 17. after Allorge 1931; 18-19. after Borge 1930; 20. after Grönblad 1942; 22-26. after Schröder 1898) (22-24. biradiate cell in frontal, lateral and apical view, respectively; 25. triradiate cell in apical view; 26. quadriradiate cell in apical view).

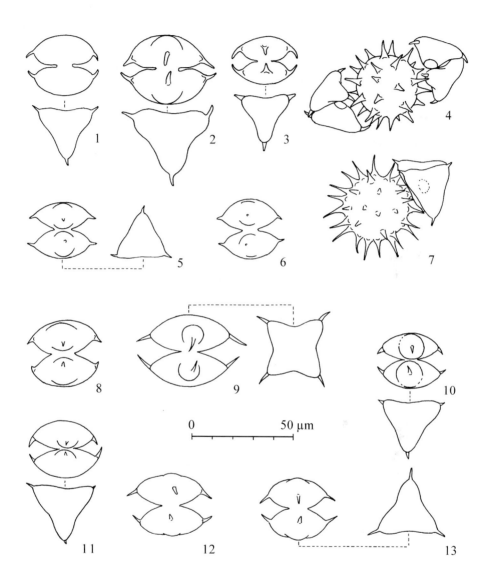

Plate 20. — 1-13. *Staurodesmus dickiei* var. *dickiei* (1. after Ralfs 1848; 2. after Lind & Brook 1980; 3-4. after Dick 1923; 5-7. after Coesel 1994; 8. after Lenzenweger 1981; 9-11. after Messikommer 1942; 12-13. after Kouwets 1987).

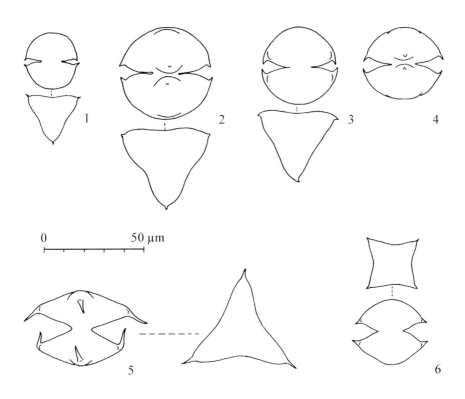

Plate 21. — 1-4. *Staurodesmus dickiei* var. *circularis* ; 5-6. *Std. dickiei* var. *rhomboides* (1. after Turner 1892; 2, 4. after Coesel 1994; 3. after Tarnogradsky 1960; 5. after West & West 1903; 6. after Borge 1911).

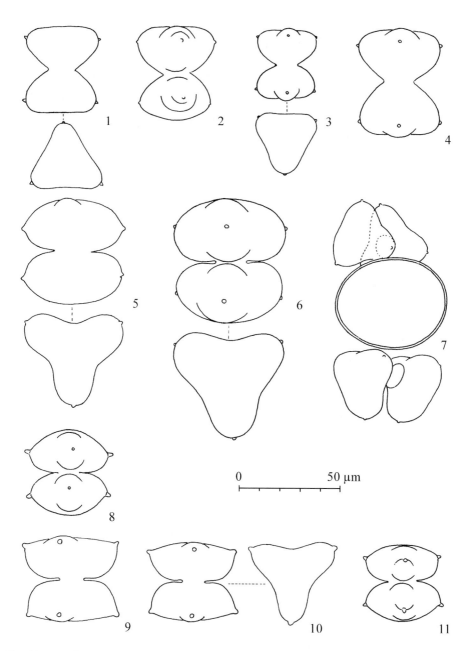

Plate 22. — 1-4. *Staurastrum aversum* ; 5-8. *S. brevispina* var. *brevispina* ; 9-11. *S. brevispina* var. *obversum* (1. after Lundell 1871 in West & West 1912; 2. after Lenzenweger 1985; 3-4, 6, 9,10. after West & West 1912; 5. after Ralfs 1848; 7. after Coesel 1994; 8, 11. after Lenzenweger 1986).

Plate 23. — 1-4. *Staurastrum brevispina* var. *boldtii* ; 5-6. *S. tumidum* (1. after Boldt 1885; 2. after Williamson 1992; 3. after Borge 1894; 4. after West & West 1912; 5. after Ralfs 1848; 6. after Lundell 1871).

Plate 24. — 1-2. *Staurastrum tumidum* ; 3. *S. julicum* ; 4-7. *S. lanceolatum* var. *lanceolatum* ; 8-11. *S. lanceolatum* var. *compressum* (1. after West & West 1912; 2. after Coesel 1994; 3. after Pevalek 1925; 4-6. after Archer 1862; 7. after Messikommer 1951; 8, 9. after West & West 1912; 10,11. after Coesel 1994).

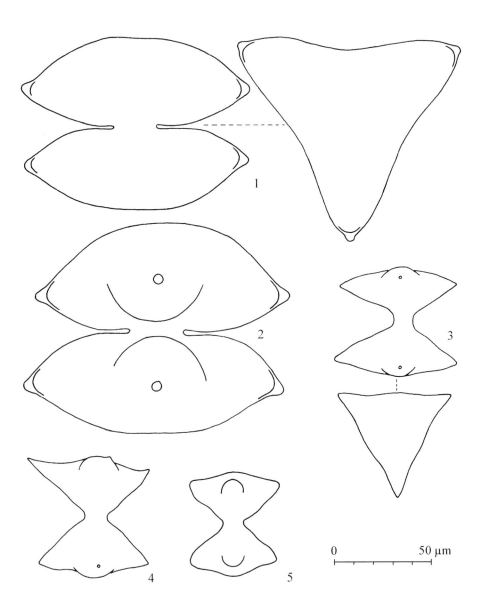

Plate 25. — 1-2. *Staurastrum conspicuum* ; 3-5. *S. inelegans* (1-3. after West & West 1912; 4-5. after Lundberg 1931).

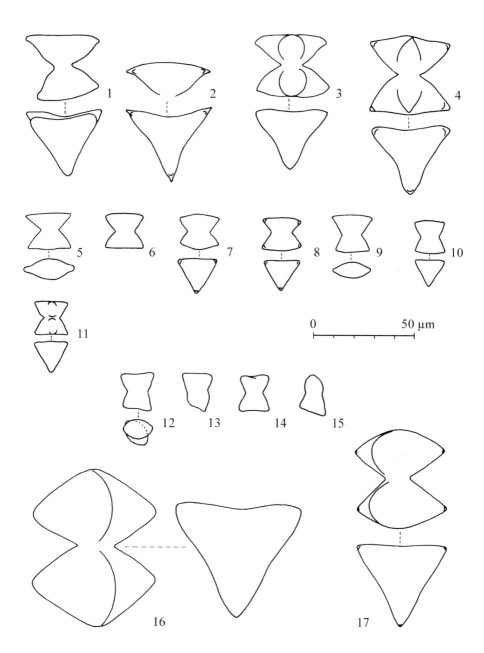

Plate 26. — 1-4. *Staurastrum clepsydra* ; 5-11. *S. sibiricum* ; 12-15. *Staurastrum tortum* ; 16. *S. angulatum* var. *angulatum* ; 17. *S. angulatum* var. *planctonicum* (1-2. after Nordstedt 1869; 3. after Messikommer 1942; 4. after Boldt 1888; 5-6. after Borge 1891; 7. after Boldt 1888; 8. after Børgesen 1894; 9-10, 16. after West & West 1912; 11. after Skuja 1964; 12-13. after Wittrock & al. 1903; 14-15. after Bijkerk, archives; 17. after West & West 1903).

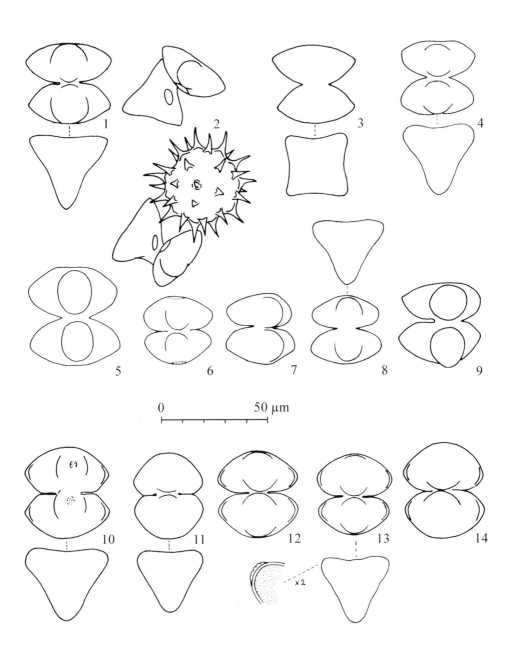

Plate 27. — 1-9. *Staurastrum bieneanum* ; 10-14. *S. crassangulatum* (1-2. after West & West 1912; 3. after Nordstedt 1875; 4. after Capdeville 1985; 5. after Nygaard 1949; 6-8. after Coesel 1997; 9. after Florin 1957; 10. after Kaiser 1919; 11. after Růžička 1972; 12-13. after Coesel 1997; 14. after Lenzenweger 1989).

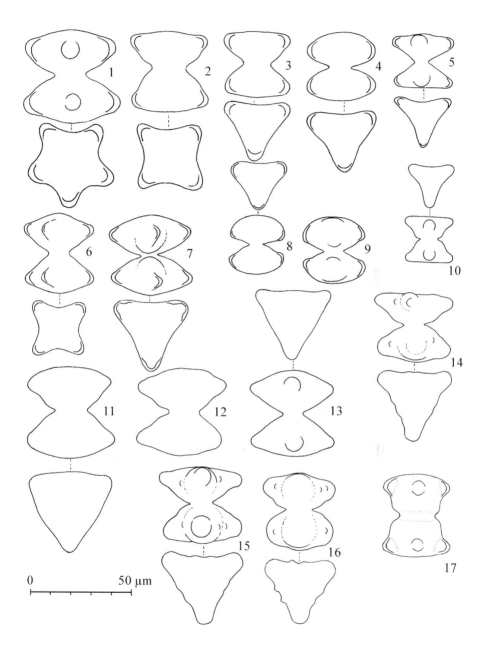

Plate 28. — 1-7. *Staurastrum pachyrhynchum* var. *pachyrhynchum* ; 8-9. *S. pachyrhynchum* var. *convergens* ; 10. *S. pachyrhynchum* var. *polonicum* ; 11-12. *S. subpygmaeum* var. *subpygmaeum* ; 13. *S. subpygmaeum* var. *subangulatum* ; 14-16. *S. subpygmaeum* var. *undulatum* ; 17. *S. bayernense* (1-3. after Nordstedt 1875; 4. after West & West 1912; 5. after Lenzenweger 1989b; 6-7. after Skuja 1964; 8. after Raciborski; 9. after Lenzenweger 1989; 10. after Eichler & Gutwiński 1895; 11-12. after W. West 1892; 13. after West & West 1912; 14-16. after John & Williamson 2009; 17. after Dick 1919).

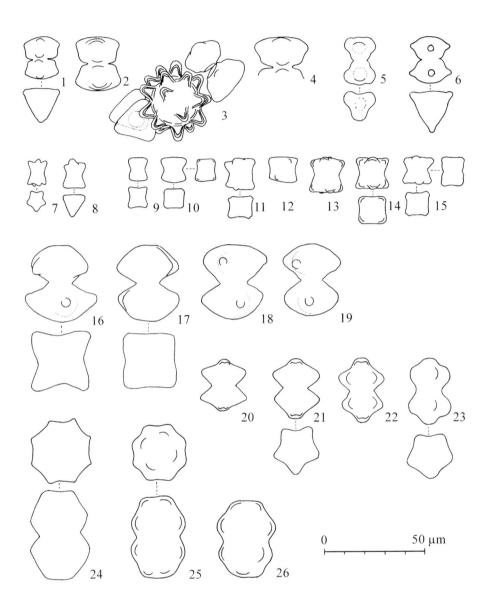

Plate 29. — 1-4. *Staurastrum groenbladii* ; 5. *S. schroederi* ; 6. *S. crassimamillatum* ; 7-8. *S. minutissimum* var. *minutissimum* ; 9-15. *S. minutissimum* var. *convexum* ; 16-19. *S. obscurum* ; 20-23. *S. insigne* ; 24-26. *S. habeebense* (1, 5 after Grönblad 1926; 2. after Lenzenweger 1997; 3-4. after Skuja 1931; 6. after Eichler & Gutwiński 1895; 7, 9. after Reinsch 1867 in West & West 1912; 8, 10. after Boldt 1888; 11-12. after Grönblad 1921; 13, 15. after Förster 1967; 14. after Krieger 1938; 16-19. after Coesel 1996; 20-21. after Lundell 1871; 22. after Peterfi 1963; 23. after Heimerl 1891; 24. after Irénéé-Marie 1949a; 25-26. after Brook & Williamson 1990).

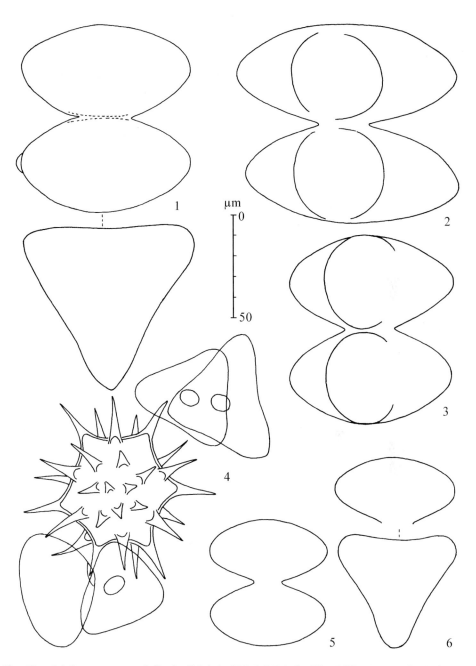

Plate 30. — 1-6. *Staurastrum grande* (1. after Bulnheim 1861; 2-3, 5-6. after West & West 1912; 4. after Cushman 1905 in West & West 1912).

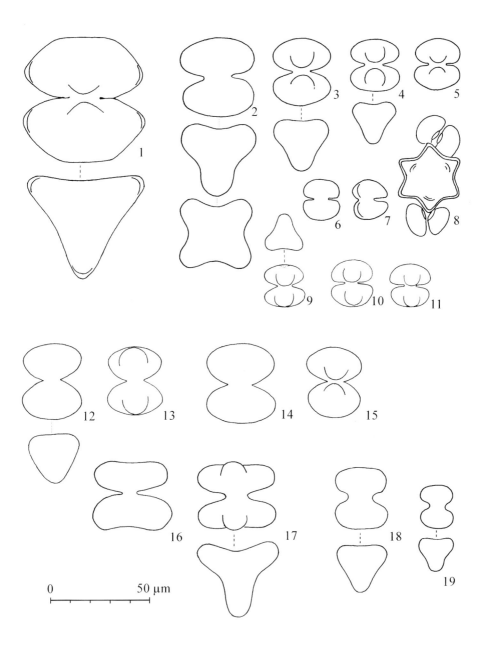

Plate 31. — 1. *Staurastrum keuruense* ; 2-11. *S. muticum* ; 12-15. *S. myrdalense* ; 16-17. *S. coarctatum* var. *coarctatum* ; 18-19. *S. coarctatum* var. *subcurtum* (1. after Grönblad 1920; 2. after Ralfs 1848; 3-5. after West & West 1912; 6-8. after Homfeld 1929; 9-11, 13. after Coesel 1997; 12. after Smith 1924; 14. after Strøm 1926; 15. after Messikommer 1956; 16. after Brébisson 1856; 17. after Hirano 1959; 18. after Nordstedt 1888; 19. after W. West 1890).

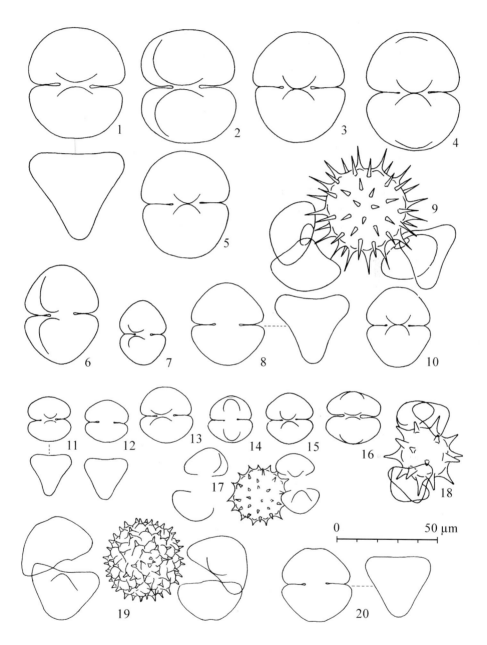

Plate 32. — 1-5. *Staurastrum orbiculare* ; 6-10. *S. ralfsii* var. *ralfsii*; 11-18. *S. ralfsii* var. *depressum*; 19-20. *S. suborbiculare* (1-2. after West & West 1912; 3. after Kouwets 1987; 4. after Lenzenweger 1997; 5, 10. after Coesel 1997; 6-9. after West & West 1912; 11, 19. after Roy & Bisset 1886; 12-13, 20. after West & West 1912; 14-15, 18. after Coesel 1997; 16. after Lind & Brook 1980; 17. after Hegde & Bharati 1983).

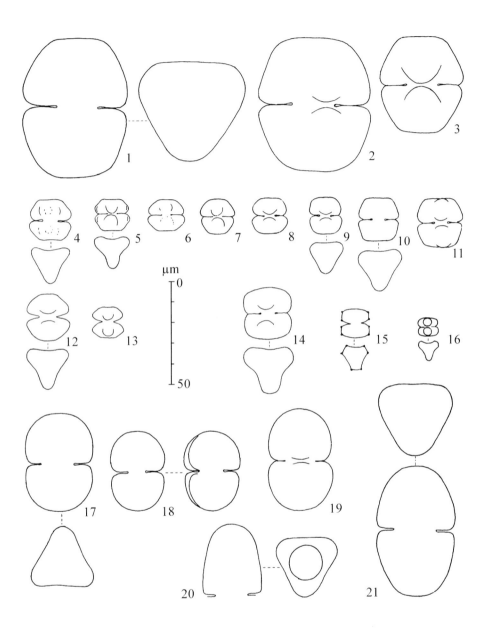

Plate 33. — 1-3. *Staurastrum hibernicum* ; 4-11. *S. retusum* var. *retusum* ; 12-13. *S. retusum* var. *hians* ; 14. *S. quadratulum* ; 15. *S. kobelianum* ; 16. *S. pokljukense* ; 17-19. *S. extensum* ; 20-21. *S. cosmarioides* (1-2, 8-11, 18, 21. after West & West 1912; 3. after Kossinskaja 1953; 4. after Turner 1892; 5-7, 13. after Eichler & Gutwiński 1895; 12. after Růžička 1972; 14. after Schmidle 1897; 15. after Schröder 1919; 16. after Pevalek 1925; 17. after Nordstedt 1873; 19. after Růžička 1957; 20. after Nordstedt 1870 in West & West 1912).

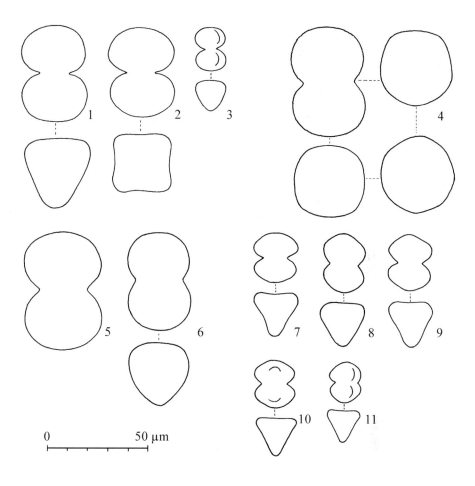

0 ——————— 50 μm

Plate 34. — 1-3. *Staurastrum ellipticum* ; 4-6. *S. subsphaericum* ; 7-11. *S. thomassonii* ; (1. after W. West 1892; 2. after Børgesen 1894; 3. after Skuja 1964; 4-6. after Nordstedt 1875; 7-11. after Nygaard 1991).

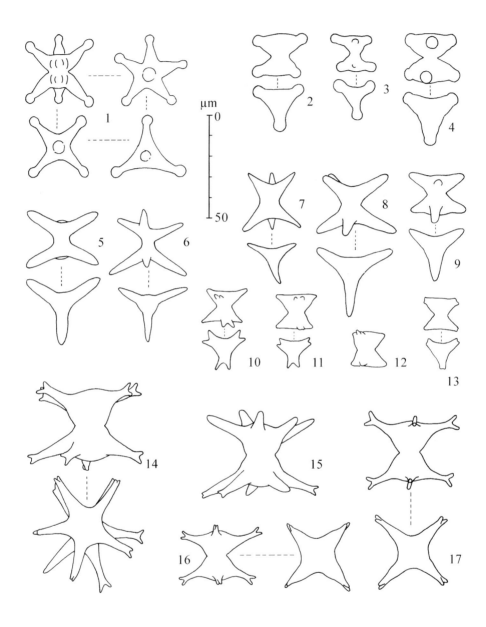

Plate 35. — 1. *Staurastrum bacillare* var. *bacillare* ; 2-4. *S. bacillare* var. *obesum* ; 5-6. *S. sublaevispinum* ; 7-9. *S. laevispinum* var. *laevispinum*; 10-13. *S. laevispinum* var. *compactum* ;14-17. *S. subnudibranchiatum* (1. after Ralfs 1848 in West & al. 1923; 2. after Lundell 1871 in West & al. 1923; 3, 8, 14-15. after West & al. 1923; 4. after Kouwets 1987; 5. after West & West 1898; 6. after Grönblad 1948; 7. after Bisset 1884 in West & al. 1923; 9. after Lenzenweger 1994; 10-13. after Grönblad 1942; 16-17. after Nygaard 1991).

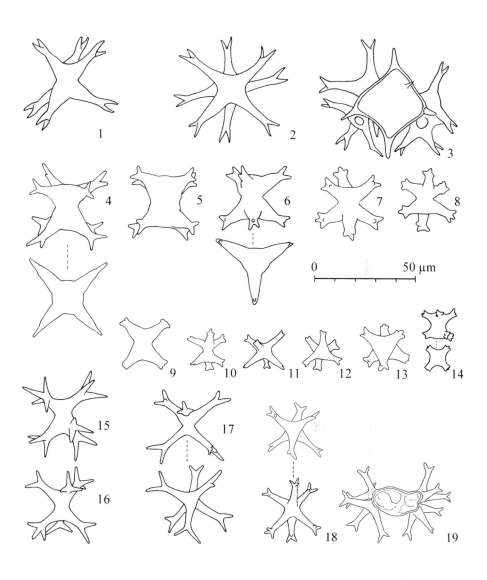

Plate 36. — 1-18. *Staurastrum brachiatum* (1-3. after Ralfs 1848; 4-5, 7-13. after Coesel 1997; 6, 14, 16. after Kouwets 1987; 15. after Coesel 1998; 17. after Williamson 1992; 18-19. after Meesters, archives 2011).

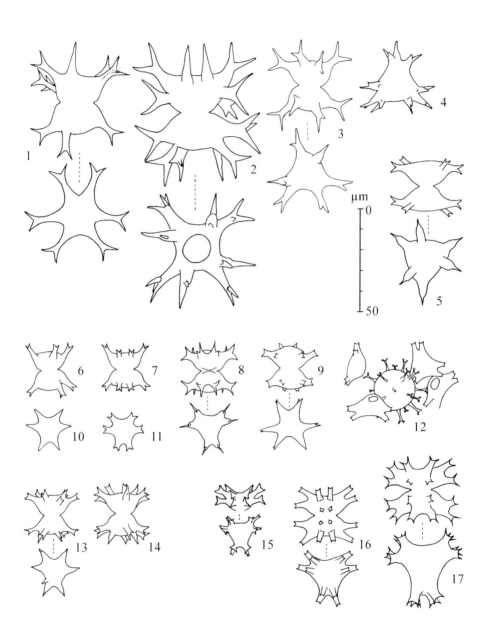

Plate 37. — 1-4. *Staurastrum clevei* var. *clevei* ; 5. *S. clevei* var. *inflatum* ; 6-14. *S. laeve* ; 15-17. *S. gemelliparum* (1. after Wittrock 1869; 2. after Ryppowa 1927; 3. after Coesel 1997; 4. after Messikommer 1960; 5. after Schmidle 1898; 6-7, 10-11. after Ralfs 1848; 8. after Lenzenweger; 9, 12. after West & al. 1923; 13-14. after Kouwets 1987; 15. after Nordstedt 1887; 16. after Tomaszewicz & Skrzeczkowska 1989; 17. after Nauwerck 1962).

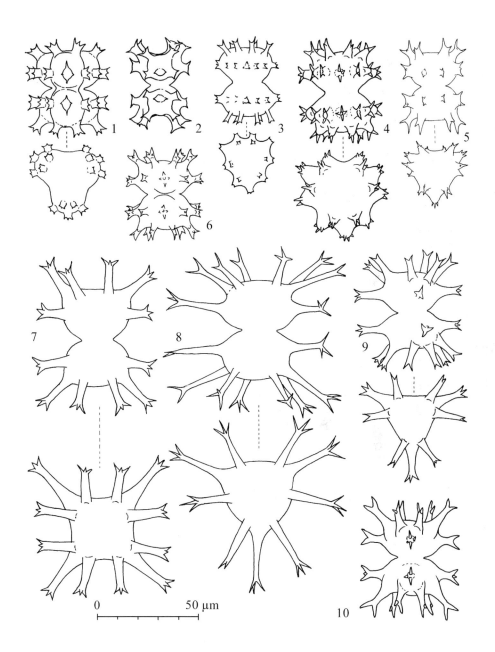

Plate 38. — 1-6. *Staurastrum hantzschii* ; 7-10. *S. tohopekaligense* (1-2. after Delponte 1878; 3. after Homfeld 1929; 4. after Dick 1919; 5. after Coesel 1997; 6. after Lenzenweger 1994; 7-9. after West & al. 1923; 10. after Dick 1923).

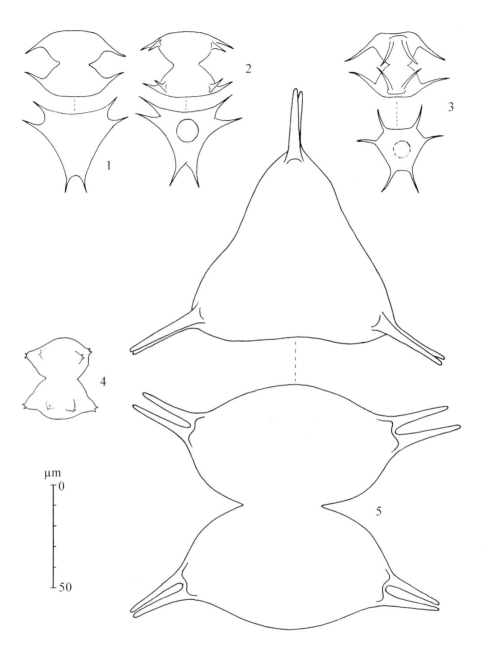

Plate 39. — 1-2. *Staurastrum bifidum* var. *bifidum* ; 3. *S. bifidum* var. *hexagonum* ; 4. *S. bispiniferum* ; 5. *S. longispinum* (1. after Lundell 1871; 2, 5. after West & al. 1923; 3. after Schaarschmidt 1883 ; 4. after Capdevielle 1978).

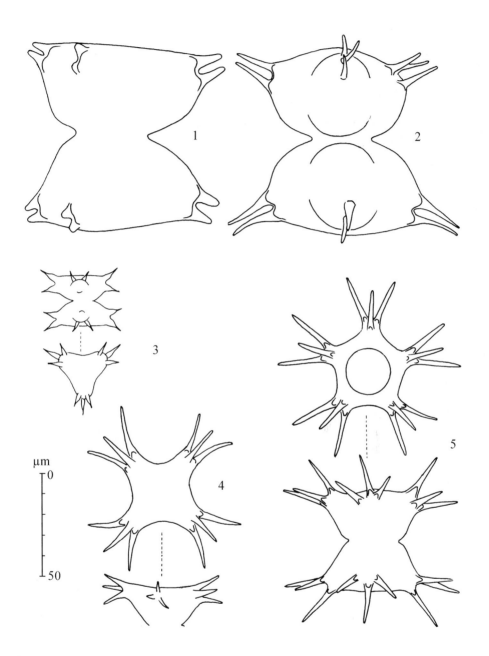

Plate 40. — 1-2. *Staurastrum longispinum* ; 3. *S. besseri* ; 4-5. *S. brasiliense* (1. after West & al. 1923; 2. after Brook 1958; 3. after Woloszynska 1921; 4. after Nordstedt 1870; 5. after Donat 1926).

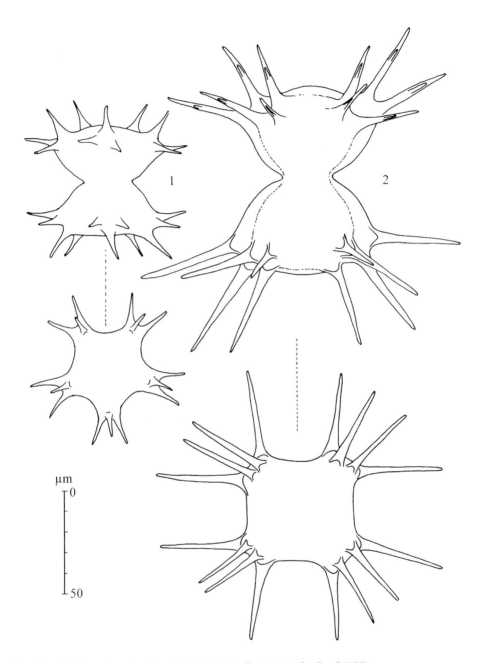

Plate 41. — 1-2. *Staurastrum brasiliense* (1. after Lütkemüller 1910; 2. after Brook 1958).

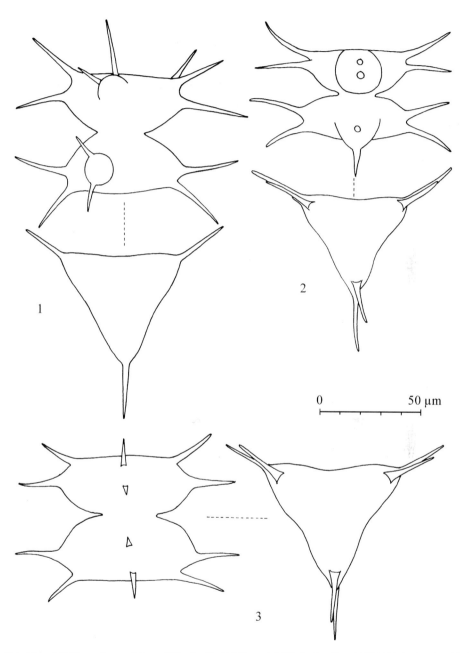

Plate 42. — 1-3. *Staurastrum wildemanii* (1. after Scott & Prescott 1956; 2. after Gutwiński 1902; 3. after Grönblad 1920).

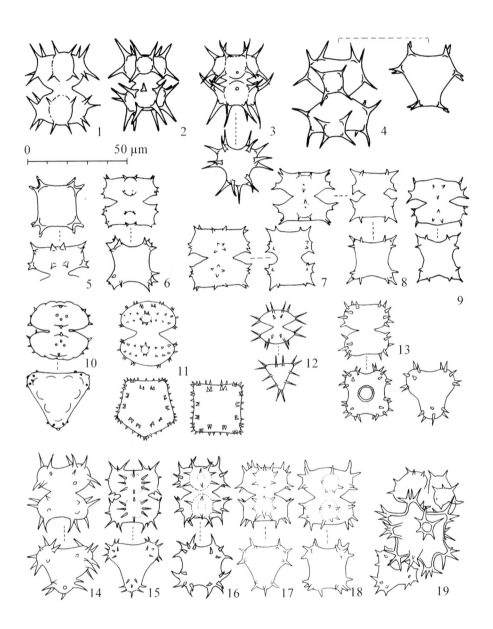

Plate 43. — 1-4. *Staurastrum quadrispinatum* ; 5-9. *S. quadrangulare* ; 10. *S. gatniense* ; 11. *S. echinodermum* ; 12. *S. kanitzii* ; 13-19. *S. hystrix* (1. after Turner 1886; 2. after Peterfi 1973; 3. after Lenzenweger 1997; 4-5, 11. after West & al. 1923; 6. after Grönblad 1960; 7. after Capdevielle 1978; 8. after Coesel 1997; 9. after Manguin 1936; 10. after West & West 1902; 12. after Schaarschmidt 1883; 13. after Ralfs 1848; 14. after Lütkemüller 1900; 15. after Allorge & Allorge 1931; 16. after Lenzenweger 1997; 17-18. after Coesel 1997; 19. after Beijerinck 1926 in Coesel 1997).

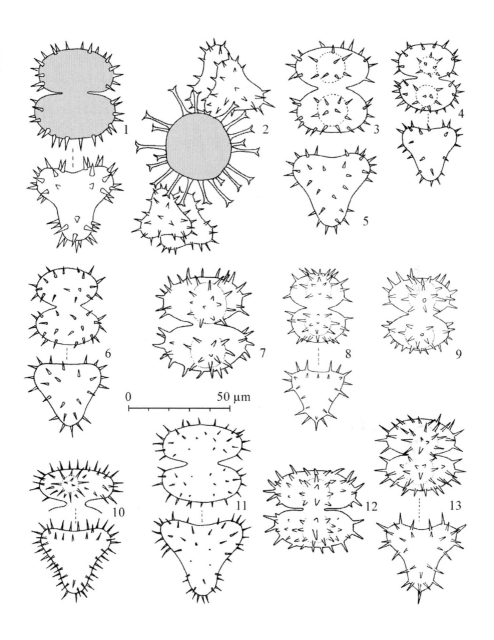

Plate 44. — 1-9. *Staurastrum teliferum* var. *teliferum* ; 10-13. *S. teliferum* var. *gladiosum* (1-2. after Ralfs 1848; 3-6, 11. after West & al. 1923; 7-9, 12-13. after Coesel 1997; 10. after Turner 1885).

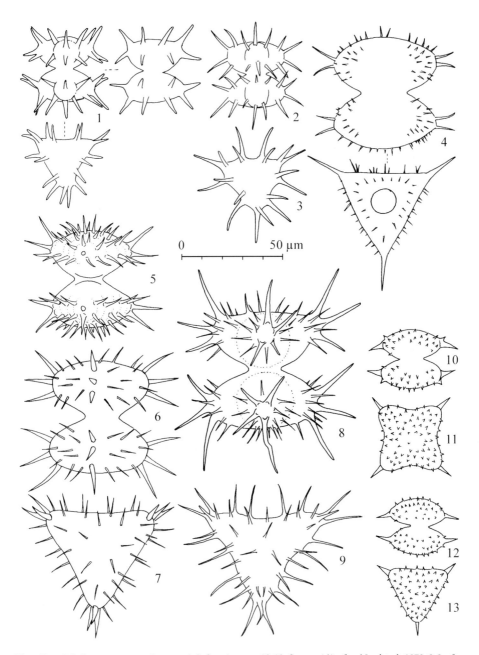

Plate 45. — 1-3. *Staurastrum geminatum* ; 4-9. *S. setigerum* ; 10-13. *S. ungeri* (1. after Nordstedt 1873; 2-3. after Printz 1915; 4. after Cleve 1864; 5. after Lenzenweger 1997; 6-7 after West & al. 1923; 8-9. after Brook 1958; 10-13. after Reinsch 1867).

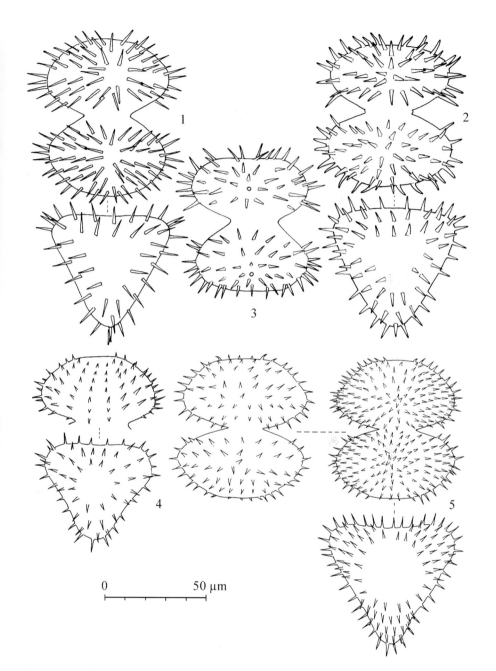

0   50 µm

Plate 46. — 1-3. *Staurastrum polytrichum* ; 4-5. *S. subbrebissonii* (1. after Roy & Bisset 1894; 2-3. after Coesel 1997; 4. after Schmidle 1894; 5. after Bourrelly 1987).

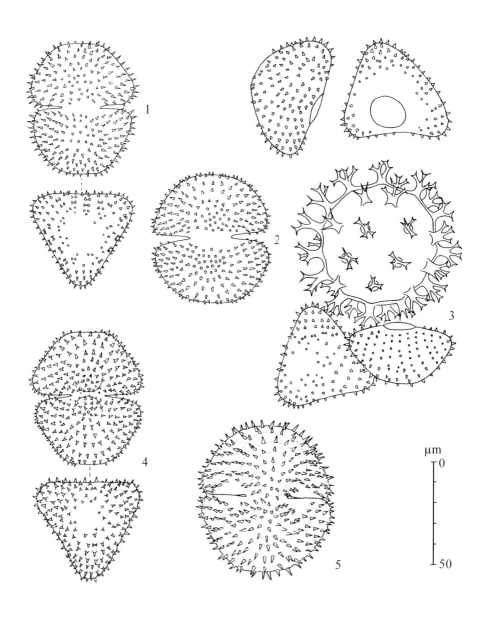

Plate 47. — 1-5. *Staurastrum pyramidatum* (1-3. after West & al. 1923; 4. after Dick 1930; 5. after Kouwets 1987).

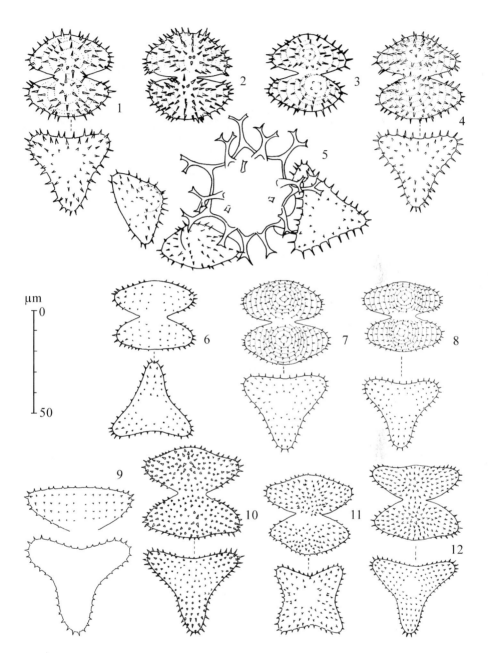

Plate 48. — 1-5. *Staurastrum kouwetsii* ; 6-8. *S. brebissonii* var. *brebissonii* ; 9-12. *S. brebissonii* var. *ordinatum* (1. after Lenzenweger 1984; 2. after Kouwets 1987; 3-5, 7-8, 12. after Coesel 1997; 6, 10. after West & al. 1923; 9. after Schmidle 1898; 11. after Messikommer 1942).

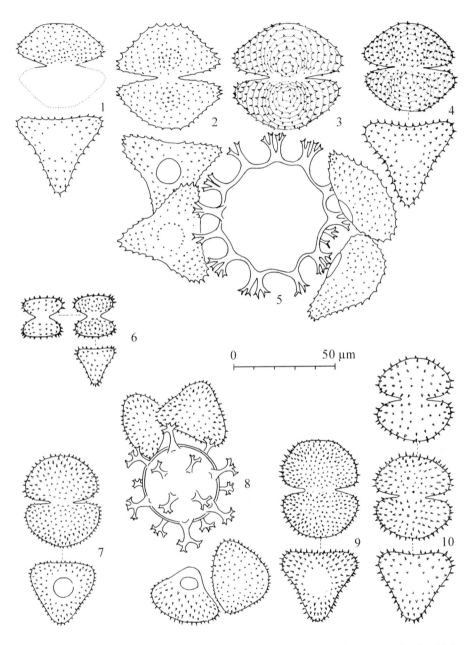

Plate 49. — 1-5. *Staurastrum trapezioides* ; 6. *S. erostellum* ; 7-10. *S. hirsutum* var. *hirsutum* (1. after Grönblad 1926; 2, 5. after Homfeld 1929; 3-4. after Coesel 1997; 6. after West & West 1897; 7-8. after Ralfs 1848; 9-10. after West & al. 1923).

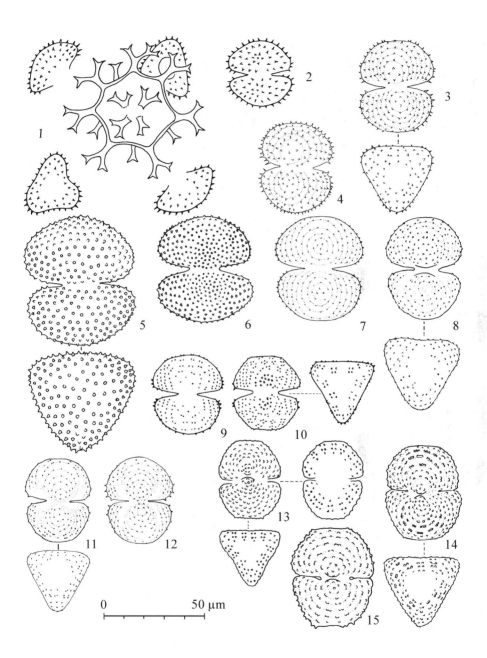

Plate 50. — 1-4. *Staurastrum hirsutum* var. *hirsutum* ; 5-8. *S. hirsutum* var. *muricatum* ; 9-12. *S. hirsutum* var. *pseudarnellii*; 13-15. *S. arnellii* (1-2. after Kossinskaja 1950; 3-4, 7, 11-12. after Coesel 1997; 5. after Ralfs 1848 and West & al. 1923; 6, 9-10. after West & al. 1923; 8 after Kouwets 1987; 13. after Boldt 1885; 14. after Lenzenweger 1997; 15. original, after Austrian material).

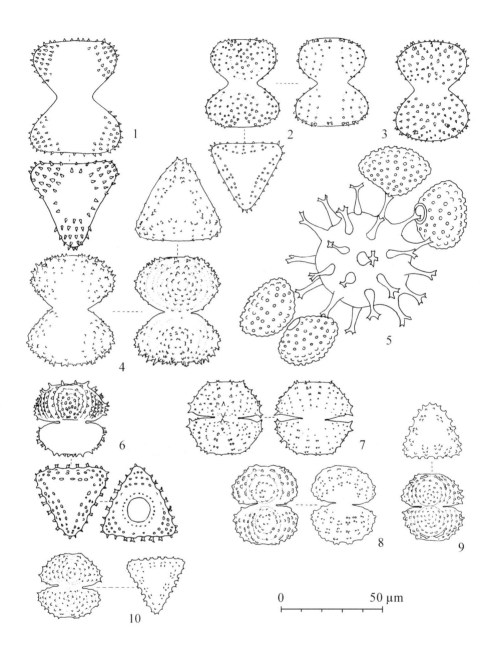

Plate 51. — 1. *Staurastrum horametrum* ; 2-5. *S. asperum* ; 6-10. *S. scabrum* (1. after Roy & Bisset 1894 in West & al. 1923; 2-3. after West & al. 1923; 4. after Williamson 1994; 5. after Ralfs 1848; 6. after Wittrock & Nordstedt 1893; 7. after Rybnicek 1960; 8-9. after Coesel 1997; 10. after Coesel & Meesters 2007).

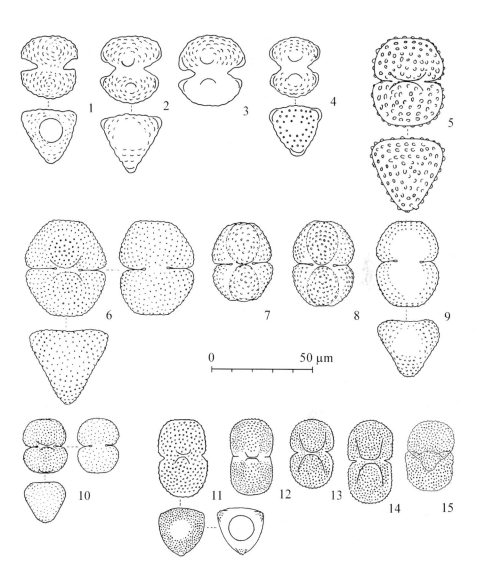

Plate 52. — 1-4. *Staurastrum novae-semliae* ; 5. *S. ricklii* ; 6-9. *S. botrophilum* ; 10. *S. donardense* ; 11-15. *S. alpinum* (1. after Wille 1879; 2-3. after Grönblad 1942; 4. after Förster 1967; 5. after Huber-Pestalozzi 1928; 6, 10. after West & West 1912; 7-9. after Brook & Williamson 1983; 11. after Raciborski 1889; 12. after Skuja 1928; 13-15. after Beck-Mannagetta 1926a, 15: cell in oblique position).

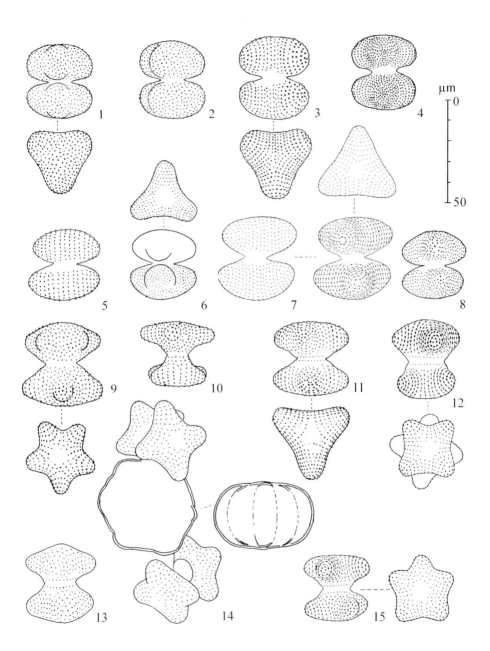

Plate 53.. — 1-4. *Staurastrum turgescens* ; 5-8. *S. lapponicum* ; 9-15. *S. dilatatum* (1-2, 9-10. after West & West 1912; 3, 11. after Růžička 1956; 4. after Lenzenweger 1994; 5. after Schmidle 1898; 6. after Grönblad 1926; 7, 12, 15. after Coesel 1997; 8. after Kouwets 1987; 13-14. after Coesel & Delfos 1986).

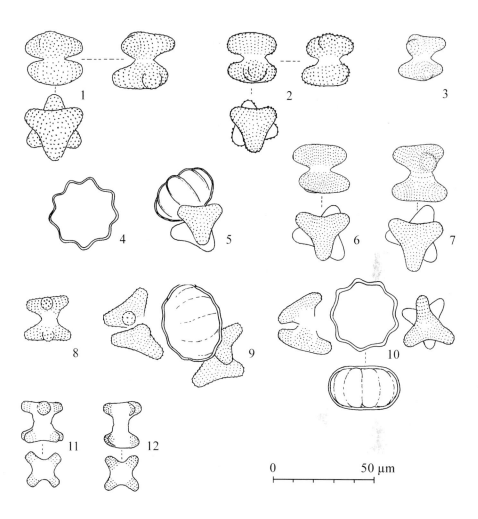

Plate 54. — 1-7. *Staurastrum alternans* ; 8-10. *S. striolatum* ; 11-12. *S. sinense* (1, 8-9. after West & West 1912; 2. after Förster 1967; 3-7, 10. after Coesel 1997; 11. after Lütkemüller 1900a; 12. after Borge 1906).

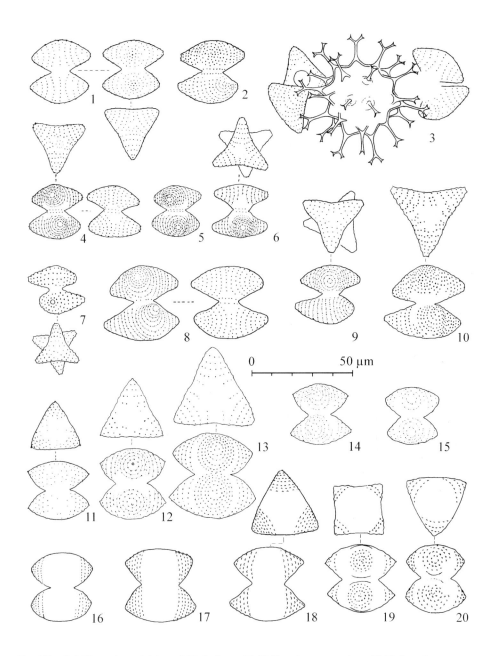

Plate 55. — 1-6. *Staurastrum striatum* ; 7-10. *S. dispar* ; 11-15. *S. acutum* var. *acutum* ; 16-20. *S. acutum* var. *varians* (1, 3, 7, 11. after West & West 1912; 2. after Růžička 1957; 4-6, 8-9, 12-15. after Coesel 1997; 10. after Kouwets 1987; 16. after Raciborski 1885; 17-18. after Schmidle 1896; 19-20. after Lenzenweger 1989).

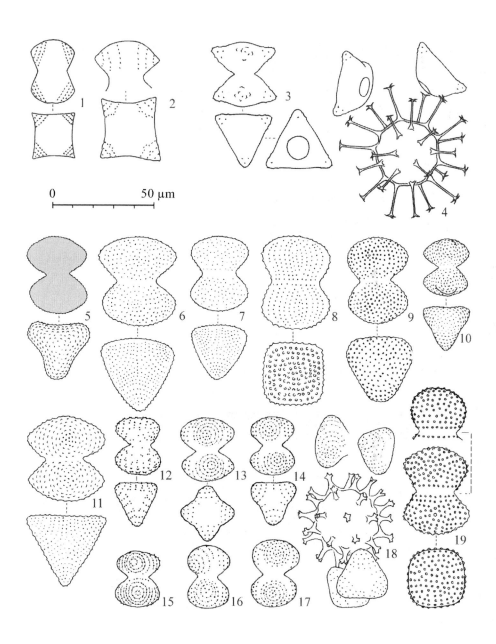

Plate 56. — 1-2. *Staurastrum tristichum* ; 3-4. *S. trachytithophorum* ; 5-19. *S. punctulatum* (1. after Elfving 1881; 2. after Grönblad 1921; 3-4, 9-10. after West & West 1912; 5. after Ralfs 1848; 6-8. after Wille 1879; 11. after Wille 1879 in West & West 1912; 12-18. after Coesel 1997; 19. after Förster 1967).

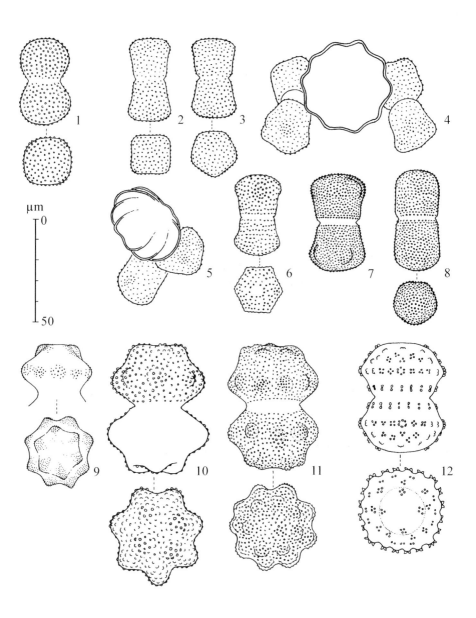

Plate 57. — 1. *Staurastrum punctulatoides* ; 2-8. *S. meriani* ; 9-11. *S. polonicum* ; 12. *S. mutilatum* (1-3 after West & West 1912; 4-5. after West & al. 1923; 6. after Růžička 1956; 7-8. after Skuja 1964; 9. after Raciborski 1884; 10. after Dick 1923; 11. after Lenzenweger 1997; 12. after Lundell 1871).

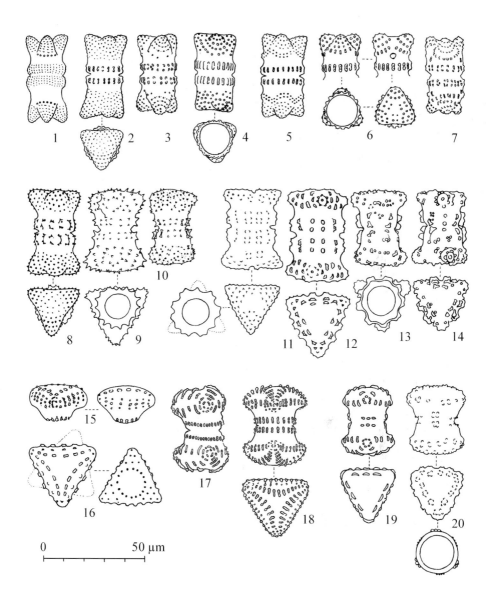

0   50 µm

Plate 58.. — 1-7. *Staurastrum pileolatum* ; 8-10. *S. capitulum* ; 11-14. *S. spetsbergense* ; 15-18. *S. bifasciatum* var. *bifasciatum* ; 19-20. *S. bifasciatum* var. *subkaiseri* (1. after Ralfs 1848; 2, 5, 8. after West & West 1912; 3, 10. after Kouwets 1987; 4. after Rybnicek 1960; 6. after Lütkemüller 1892; 7 after Lenzenweger 1984; 9. after Peterfi 1963; 11. after Nordstedt 1872; 12. after Krieger 1938; 13, 14. after Förster 1967; 15-16. after Lütkemüller 1900; 17. after Pevalek 1925; 18. after Dick 1919; 19. after Messikommer 1956; 20. after Bourrelly 1987).

275

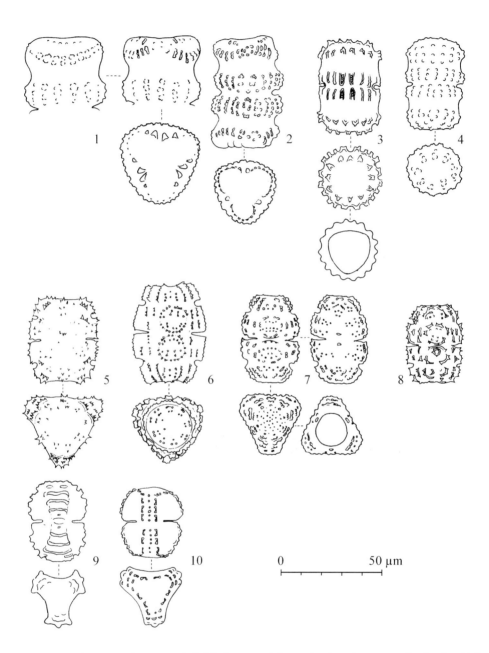

Plate 59. — 1-2. *Staurastrum borgei* ; 3-4. *S. rhabdophorum* ; 5-8. *S. acarides* ; 9-10. *S. maamense* (1. after Grönblad 1933; 2. after Lenzenweger 1977; 3. after Nordstedt 1875; 4. after Grönblad 1963; 5. after Nordstedt 1872; 6, 10. after West & al. 1923; 7. after Růžička 1956; 8. after Messikommer 1942; 9. after Cooke 1887).

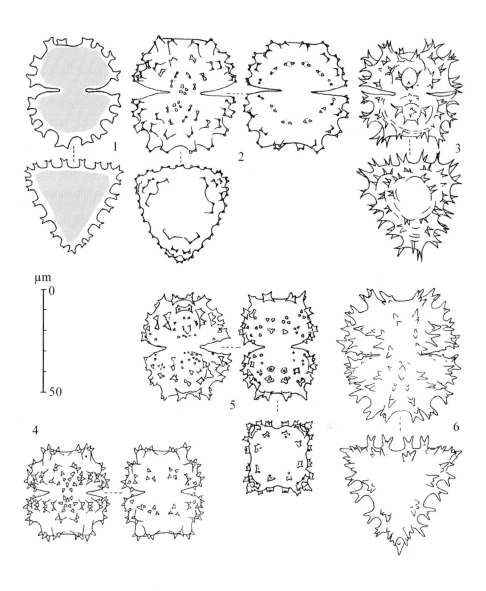

µm

0

50

Plate 60. — 1-6. *Staurastrum spongiosum* (1. after Ralfs 1848; 2. after West & al. 1923; 3. after W. West 1892a; 4. after Dick 1923; 5. after Förster 1967; 6. after Coesel 1997).

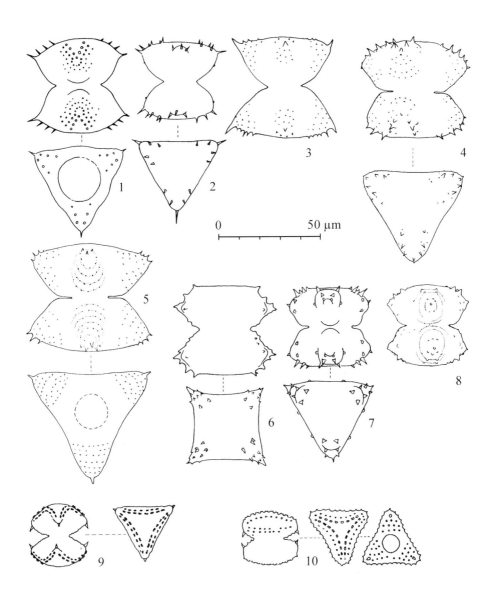

Plate 61. — 1-3. *Staurastrum cristatum* var. *cristatum* ; 4-5. *S. cristatum* var. *cuneatum* ; 6-8. *S. cristatum* var. *oligacanthum* ; 9. *S. oxyrhynchum* var. *oxyrhynchum* ; 10. *S. oxyrhynchum* var. *truncatum* (1. after Nägeli 1849; 2. after West & al. 1923; 3. after Coesel & Meesters 2007; 4-5, 8. after Coesel 1997; 6. after Nordstedt 1875 in West & al. 1923; 7. after Förster 1970; 9. after Roy & Bisset 1886; 10. after Lütkemüller).

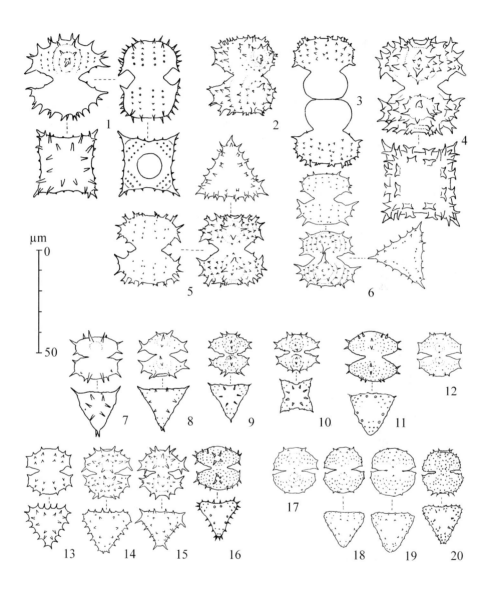

Plate 62. — 1-6. *Staurastrum echinatum* ; 7-12. *S. simonyi* var. *simonyi* ; 13-16. *S. simonyi* var. *sparsiaculeatum* ; 17-20. *S. simonyi* var. *semicirculare* (1. after Børgesen 1889; 2-3. after Heimans 1926; 4. after Messikommer 1962; 5-6, 8, 12, 14-15, 17-19. after Coesel 1997; 7. after Heimerl 1891; 9-10, 20. after Kouwets 1988; 11. after West & al. 1923; 13. after Schmidle 1896).

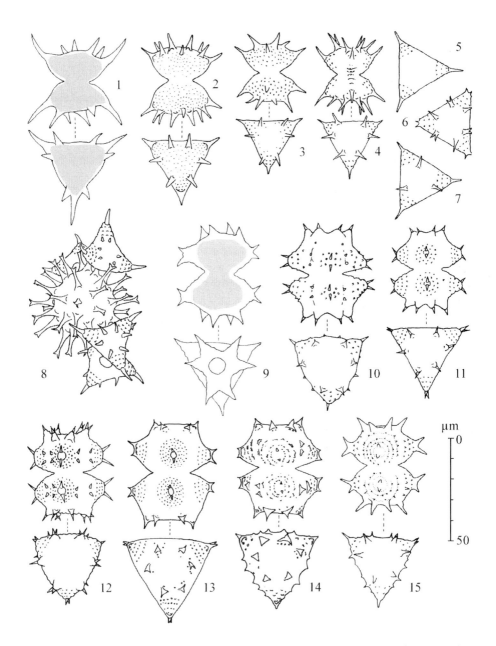

Plate 63. — 1-8. *Staurastrum pungens* ; 9-15. *S. monticulosum* (1, 9. after Ralfs 1848; 2, 10-12. after West & al. 1923; 3, 15. after Coesel 1997; 4-7. after Růžička 1972; 8. after Dick 1923; 13-14. after Grönblad 1920).

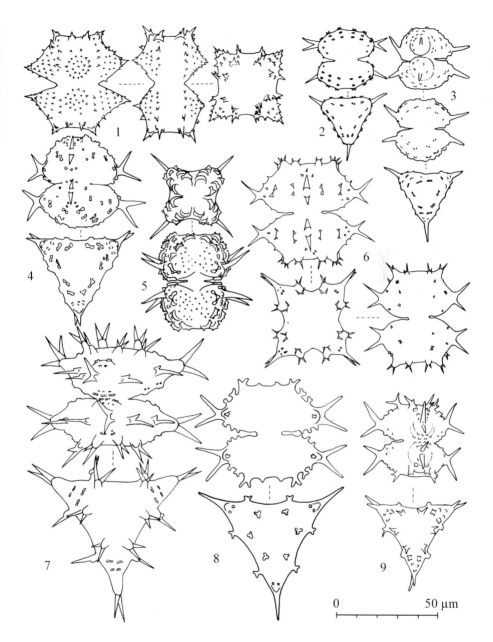

Plate 64. — 1. *Staurastrum megalonotum* ; 2-4. *S. cornutum* var. *cornutum* ; 5. *S. cornutum* var. *skujae* ; 6-9. *Staurastrum forficulatum* var. *forficulatum* (1. after Nordstedt 1875; 2. after Roy & Bisset 1894; 3. after John & Williamson 2009; 4. after Grönblad 1920; 5. after Skuja 1948; 6. after West & West 1905; 7. after West & al. 1923; 8. after Lundell 1871; 9. after Coesel 1997).

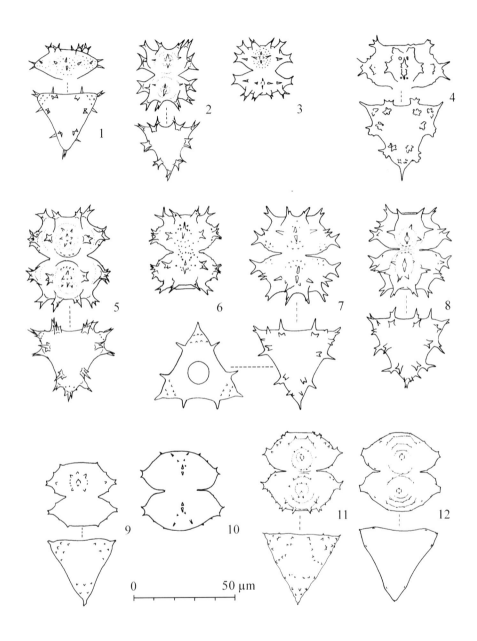

Plate 65. — 1-3. *Staurastrum forficulatum* var. *subsenarium* ; 4-8. *S. forficulatum* var. *verrucosum* ; 9-12. *S. podlachicum* (1. after West & al. 1923; 2. after Lenzenweger 1986b; 3. after Sieminska 1967; 4. after Grönblad 1920; 5. after Lenzenweger 1981; 6. after Kouwets 1987; 7-8. after Coesel 1997; 9. after Eichler & Gutwiński 1895; 10. after Grönblad 1920; 11-12. after Coesel 1997).

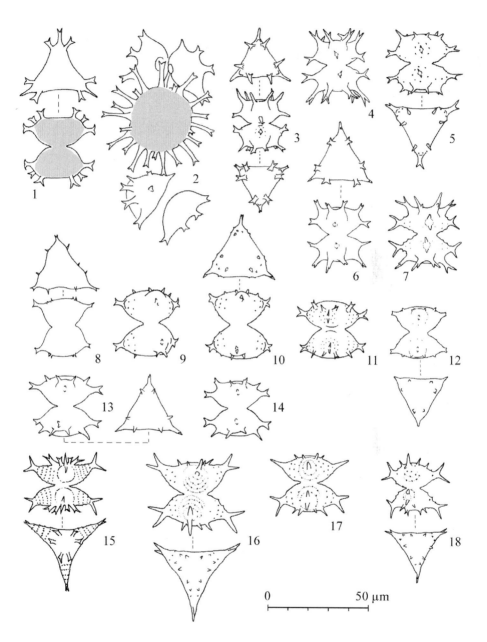

Plate 66. — 1-7. *Staurastrum furcatum* var. *furcatum* ; 8-14. *S. furcatum* var. *aciculiferum* ; 15-18. *S. arcuatum* var. *arcuatum* (1-2. after Ralfs 1848; 3, 5. after Kouwets 1988; 4. after Kouwets 1987; 6-7, 12-14, 16-18. after Coesel 1997; 8. after W. West 1889; 9-10. after West & al. 1923; 11. after Lenzenweger 1989d; 15. after Nordstedt 1873).

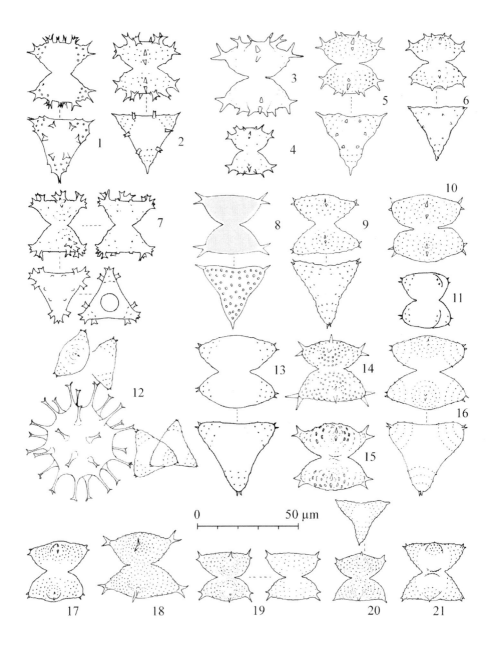

Plate 67. — 1-6. *Staurastrum arcuatum* var. *subavicula* ; 7. *S. dicroceros* ; 8-21. *S. avicula* var. *avicula* (1. after W. West 1892; 2. after Förster 1978; 3-6, 19-20. after Coesel 1997; 7. after Růžička 1963; 8. after Ralfs 1848; 9-13. after West & al. 1923; 14. after Messikommer 1943; 15, 18. after Lenzenweger 1989; 16. after Nygaard 1991; 17. after Manguin 1935; 21. after Lenzenweger 1988).

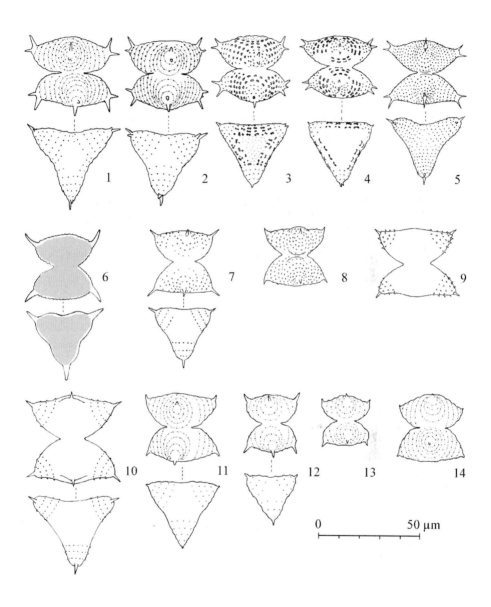

Plate 68. — 1-5. *Staurastrum avicula* var. *avicula* ; 6-14. *S. avicula* var. *lunatum* (1-3, 11-14. after Coesel 1997; 4. after Messikommer 1943; 5. after Skuja 1956; 6. after Ralfs 1848; 7. after Růžička 1972; 8. after Lenzenweger 1989; 9-10. after Nygaard 1991).

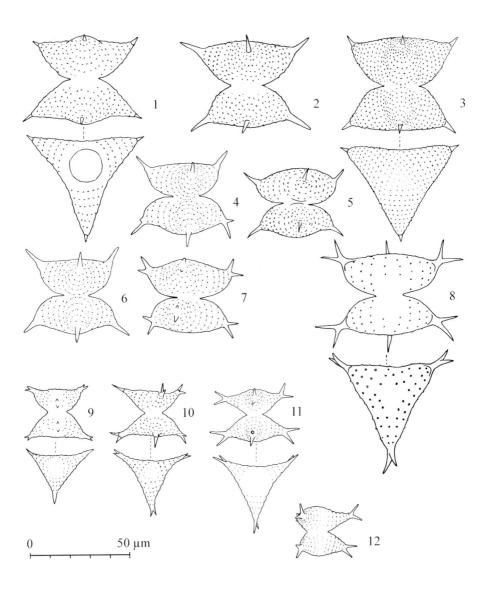

Plate 69. — 1-7. *Staurastrum avicula* var. *planctonicum* ; 8. *S. pelagicum* ; 9-12. *S. subcruciatum* (1-2, 8, 10. after West & al. 1923; 3. after Skuja 1953; 4, 6, 7, 11, 12. after Coesel 1997; 5. after Lenzenweger 1989; 9. after Cooke 1887).

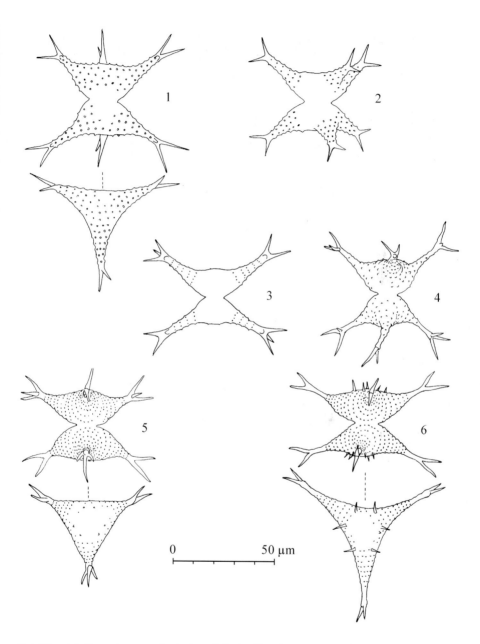

Plate 70. — 1-6. *Staurastrum pseudopelagicum* (1-2. after West & West 1903; 3. after Grönblad 1938; 4-6. after Brook 1957).

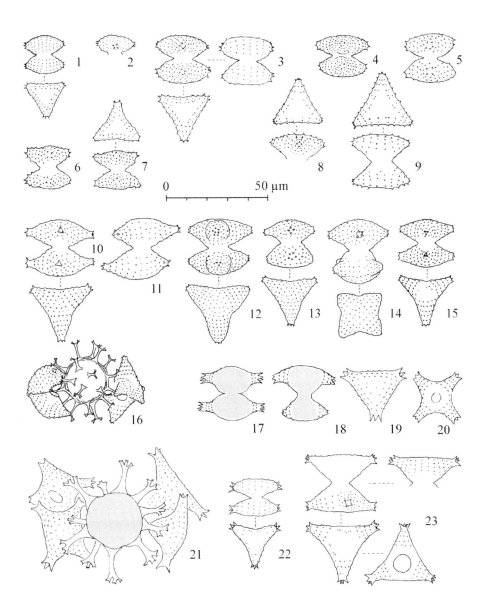

Plate 71. — 1-7. *Staurastrum bohlinianum* var. *bohlinianum* ; 8-9. *S. bohlinianum* var. *subpygmaeum* ; 10-16. *S. hexacerum* ; 17-23. *S. polymorphum* (1-2. after Schmidle 1898; 3. after Van Westen archives; 4. after Lenzenweger 1989; 5-7. after Coesel 2008; 8-9. after Růžička 1972; 10-11. after West & al. 1923; 12-14. after Förster 1970; 15. after Tomaszewicz 1988; 16. after Skuja 1931; 17-21. after Ralfs 1848; 22-23. after Grönblad 1921).

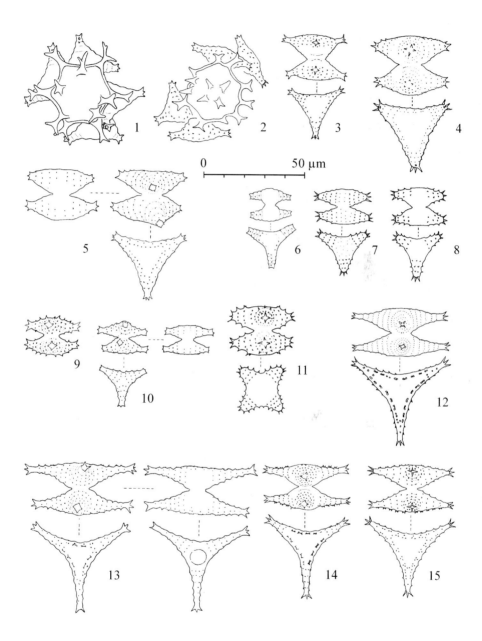

Plate 72. — 1-5. *Staurastrum polymorphum* ; 6-11. *S. haaboeliense* ; 12-15. *S. dybowskii* (1, 7, 8. after West & al. 1923; 2. after Kossinskaja 1950; 3. after Lenzenweger 1994; 4. after Lenzenweger 1986; 5, 9, 10, 13. after Coesel 1997; 6. after Wille 1881; 11. after Lenzenweger 1987; 12. after Woloszynska 1919; 14. after Lenzenweger 1989; 15. after Skuja 1948).

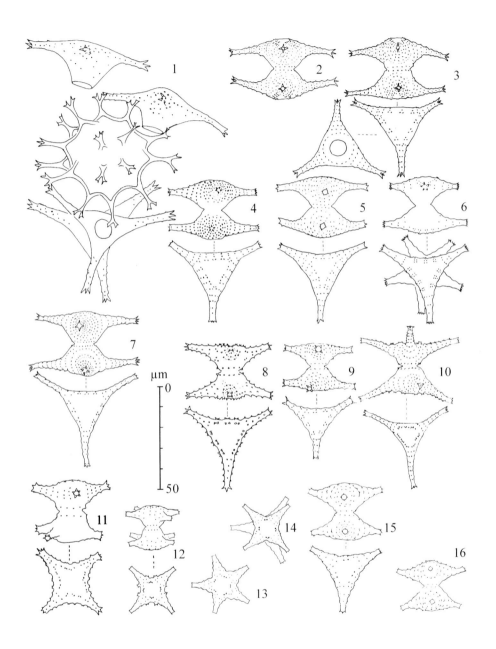

Plate 73. — 1-7. *Staurastrum gracile* ; 8-16. *S. boreale* (1. after Homfeld 1929; 2-3. after Brook 1959; 4. after Růžička 1972; 5. after Coesel 1997; 6. after Nygaard 1949; 7. after Lenzenweger 1989; 8. after West & al. 1923; 9. after Brook 1959c; 10. after Lenzenweger 1989e; 11. after Cedercreutz & Grönblad 1936; 12-16. after Coesel 1997).

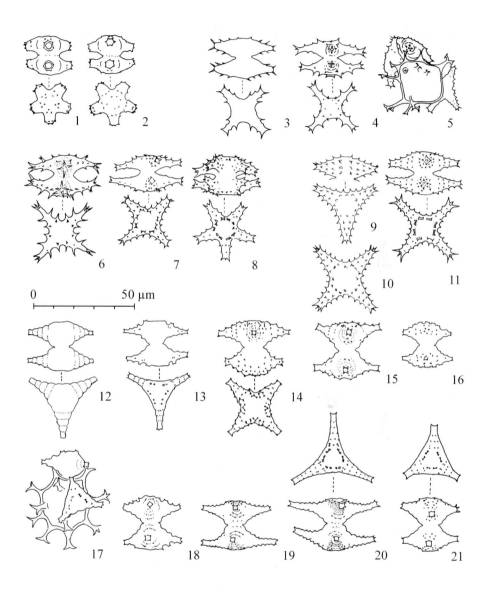

0   50 µm

Plate 74. — 1-2. *Staurastrum glaronense* ; 3-8. *S. heimerlianum* var. *heimerlianum* ; 9-11. *S. heimerlianum* var. *spinulosum* ; 12-21. *S. crenulatum* (1-2. after Messikommer 1951; 3. after Heimerl 1891; 4–5, 13, 16. after West & al. 1923; 6. after Lenzenweger 1986; 7. after Lenzenweger 1987; 8. after Lenzenweger 1991; 9-10. after Lütkemüller 1893; 11. after Lenzenweger 1994; 12. after Nägeli 1849; 14. after Dick 1980; 15. after Brook 1959; 17-18. after Coesel archives; 19-21. after Coesel 1997).

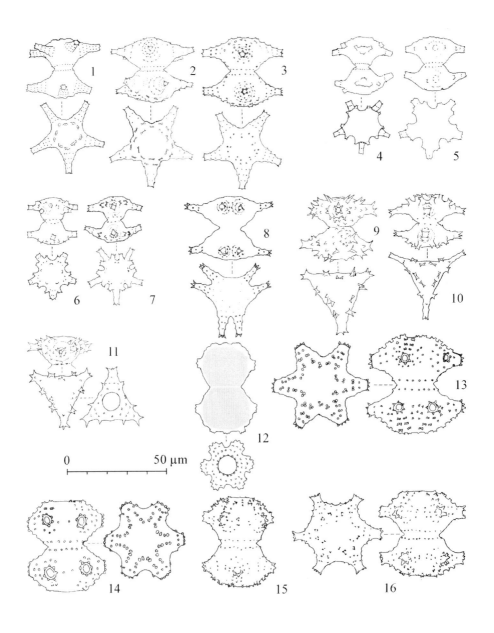

Plate 75. — 1-3. *Staurastrum pentasterias* ; 4-7. *S. pertyanum* ; 8. *S. barbaricum* ; 9-11. *S. suchlandtianum* ; 12-16. *S. sexcostatum* (1. after Grönblad 1963; 2. after Lenzenweger 1989; 3. after Růžička 1972; 4. after Schmidle 1896; 5. after Laporte 1931; 6-7. after Lenzenweger 1988; 8, 13, 14. after West & al. 1923; 9, 11. after Messikommer 1942, 10. after Lenzenweger 1986b; 12. after Ralfs 1848; 15-16. after Coesel 1997).

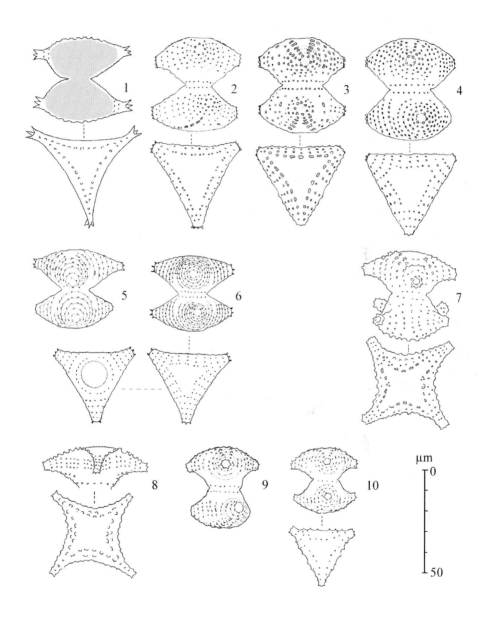

Plate 76. — 1-6. *Staurastrum proboscideum* ; 7-10. *S. borgeanum* (1. after Ralfs 1848; 2. after West & al. 1923; 3-4. after Růžička 1972; 5-6, 9-10. after Coesel 1997; 7. after Förster 1967; 8. after Schmidle 1898).

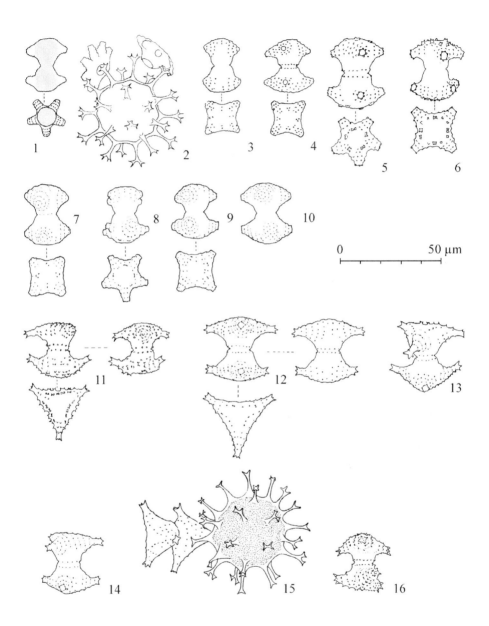

Plate 77. — 1-10. *Staurastrum margaritaceum* ; 11-16. *S. subnivale* (1. after Ralfs 1848; 2-5. after West & al. 1923; 6. after Dick 1930; 7-10, 12-15. after Coesel 1997; 11. after Messikommer 1942; 16. after Lenzenweger 1997).

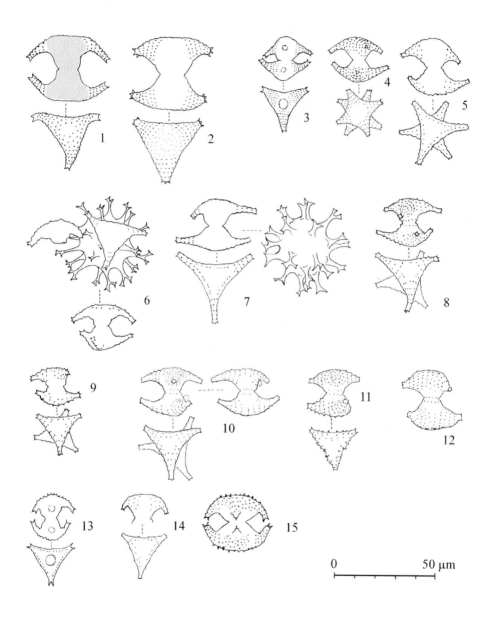

Plate 78. — 1-2. *Staurastrum cyrtocerum* var. *cyrtocerum* ; 3-12. *S. cyrtocerum* var. *inflexum* ; 13-15. *S. cyrtocerum* var. *brachycerum* (1. after Ralfs 1848; 2, 5-6, 15. after West & al. 1923; 3, 13. after Brebisson 1856; 4. after Tomaszewicz 1988; 7. after Villeret 1955; 8-9. after Kouwets 1987; 10-12. after Coesel 1997; 14. after Grönblad 1934).

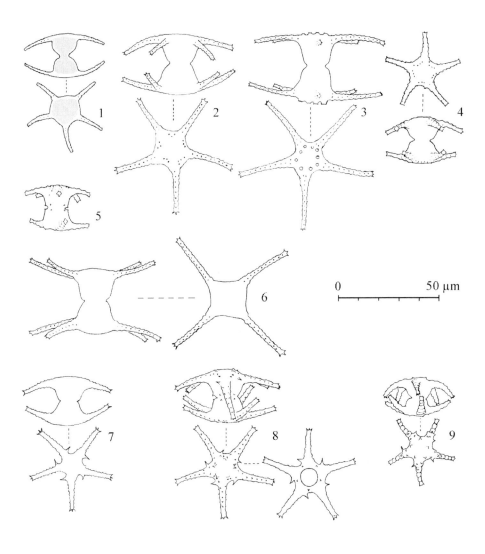

Plate 79. — 1-5. *Staurastrum arachne* var. *arachne* ; 6. *S. arachne* var. *curvatum* ; 7-9. *S. arachne* var. *gyrans* (1. after Ralfs 1848; 2-3, 6. after West & al. 1923; 4-5. after Coesel 1997; 7. after Johnson 1894; 8. after Capdevielle & Couté 1980; 9. after Messikommer 1942).

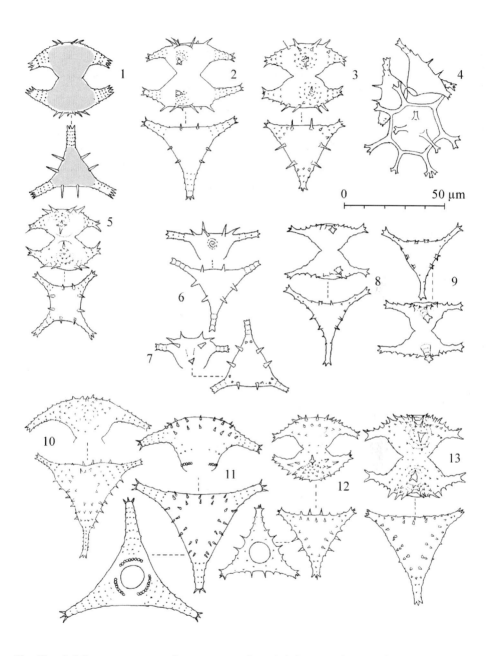

Plate 80. — 1-5. *Staurastrum oxyacanthum* var. *oxyacanthum* ; 6- 9. *S. oxyacanthum* var. *sibiricum* ; 10-13. *S. oxyacanthum* var. *polyacanthum* (1. after Archer 1860; 2, 11. after West & al. 1923; 3, 12. after Růžička 1972; 4. after Homfeld 1929; 5. after Lenzenweger 1993; 6-7. after Boldt 1885; 8-9, 13. after Coesel 1997; 10. after Nordstedt 1885).

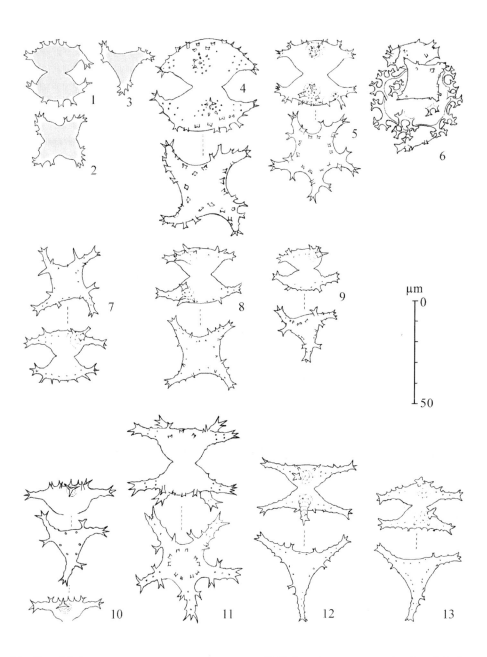

Plate 81. — 1-9. *Staurastrum controversum* var. *controversum*; 10-13. *S. controversum* var. *semivestitum* (1-3. after Ralfs 1848; 4. after West & al. 1923; 5. after Lenzenweger 1986; 6. after Lütkemüller 1900 in West & al. 1923; 7-9. after Coesel 1997; 10. after West 1892; 11. after Dubois-Tylski 1969; 12-13. after Coesel 1997).

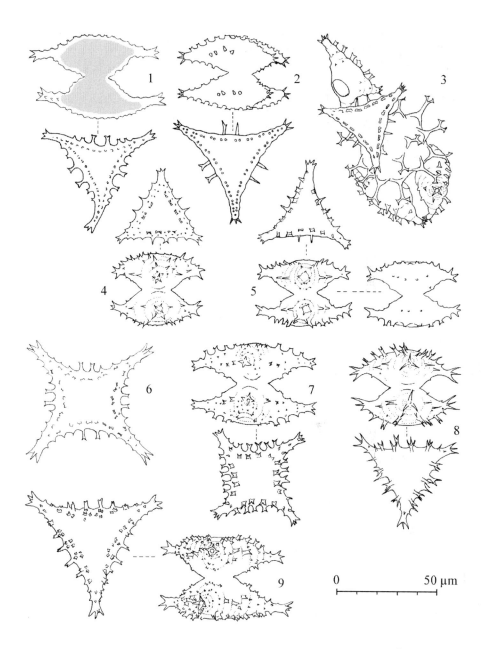

Plate 82. — 1-5. *Staurastrum vestitum* var. *vestitum* ; 6-9. *S. vestitum* var. *splendidum* (1. after Ralfs 1848; 2. after West & al. 1923; 3. after Skuja 1934; 4, 8. after Lenzenweger 1989c; 5. after Coesel 1997; 6. after Grönblad 1920; 7. after Lenzenweger 1989d; 9. after Kouwets 1989).

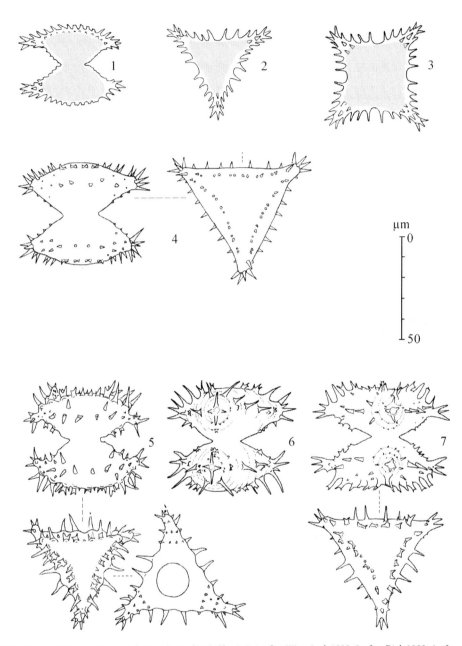

Plate 83. — 1-7. *Staurastrum aculeatum* (1-3. after Ralfs 1848; 4. after West & al. 1923; 5. after Dick 1923; 6. after Brook 1959; 7. after Coesel 1997).

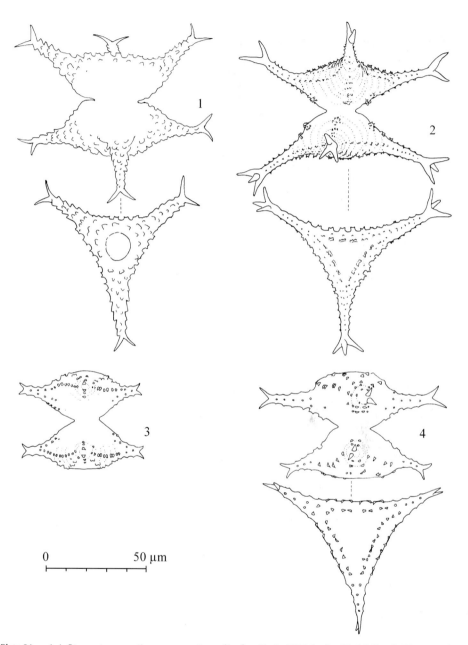

0   50 μm

Plate 84. — 1-4. *Staurastrum anatinum* var. *anatinum* (1. after Cooke 1880; 2. after Lind & Brook 1980; 3-4. after West & al. 1923).

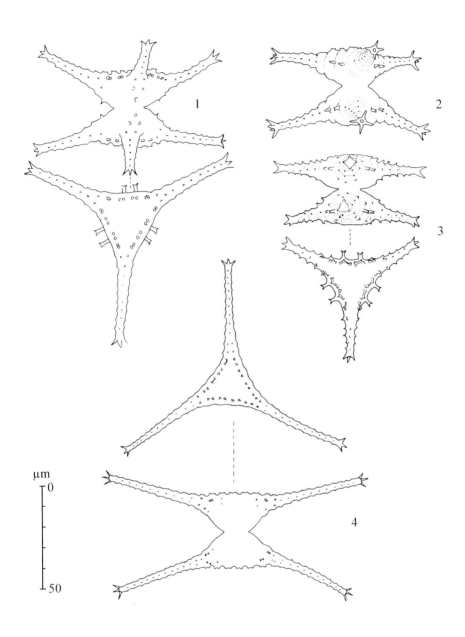

Plate 85. — 1-3. *Staurastrum anatinum* var. *subanatinum* ; 4. *S. anatinum* var. *longibrachyatum* (1, 4. after West & al. 1923; 2. after Kouwets 1987; 3. after Coesel).

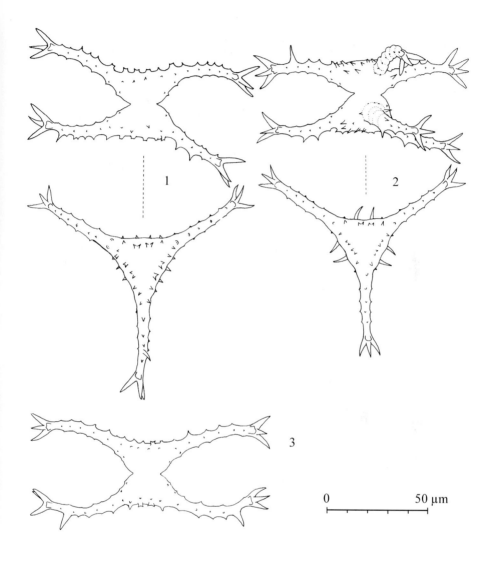

Plate 86. — 1-3. *Staurastrum anatinum* var. a*rmatum* (1-3. after Capdevielle 1978).

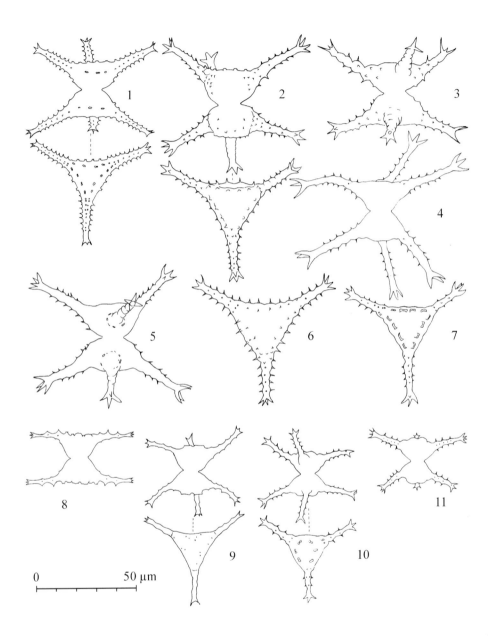

Plate 87. — 1-7. *Staurastrum anatinum* var. *denticulatum* ; 8-11. *S. floriferum* (1. after Smith 1924; 2, 4-7, 10-11. after Florin 1957; 3. after Scharf 1986; 8. after West & West 1896; 9. after Grönblad 1938).

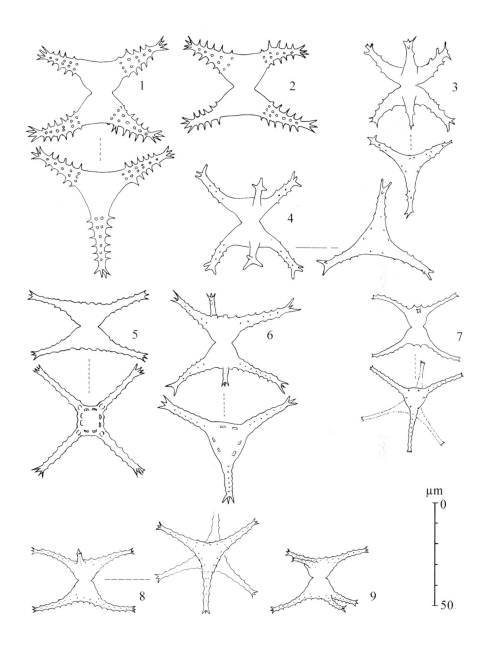

Plate 88. — 1-2. *Staurastrum magdalenae* ; 3-4. *S. informe* ; 5-6. *S. subosceolense* ; 7. *S. saltator* var. *saltator* ; 8-9. *S. saltator* var. *pendulum* (1-2. after Børgesen & Ostenfeld 1903; 3, 5. after Grönblad 1920; 4, 7. after Grönblad 1938; 6. after Borge 1939; 8-9. after Nygaard 1949).

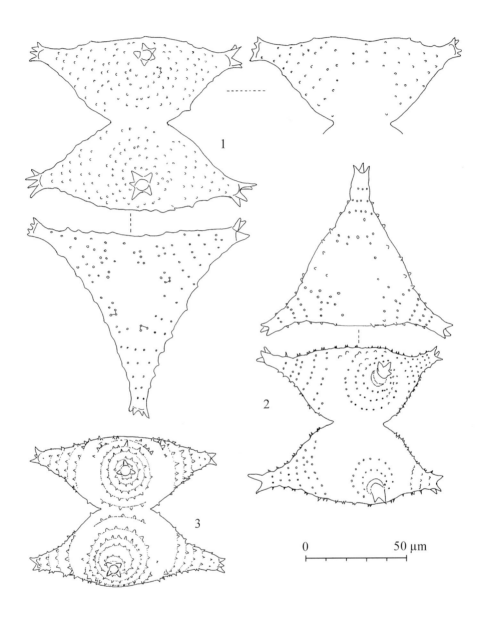

Plate 89. — 1-3. *Staurastrum petsamoense* (1. after Järnefeld 1934; 2. after Thomasson 1957; 3. after Lenzenweger 1989).

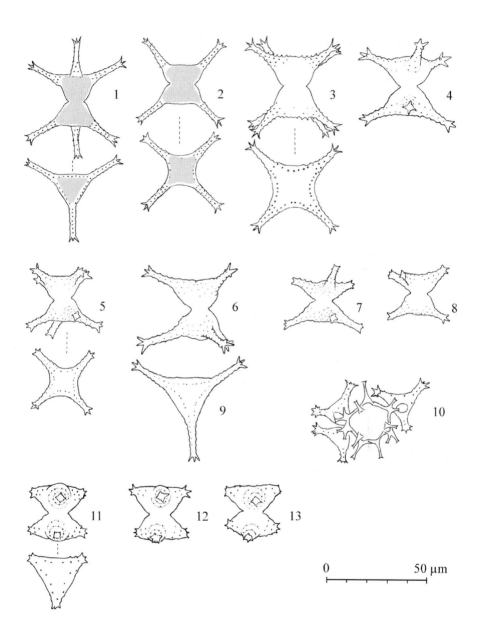

Plate 90. — 1-10. *Staurastrum paradoxum* ; 11-13. *S. paradoxoides* (1-2. after Ralfs 1848; 3. after West & al. 1923; 4-13. after Coesel 1997).

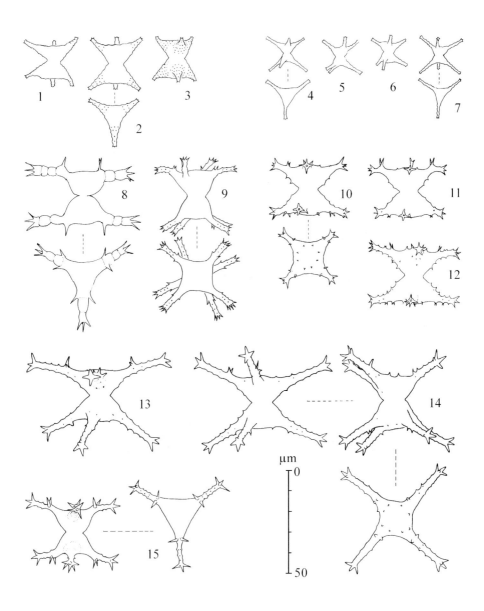

Plate 91. — 1-3. *Staurastrum pseudotetracerum* ; 4-7. *S. minimum* ; 8-15. *S. diacanthum* (1. after Nordstedt 1888; 2. after West & al. 1923; 3. after Dick 1923; 4-6. after Coesel 1996; 7. after Kouwets 2001; 8. after Lemaire 1890; 9. after Homfeld 1929; 10-12. after Kouwets 1987; 13-14. after Coesel 1997; 15 after Lenzenweger 2000a).

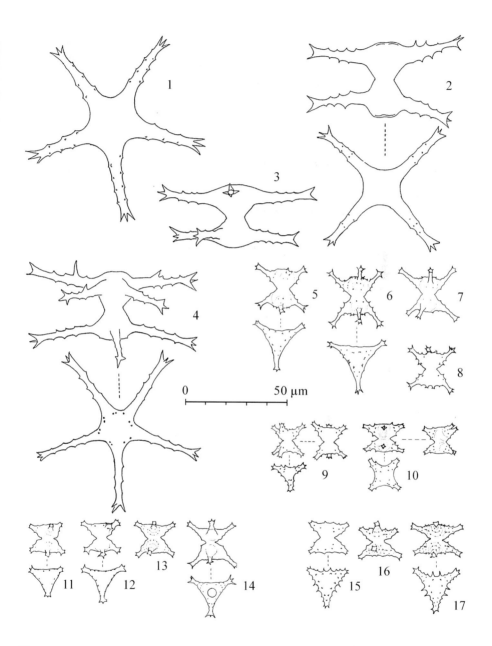

Plate 92. — 1-4. *Staurastrum platycerum* ; 5-8. *S. micronoides* ; 9-14. *S. micron* var. *micron* ; 15-17. *S. micron* var. *spinulosum* (1. after Joshua 1886; 2-4. after Grönblad 1938; 5-8, 11-14, 16-17. after Coesel 1997; 9. after West & al. 1923; 10. after Lenzenweger 1991; 15. after Grönblad 1942).

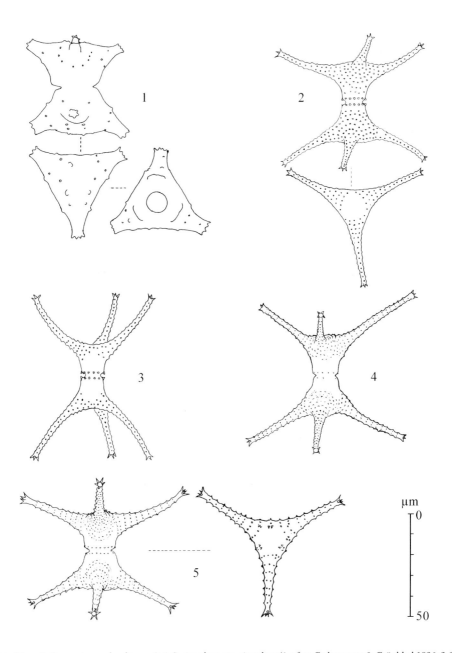

Plate 93. — 1. *Staurastrum alandicum* ; 2-5. *S. cingulum* var. *cingulum* (1. after Cedercreutz & Grönblad 1936; 2-3. after West & West 1903; 4-5. after Brook 1959).

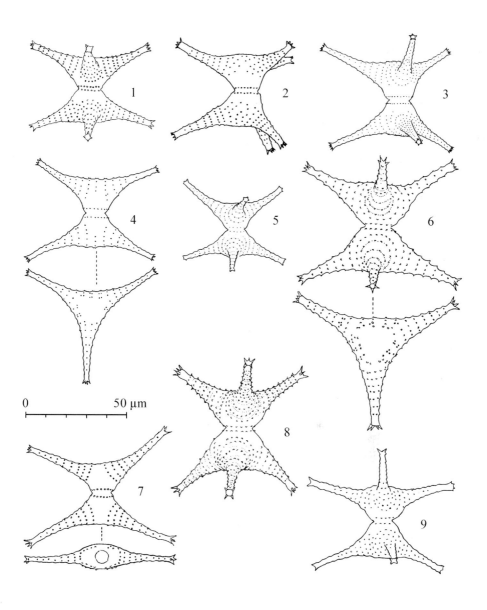

Plate 94. — 1-9. *Staurastrum cingulum* var. *obesum* (1-2. after Smith 1922; 3-4. after Nygaard 1949; 5. after Coesel 1997; 6, 8. after Brook 1959; 7. after Teiling 1946; 9. after Coesel & Wardenaar 1990).

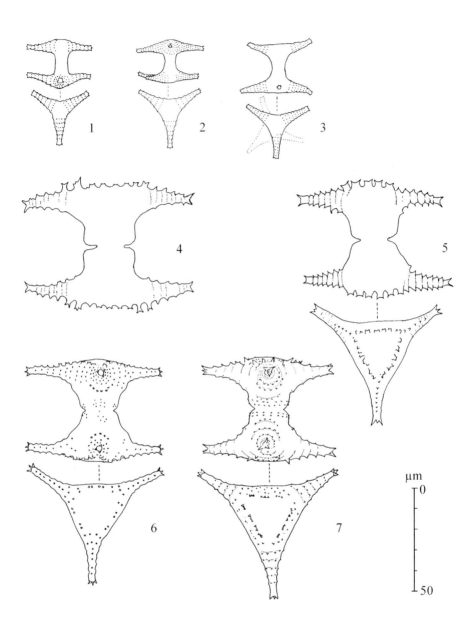

Plate 95. — 1-3. *Staurastrum neglectum* ; 4-7. *S. manfeldtii* var. *manfeldtii* (1. after G.S. West 1909; 2, 6. after West & al. 1923; 3. after Grönblad 1947; 4-5. after Delponte 1878; 7. after Lenzenweger 1989d).

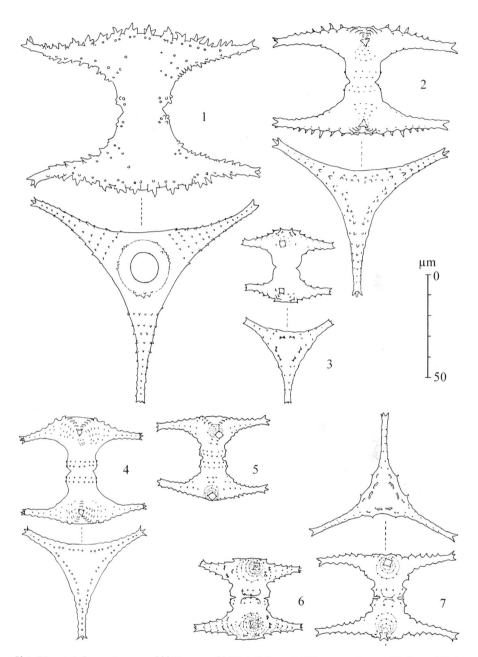

Plate 96. — 1-3. *Staurastrum manfeldtii* var. *manfeldtii* ; 4-5. *S. manfeldtii* var. *annulatum* ; 6-7. *S. manfeldtii* var. *splendidum* (1. after Nordsted 1873; 2-3, 7. after Coesel 1997; 4-5. after West & al. 1923; 6. after Lenzenweger 1989d).

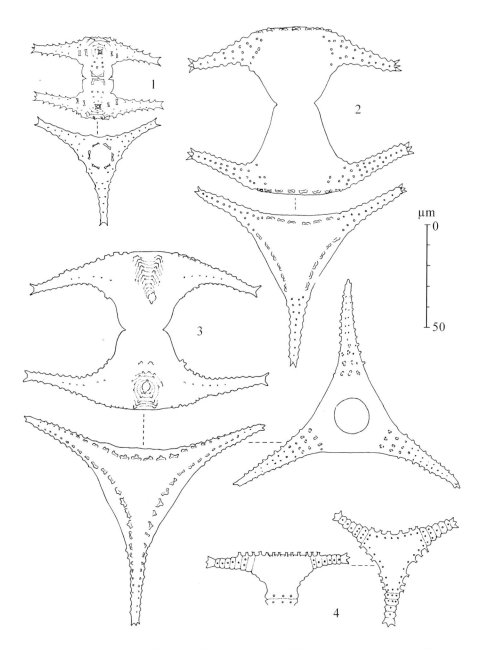

Plate 97. — 1. *Staurastrum manfeldtii* var. *splendidum* ; 2-3. *S. manfeldtii* var. *productum* ; 4. *S. manfeldtii* var. *pseudosebaldi* (1. after Messikommer 1928; 2. after West & al. 1923; 3. after Coesel 1997; 4. after Wille 1881).

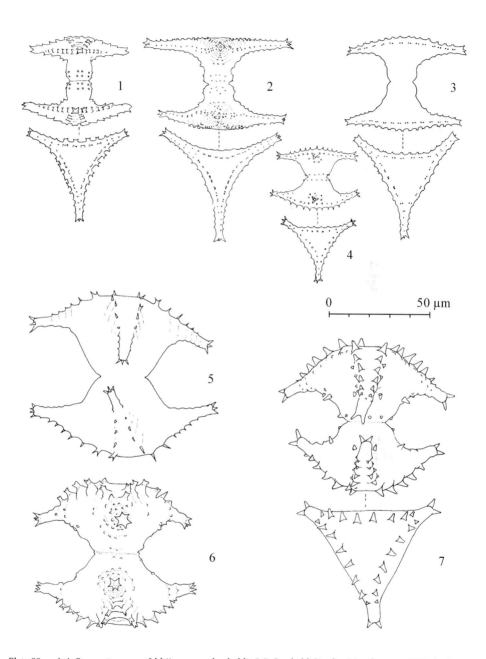

Plate 98. — 1-4. *Staurastrum manfeldtii* var. *pseudosebaldi* ; 5-7. *S. sebaldi* (1. after Messikommer 1957; 2. after Lenzenweger 1989d; 3. after Grönblad 1947; 4, 5. after West & al. 1923; 6. after Lenzenweger 1994; 7. after Dick 1923).

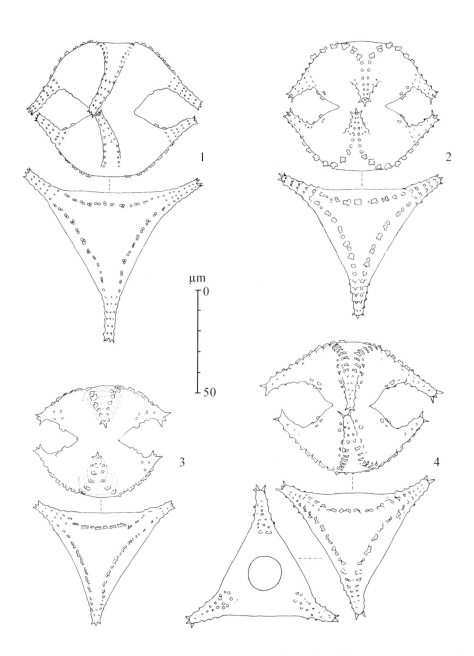

Plate 99. — 1-4. *Staurastrum traunsteineri* (1. after Hustedt 1911; 2. after Dick 1919; 3. after Lenzenweger 1989a; 4. after Dick 1923).

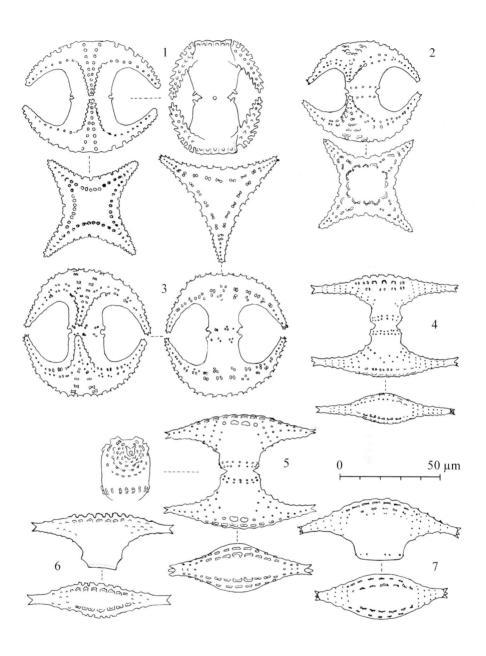

Plate 100. — 1-3. *Staurastrum cerastes* ; 4-7. *S. bicorne* (1. after Lundell 1871; 2-3, 6. after West & al. 1923; 4. after Børgesen 1889; 5. after Förster 1967; 7. after Lenzenweger 1986).

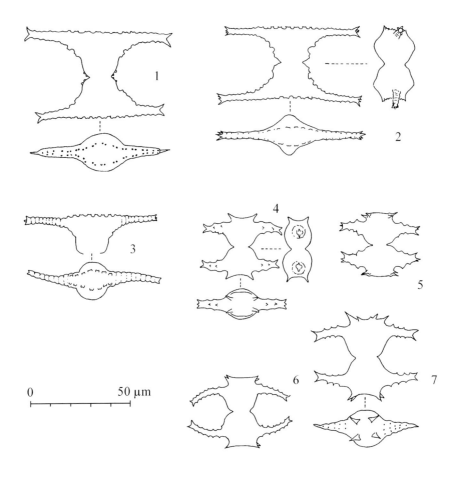

Plate 101. — 1-3. *Staurastrum duacense* ; 4-8. *S. miedzyrzecense* (1. after West & al. 1923; 2. after Grönblad 1945; 3. after Boldt 1885; 4. after Eichler 1896; 5. after Kouwets archives; 6-7. after Cedercreutz & Grönblad 1936).

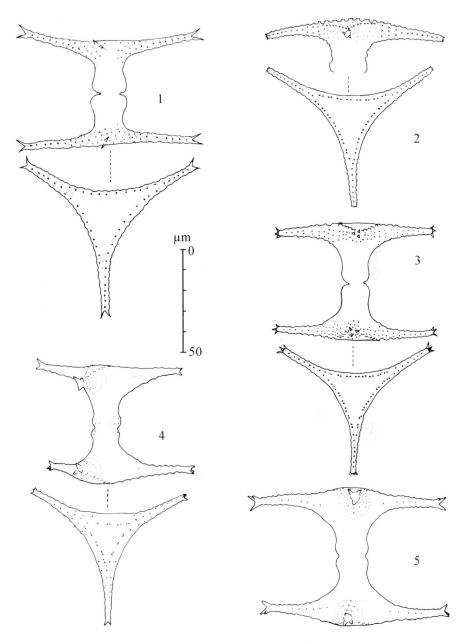

Plate 102. — 1. *Staurastrum bulbosum* var. *bulbosum* ; 2-5. *S. bulbosum* var. *cyathiforme* (1. after W. West 1892; 2. after W. & G.S. West 1895; 3. after West & al. 1923; 4-5. after Coesel 1997).

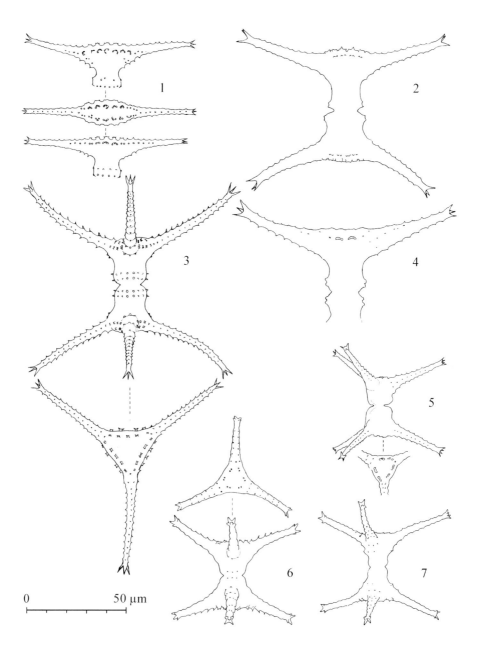

Plate 103. — 1-4. *Staurastrum johnsonii* ; 5-7. *Staurastrum pingue* var. *pingue* (1. after W. & G.S. West 1896; 2. after Strøm 1926; 3. after Skuja 1948; 4. after Grönblad 1920; 5. after Teiling 1942; 6-7. after Förster 1967).

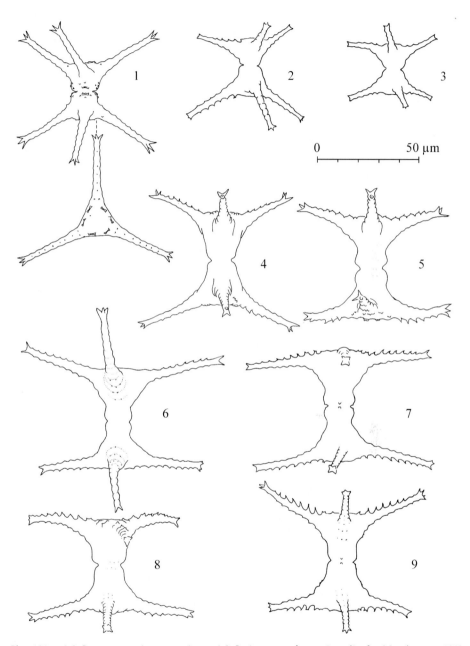

Plate 104. — 1-3. *Staurastrum pingue* var. *pingue* ; 4-9. *S. pingue* var. *planctonicum* (1. after Messikommer 1942; 2-3, 6-9. after Coesel 1997; 4-5. after Teiling 1946).

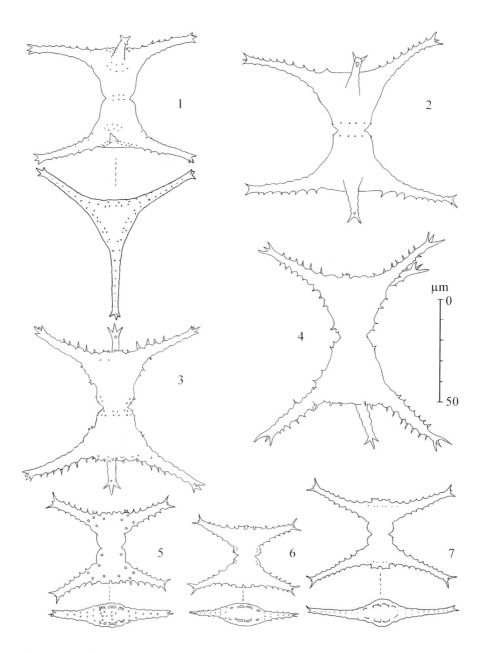

Plate 105. — 1-4. *Staurastrum pingue* var. *planctonicum* ; 5-7. *S. reductum* (1, 7. after Coesel 1997; 2. after Grönblad 1938; 3-4. after Florin 1957; 5. after Messikommer 1927a; 6. after Lenzenweger 2000).

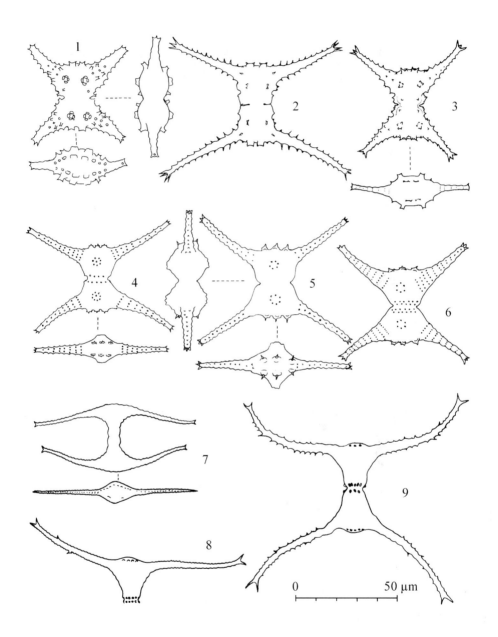

Plate 106. — 1-3. *Staurastrum dimazum* ; 4-6. *S. natator* ; 7-9. *S. leptocladum* var. *leptocladum* (1. after Lütkemüller 1910; 2. after Skuja 1948; 3. after Coesel 1997; 4. after Smith 1924; 5. after West & al. 1923; 6. after Grönblad 1920; 7. after Nordstedt 1869; 8. after Teiling 1944; 9. after Florin 1957).

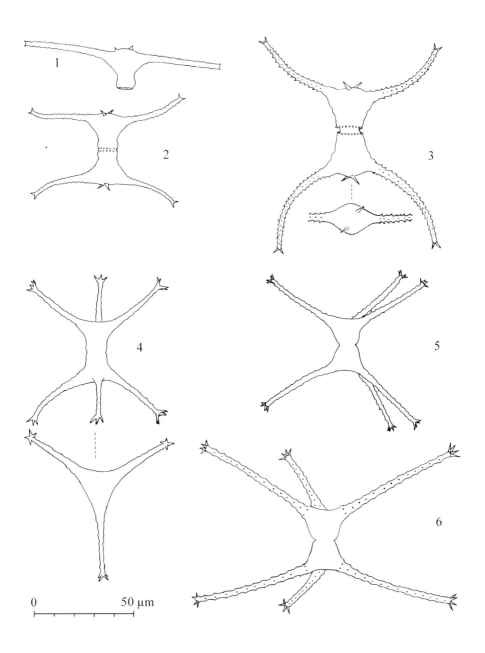

Plate 107. — 1-3. *Staurastrum leptocladum* var. *cornutum* ; 4-6. *S. longipes* var. *longipes* (1. after Wille 1884; 2. after Grönblad 1926; 3. after Skuja 1934; 4. after Nordstedt 1873; 5-6. after West & al. 1923).

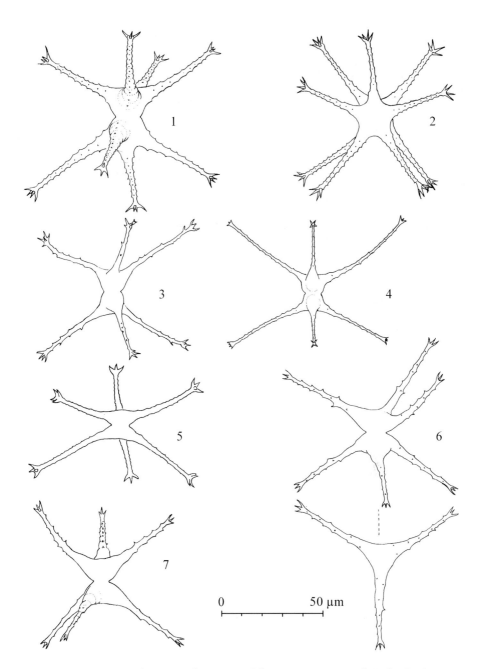

Plate 108. — 1-4. *Staurastrum longipes* var. *longipes* ; 5-7. *S. longipes* var. *contractum* (1-2. after Brook 1959; 3-4, 6-7. after Brook 1958; 5. after Teiling 1946).

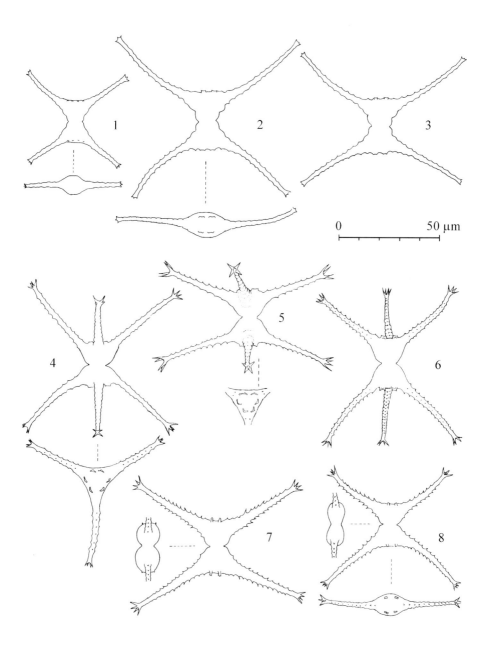

Plate 109. — 1-3. *Staurastrum multinodulosum* ; 4-8. *S. bullardii* (1. after Grönblad 1926; 2-3. after Coesel & Alfinito 2006; 4. after Smith 1924; 5-6. after Skuja 1948; 7-8 after Teiling 1942a).

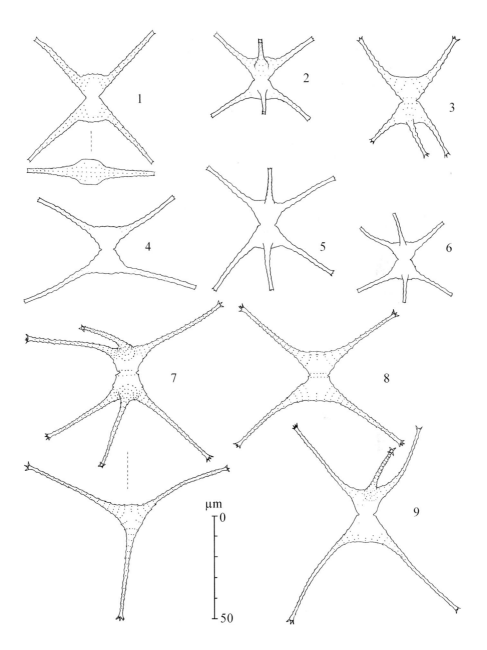

Plate 110. — 1-9. *Staurastrum chaetoceras* (1. after Schröder 1898; 2-6. after Coesel 1997; 7-9. after Brook 1959).

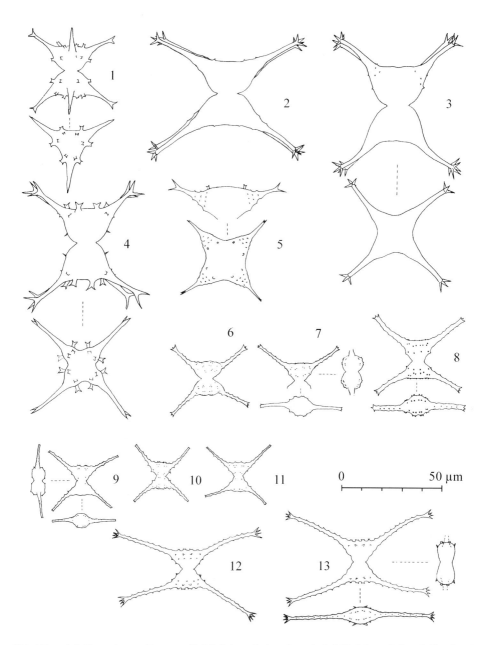

Plate 111. — 1-5. *Staurastrum subboergesenii* ; 6-8. *S. levanderi* var. *levanderi* ; 9-11. *S. levanderi* var. *hollandicum* ; 12-13. *S. iversenii* (1, 6-7. after Grönblad 1938; 2-4. after Skuja 1964; 5. after Borge 1913; 8. after Capdevielle 1982; 9-11. after Coesel & Joosten 1996; 12-13. after Nygaard 1949).

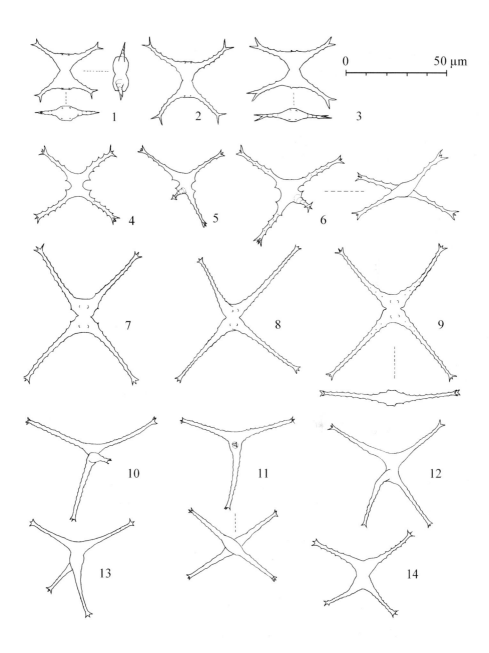

Plate 112. — 1-3. *Staurastrum bloklandiae* ; 4-6. *S. lenzenwegeri* ; 7-9. *S. nygaardii* ; 10-14. *S. smithii* (1-2. after Coesel & Joosten 1996; 3. after Lenzenweger 2003; 4-6. after Lenzenweger 1999; 7-9. after Nygaard 1949; 10-11. after Smith 1924; 12. after Florin 1957; 13-14. after Coesel 1997).

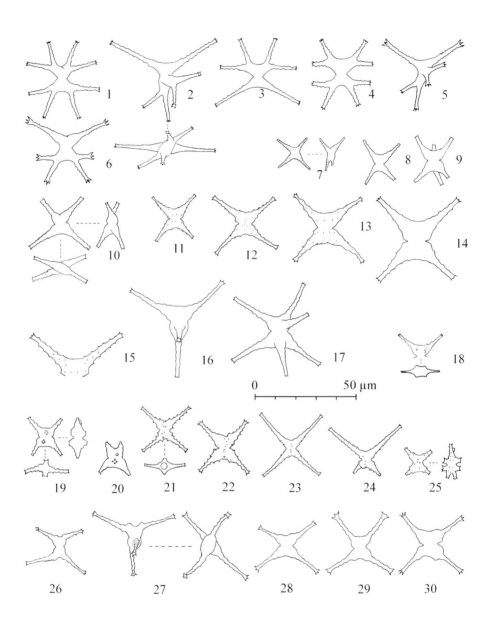

Plate 113. — 1-6. *Staurastrum bibrachiatum* ; 7-17. *S. tetracerum* var. *tetracerum* ; 18. *S. tetracerum* var. *biverruciferum* ; 19-25. *S. tetracerum* var. *irregulare* ; 26-30. *S. tetracerum* var. *cameloides* (1-3. after Grönblad & Scott 1955; 4-6. after Lenzenweger 2001; 7. after Ralfs 1848; 8-10, 19-20. after West & al. 1923; 11-12, 23-25. after Kouwets 1987; 13. after Brook 1982; 14-15. after Růžička 1972; 16-17, 21-22. after Coesel 1972; 18. after Grönblad 1921; 26-27. after Florin 1957; 28. after Grönblad 1960; 29. after Coesel 1997; 30. after Georgevitch 1910.

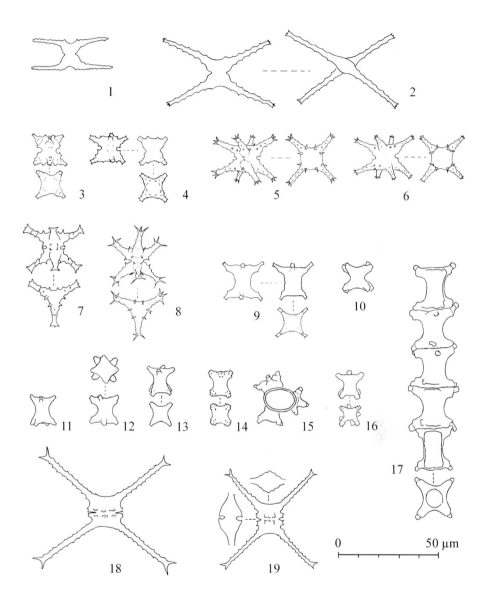

Plate 114. — 1-2. *Staurastrum subexcavatum* ; 3-4. *S. chavesii* var. *chavesii* ; 5-6. *S. chavesii* var. *latiusculum* ; 7-8. *S. dentatum* ; 9-17. *S. inconspicuum* ; 18-19. *S. uhtuense* (1. after Allorge & Allorge 1931; 2, 18. after Grönblad 1921; 3. after Bohlin 1901; 4, 11-12. after Coesel 1997; 5. after West & West 1902; 6. after West & West 1923; 7. after Krieger 1932; 8. after Lenzenweger 1997; 9. after Nordstedt 1873; 10. after Tomaszewicz & Kowalski 1993; 13-14. after Kouwets 1987; 15. after Lütkemüller 1900; 16. after Gay 1884; 17. after Børgesen 1901; 19. after Grönblad 1938).

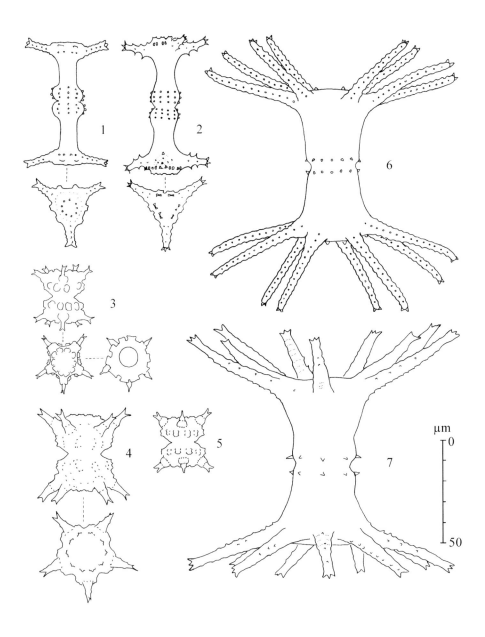

Plate 115. — 1-2. *Staurastrum elongatum* ; 3-5. *S. eichleri* ; 6-7. *S. verticillatum* (1. after Cooke 1887; 2. after West & al. 1923; 3. after Eichler & Raciborski 1893; 4. after Grönblad 1938; 5. after Eichler 1896; 6. after West & West 1903; 7. after John & Williamson 2009).

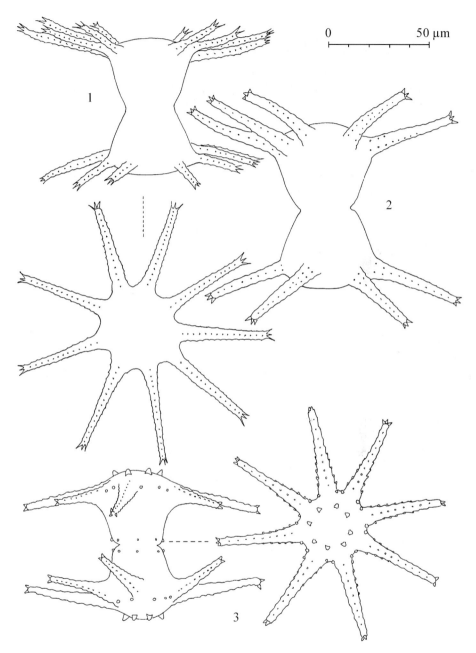

0   50 μm

Plate 116. — 1-2. *Staurastrum archeri* ; 3. *S. ophiura* var. *ophiura* (1. after W. West 1892a; 2-3. after West & al. 1923).

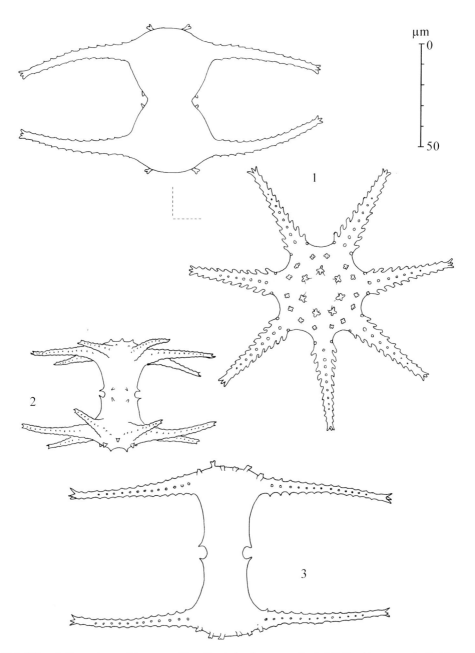

Plate 117. — 1. *Staurastrum ophiura* var. *ophiura* [front view in cross section] ; 2-3. *S. ophiura* var. *subcylindricum* (1. after Lundell 1871; 2. after Donat 1926; 3. after Borge 1939).

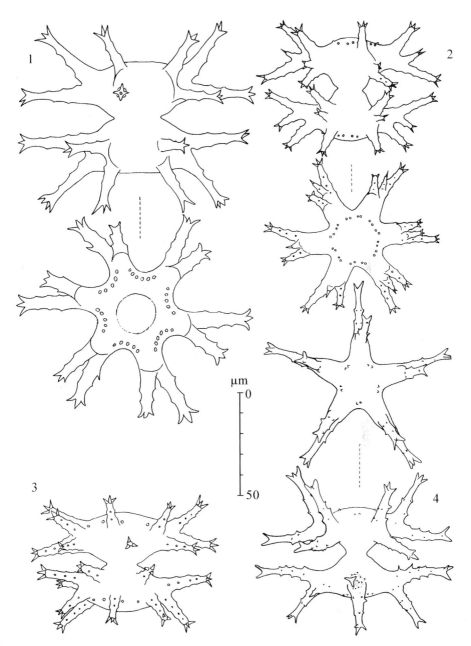

Plate 118. — 1-4. *Staurastrum sexangulare* (1. after Lundell 1871; 2, 3. after West & al. 1923; 4. after Williamson 1992).

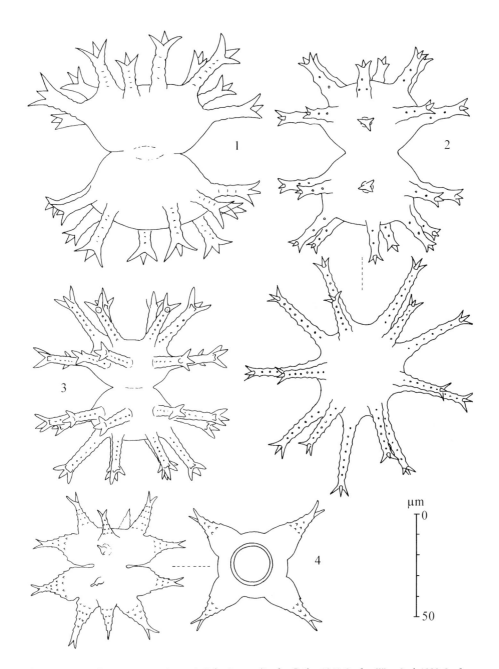

µm
0

50

Plate 119. — 1-3. *Staurastrum arctiscon* ; 4. *S. furcigerum* (1. after Bailey 1841; 2. after West & al. 1923; 3. after Heimans 1960; 4. after Ralfs 1948).

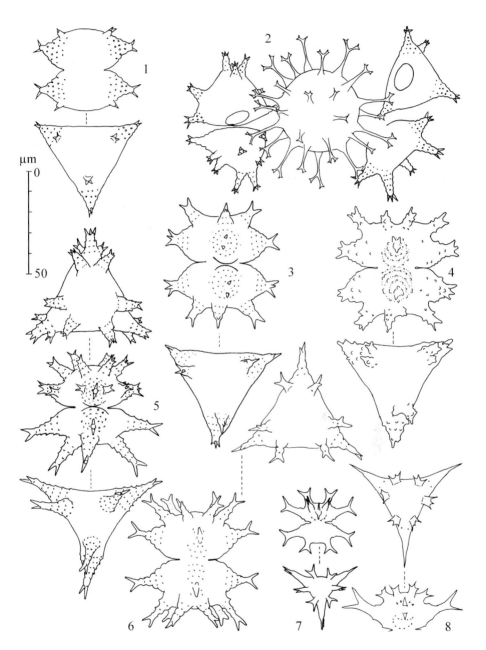

Plate 120. — 1-6. *Staurastrum furcigerum* ; 7-8. *S. pseudopisciforme* (1, 2. after West & al. 1923; 3. after Coesel 1997; 4, 6. after Coesel archives; 5. after Brook 1958; 7. after Eichler & Gutwiński 1895; 8. after Grönblad 1920).

Index of species

(For excluded species, see page 166)

Arthrodesmus

Cosmarium

Desmidium

Didymocladon

Micrasterias

Pentasterias

Phycastrum

Colophon

Authors
P.F.M. Coesel, PhD
J. Meesters, BSc

Graphic design and layout
Erik de Bruin, Varwig Design, Hengelo

Cover photography
Photograph A.M.T. Joosten (photos stacked by J. Meesters)
Staurastrum bicorne

This publication was made possible by a contribution of Stichting Hugo de Vries fonds

© KNNV Publishing, Zeist, the Netherlands, 2013.
ISBN 978-90-5011-458-5
NUR 940
www.knnvpublishing.nl

Printed in the United States
By Bookmasters